Modern Certificate Chemistry

Arthur Atkinson
B.Sc., A.R.I.C., B.Sc.(Econ.), Dip. Ed.
Formerly in the Education Services of Malaysia, Hong Kong and Uganda

OXFORD UNIVERSITY PRESS

Oxford University Press, Walton Street, Oxford OX2 6DP

OXFORD LONDON GLASGOW
NEW YORK TORONTO MELBOURNE WELLINGTON
IBADAN NAIROBI DAR ES SALAAM CAPE TOWN
KUALA LUMPUR SINGAPORE JAKARTA HONG KONG TOKYO
DELHI BOMBAY CALCUTTA MADRAS KARACHI

First Edition 1971
Reprinted 1973 (with corrections), 1975, 1978, 1979

Printed in Great Britain at The Pitman Press, Bath

Relative Atomic Masses of the Elements (based on carbon-12)

Atomic number	Element	Symbol	Relative atomic mass	Atomic number	Element	Symbol	Relative atomic mass	Atomic number	Element	Symbol	Relative atomic mass
1	hydrogen	H	1·008 0	36	krypton	Kr	83·80	71	lutetium	Lu	174·97
2	helium	He	4·002 60	37	rubidium	Rb	85·47	72	hafnium	Hf	178·49
3	lithium	Li	6·941	38	strontium	Sr	87·62	73	tantalum	Ta	180·948
4	beryllium	Be	9·012 2	39	yttrium	Y	88·906	74	tungsten	W	183·85
5	boron	B	10·81	40	zirconium	Zr	91·22	75	rhenium	Re	186·2
6	carbon	C	12·011 15	41	niobium	Nb	92·906	76	osmium	Os	190·2
7	nitrogen	N	14·006 7	42	molybdenum	Mo	95·94	77	iridium	Ir	192·2
8	oxygen	O	15·999 4	43	technetium	Tc	98·906	78	platinum	Pt	195·09
9	fluorine	F	18·998 4	44	ruthenium	Ru	101·07	79	gold	Au	196·967
10	neon	Ne	20·179	45	rhodium	Rh	102·906	80	mercury	Hg	200·59
11	sodium	Na	22·989 8	46	palladium	Pd	106·4	81	thallium	Tl	204·37
12	magnesium	Mg	24·305	47	silver	Ag	107·868	82	lead	Pb	207·2
13	aluminium	Al	26·981 5	48	cadmium	Cd	112·40	83	bismuth	Bi	208·981
14	silicon	Si	28·086	49	indium	In	114·82	84	polonium	Po	210*
15	phosphorus	P	30·973 8	50	tin	Sn	118·69	85	astatine	At	210*
16	sulphur	S	32·06	51	antimony	Sb	121·75	86	radon	Rn	222*
17	chlorine	Cl	35·453	52	tellurium	Te	127·60	87	francium	Fr	223*
18	argon	Ar	39·948	53	iodine	I	126·904 5	88	radium	Ra	226·025
19	potassium	K	39·102	54	xenon	Xe	131·30	89	actinium	Ac	227*
20	calcium	Ca	40·08	55	caesium	Cs	132·906	90	thorium	Th	232·038
21	scandium	Sc	44·956	56	barium	Ba	137·34	91	protoactinium	Pa	231·036
22	titanium	Ti	47·90	57	lanthanum	La	138·91	92	uranium	U	238·029
23	vanadium	V	50·941	58	cerium	Ce	140·12	93	neptunium	Np	237·048
24	chromium	Cr	51·996	59	praseodymium	Pr	140·908	94	plutonium	Pu	242*
25	manganese	Mn	54·938 0	60	neodymium	Nd	144·24	95	americium	Am	243*
26	iron	Fe	55·847	61	promethium	Pm	147*	96	curium	Cm	247*
27	cobalt	Co	58·933 2	62	samarium	Sm	150·4	97	berkelium	Bk	249*
28	nickel	Ni	58·71	63	europium	Eu	151·96	98	californium	Cf	251*
29	copper	Cu	63·546	64	gadolinium	Gd	157·25	99	einsteinium	Es	254*
30	zinc	Zn	65·37	65	terbium	Tb	158·925	100	fermium	Fm	253*
31	gallium	Ga	69·72	66	dysprosium	Dy	162·50	101	mendelevium	Md	256*
32	germanium	Ge	72·59	67	holmium	Ho	164·930	102	nobelium	No	254*
33	arsenic	As	74·921 6	68	erbium	Er	167·26	103	lawrencium	Lr	257*
34	selenium	Se	78·96	69	thulium	Tm	168·934				
35	bromine	Br	79·904	70	ytterbium	Yb	173·04				

* An asterisk means that the value is the mass number of one isotope of the element. The values of the relative atomic masses are those recommended by the International Union of Pure and Applied Chemistry (IUPAC) in 1970.

Preface

This book covers the modern 'O' level syllabus of the University of London, the Cambridge School Certificate syllabus, and the syllabus of the West African Examinations Council School Certificate and various other regional examinations. It should prove most useful during the final two or three years of a chemistry course. The treatment in the first seven chapters closely follows the Nuffield approach as recommended by London University although some changes have been made in order to make the subject matter more suitable for average pupils.

The metal reactivity series, electrochemical series and electronic theory have been used wherever possible to allow facts which are apparently unrelated to fit into a logical system. Modern views of acids, bases, salts, oxidation, valency and so on are stressed, and outdated ideas have been omitted.

The work is firmly based on experiments, most of which can be done by pupils. Each experiment is described in a step-by-step manner so that the procedure is clear and teachers can readily refer to any stage in the experiment. Clear diagrams, which have been labelled more than usual, help to reduce descriptive matter to the essential minimum.

SI units, signs, symbols and abbreviations are used throughout the book. Modern chemical nomenclature is used, but the old names are also given to avoid confusion.

I am grateful for the permission given by the Cambridge Local Examinations Syndicate (C.) and the University of London (L.) to use questions from various past examination papers.

The wording of some of the questions has been changed where necessary by making use of SI units and modern chemical names; such changes are the responsibility of the author and not of the examination authorities.

1978 A.A.

The International System of Units (SI)

In this system, first adopted in 1960 and now approved for use in many countries, there are seven basic units. Those used in Chemistry are:

Physical quantity	*Name of SI unit and symbol*	
Length	Metre	m
Mass	Kilogram	kg
Time	Second	s
Temperature	Kelvin	K
Amount of a substance	Mole	mol
Electric Current	Ampere	A

Units for all other physical quantities are based on the seven units. For example, volume is cubic metre (m^3), cubic centimetre (cm^3), cubic decimetre (dm^3), and so on, velocity or speed is metre per second (m/s or $m\ s^{-1}$), and concentration is moles per cubic decimetre (mol/dm^3 or $mol\ dm^{-3}$).

In 1964, the litre was re-defined to mean one cubic decimetre or 1000 cm^3 and therefore 1 millilitre (1 ml) is exactly 1 cubic centimetre (1 cm^3). The kelvin, symbol K, is used both for temperature and temperature interval; for example, a temperature of 0 °C = 273 K (more accurately, 273·15 K) and a temperature rise of 12 °C is the same as a rise of 12 kelvin.

Prefixes are used to indicate decimal sub-multiples and multiples of the basic units:

Sub-multiple	*Prefix*	*Symbol*	*Multiple*	*Prefix*	*Symbol*
10^{-1} or 0·1	deci	d	10^3	kilo	k
10^{-2} or 0·01	centi	c	10^6	mega	M
10^{-3} or 0·001	milli	m			
10^{-6}	micro	μ			
10^{-9}	nano	n			

Examples of the use of prefixes are:

1 centimetre = 0·01 metre
(1 cm = 0·01 m)
1 millimetre = 0·001 metre
(1 mm = 0·001 m = 0·1 cm)
1 milligram = 0·001 gram
(1 mg = 0·001 g = 0·000 001 kg)
1 megagram = 1 000 000 gram
(1 Mg = 1 000 000 g = 1000 kg)
1 microsecond = 0·000 001 second
(1 μs = 0·000 001 s)

Contents

1 Flames. Solutions and Mixtures

THE BUNSEN BURNER

The flames of a candle, kerosine (paraffin oil), petrol, oil, coal or other fuel give out light and therefore are called luminous flames. We rarely use luminous flames in science laboratories because they are not hot enough and they usually cover anything placed in them with a film of soot (carbon). The ordinary flames of gaseous fuels such as town gas, coal gas, petrol gas or bottled gas are also luminous, smoky and not very hot.

In 1855 a German scientist named Bunsen invented a burner which is used even today. In the burner the gas is mixed with air before it burns and the flame is then hotter and not smoky.

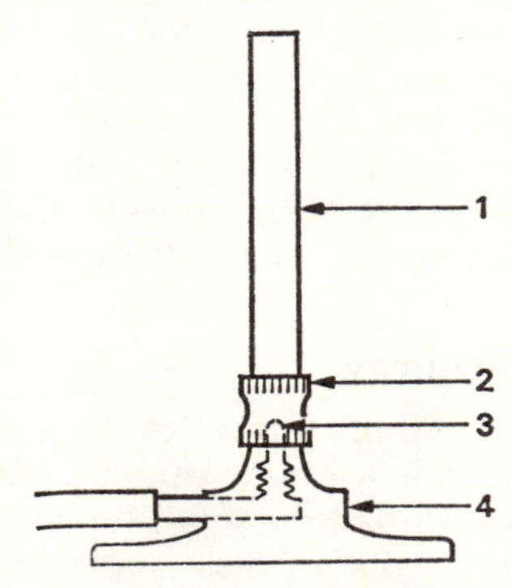

Fig. 1.1 The Bunsen burner

Note the various parts of the burner. They are:

1. The chimney, barrel or tube, which has either one or two holes near the bottom.
2. The collar, which is a metal ring with one or two holes of the same size as those in the chimney, through which air can enter when required.
3. A small jet through which gas enters the chimney.
4. A heavy base with an inlet tube for gas.

To study the Bunsen flame

1. Close the air-holes, turn the gas full on and light it. Note the size, shape and colours of the luminous flame.
2. Slowly open the holes so that air can enter and mix with the gas. Note how the flame changes. It is now non-luminous.
3. *Which parts of a flame are hot?* Use tongs to hold a piece of asbestos paper flat at the top of the non-luminous flame when the air-holes are half closed. Look at the paper from underneath and note which part becomes red-hot. Lower the paper slowly. How do the size and shape of the red hot part change?

The test can also be done with a piece of glossy paper or white cardboard. Hold it flat and steady until it just starts to turn brown and then remove it at once. Note the shape of the scorch mark.

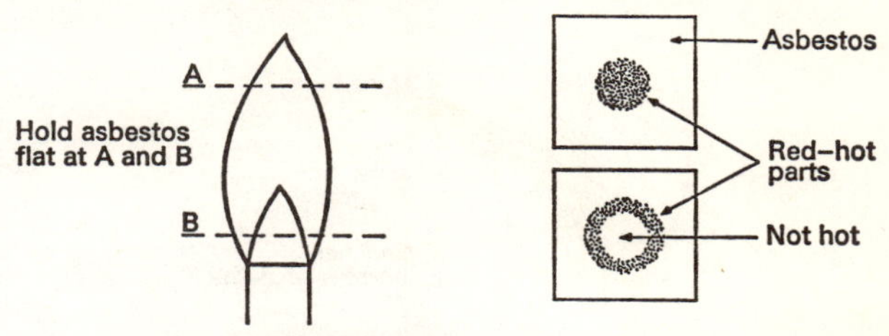

Fig. 1.2 Which parts of a flame are hot?

4. Now hold another piece of asbestos or cardboard vertically in the flame and note which part becomes red hot or scorches first.

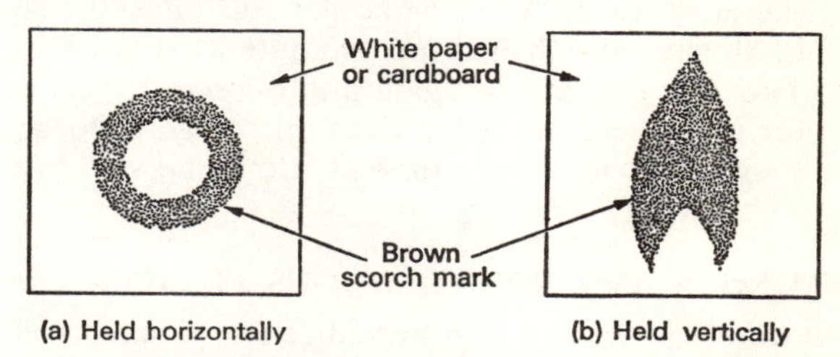

Fig. 1.3 Scorch marks on glossy paper or cardboard

5. Repeat both tests with the luminous flame when the air-holes are closed.
6. *Which flame is hotter?* Use tongs to hold a piece of broken porcelain in a luminous flame. What happens?
7. Open the air-hole until the flame is non-luminous and roaring. What happens to the porcelain?
8. *Is the inside of a flame hot?* Turn off the gas. Place a match as in Fig. 1.4. Turn on and light the gas. Does the match head burn? Do the test with both flames. Use a screen to prevent draughts blowing the flame about.

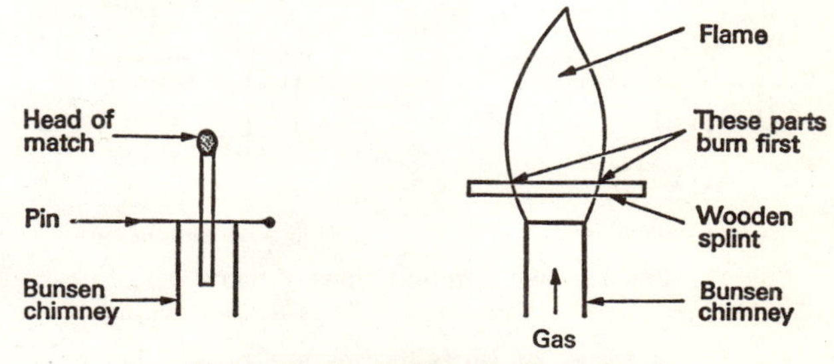

Fig. 1.4 Is the inside of a flame hot?

9. Hold a wooden splint horizontal and near the bottom of a flame. Which part of the wood scorches first? Do the test with both flames, and again use a large screen.

10. *Does a flame contain unburnt gas?* Hold a hard-glass or silica tube as in Fig. 1.5. Put a flame near the top of the tube. Note if the gas in the tube burns. Do the test with both flames.

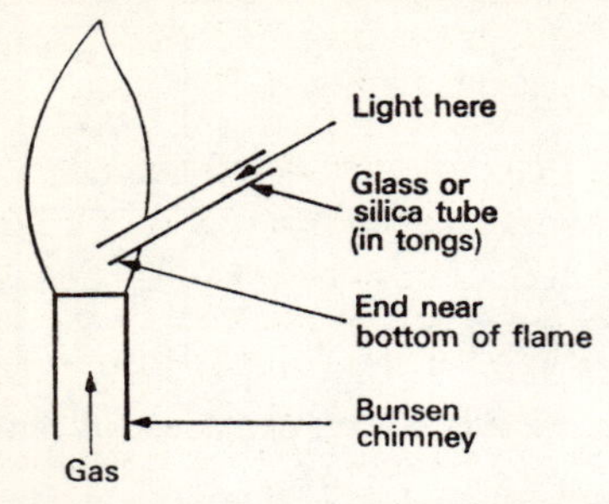

Fig. 1.5 Does a flame contain unburnt gas?

11. *To demonstrate burning back (striking back).* Open the air-holes fully, turn the gas full on and light it. Slowly close the gas tap and note what happens. Turn the gas full on again and observe the flame for a few seconds only. Turn off the gas. Do not touch the collar for a time as it may be very hot.

LUMINOUS AND NON-LUMINOUS FLAMES

A flame is a space in which burning gases give out heat and usually light. A luminous flame gives out light because tiny particles of hot carbon (soot) give out a yellow light. When the gas is mixed with air before burning, it burns more quickly and completely. No carbon particles form; therefore it gives out only a little light and it is called non-luminous. The hottest point in a non-luminous flame is just above the top of the middle zone, and its temperature is about 1000 °C or about 1300 K.

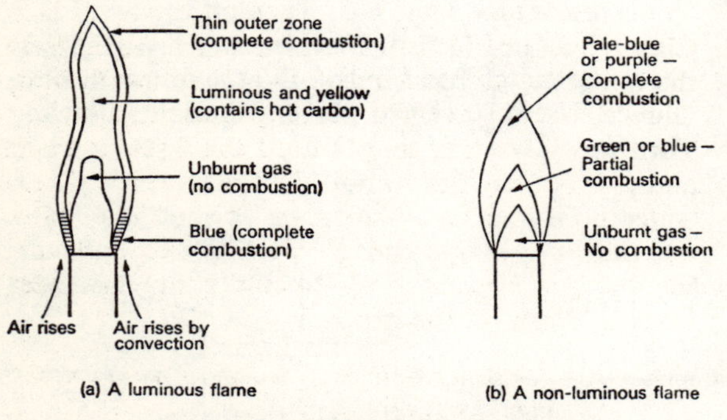

Fig. 1.6 Two different types of flame

The inner zones of both flames are cooler because they contain only gas (in the luminous flame) or a mixture of gas and air (in the non-luminous flame). Therefore when a substance is being heated it should not be placed near the bottom of a flame.

The lists below show the differences between a luminous flame and a non-luminous flame.

Luminous flame	*Non-luminous flame*
Larger; unsteady	Smaller; steady
Yellow	Greenish-blue or pale blue
Four zones	Three zones
Contains soot	No soot (carbon)
Not so hot	Hotter (about 1000 °C)
Always quiet	Can be noisy
Never burns back	Can burn back to the jet

The flames of a candle, a kerosine lamp and a coal or wood fire are luminous. The flames of gas-cookers, gas-fires, and gas water-heaters are non-luminous.

Burning back or striking back When this happens the flame travels down the chimney and burns noisily at the jet. A greenish noisy flame, which causes an unpleasant smell, may burn at the top of the chimney. Burning back occurs when the gas supply is too low and the air-holes are open too wide; the flame then passes down the chimney faster than the gas-air mixture passes up. Turn off the gas as soon as burning back occurs, close the hole taking care not to burn your fingers on the hot collar, and re-light the gas. Burners that use petrol gas, calor gas, Afrigas, etc. do not burn back; instead the flame floats off into the air when the air-holes are open too wide. Certain burners have wire gauze across the top to stop burning back.

STATES OF MATTER

Water can exist in three forms. Ice is a solid, water is a liquid, and steam is a gas or vapour. Water exists in the three states of matter—solid, liquid, and gas. All substances exist in one of these states and many, but not all, can exist in all three states. A gas has neither definite shape not definite size; a liquid has a definite size, but its shape can change; a solid has both definite shape and size. Usually a change in temperature can change the state of a substance:

$$\text{Solid} \underset{\text{freezing}}{\overset{\text{melting}}{\rightleftharpoons}} \text{Liquid} \underset{\text{condensing}}{\overset{\text{evaporating, boiling}}{\rightleftharpoons}} \text{Gas (vapour)}$$

Some substances exist in only one or two states. For example, wood and chalk are solids and cannot change to liquids or gases.

SOLUTIONS

Add a little common salt (sodium chloride) to water in a test-tube and shake. The salt seems to disappear. Taste the liquid formed. Clearly the salt is still present. The salt dissolves in the water, i.e. it mixes and the mixture is homogeneous—the same throughout. We say that the salt is soluble in water. The salt is a *solute*, the water is a *solvent*, and the mixture of salt and water is a *solution*. Solutions in which water is the solvent are called *aqueous* solutions.

Any liquid can dissolve some substances and therefore can act as a solvent. Water petrol, and ethanol—sometimes called ethyl alcohol or alcohol, are common solvents. Some substances, for example sand and glass, do not dissolve in water. These substances are *insoluble* in water.

A solute can be either a solid, a liquid, or a gas. Water, for example, can dissolve sugar (a solid), ethanol (a liquid), and air (a gas). Fizzy drinks are solutions in water of a gas together with solids and sometimes liquids. Some useful solutions are given below.

Solvent	*Solute*	*Use of solution*
Ethanol	Iodine	On cuts and wounds
Ethanol	Shellac	Lacquer or varnish
Trichloroethene	Grease	Dry-cleaning of clothes
Linseed oil	Pigments	Paint
Petrol	Rubber	Repair of bicycle tubes
Petrol	Grease	Cleaning clothes

SUSPENSIONS

Add a little clay, sand, chalk, powdered rice or bread to water in a test-tube and shake. Unlike common salt, these substances do not seem to disappear when the mixture is shaken. The mixture of the water with each solid is called a suspension; the particles of solid can be seen and they settle to the bottom on standing. The particles in a solution cannot be seen and do not settle on standing.

To find if tap-water is a solution of solids in water

1. Add water to a beaker until it is about one-quarter full. Place on a tripod and gauze, and boil.
2. Place a large clock-glass on the rim of the beaker and add tap-water to the clock-glass until it is about three-quarters full. Are there any solids suspended in the tap-water?
3. Steam from the boiling water evaporates the tap-water without breaking the clock-glass. When all the tap-water has evaporated, remove the flame, allow the clock-glass to cool, and then remove it and wipe the underside dry.

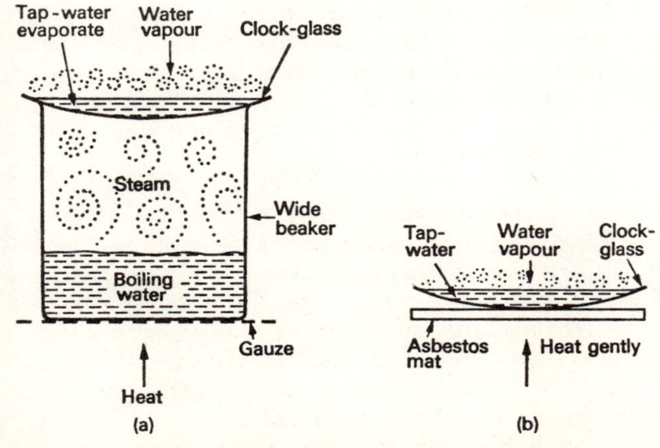

Fig. 1.7 Does tap-water contain dissolved solids?

4. Hold up the dry clock-glass to the light and look through it. Is there any solid residue?

Repeat this test with distilled water, rain-water, sea-water, river-water, etc.

If a pyrex clock-glass is used, it can be heated directly with a flame provided the glass is on an asbestos mat. Refer to Fig. 1.7b.

To find if tap-water contains dissolved gases

1. Place a short-stemmed funnel in a rubber stopper in the mouth of a large flask, at least 1000 cm^3 in in size. Make sure that the stem of the funnel is not below the base of the stopper so that gas can pass through the stem. Fill with tap-water to near the top of the funnel, and invert a test-tube full of tap-water in the funnel.

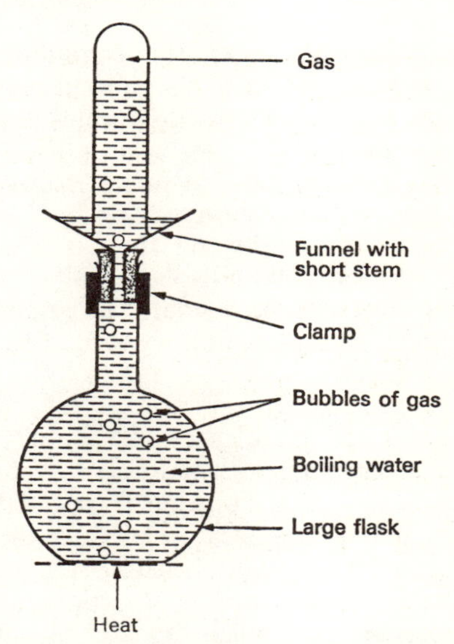

Fig. 1.8 Does tap-water contain dissolved gases?

2. Clamp the flask firmly on a tripod and gauze. Heat the water with a hot flame until it is almost boiling. Observe all the changes inside the flask.
3. Turn down the gas supply until the flame is small and just hot enough to boil the water gently. If the water boils too fast, the steam formed may blow out the stopper and funnel. Continue to boil until the steam pushes out any gas bubbles that have been formed. Turn off the gas.

To obtain pure common salt from a mixture of sand and salt

1. Add a mixture of sand and either common salt or rock salt (impure common salt) to a beaker. Add cold or warm water and stir until all the salt dissolves. The sand remains at the bottom of the beaker.
2. Fold a filter paper to form a cone, fit it into a funnel and wet it with water so that it sticks to the funnel.
3. Pour the salt solution through the paper. The level of the liquid must not rise to the top of the paper

because some solution then passes down between the paper and the funnel. The filtrate is the solution

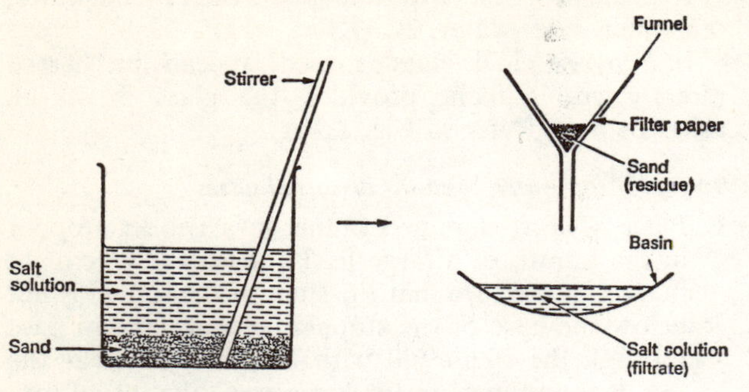

Fig. 1.9 Separating sand and salt

passing through the paper; the residue is the solid left on the paper. Most of the sand usually remains in the beaker unless the mixture is stirred well before filtering.

4. *Obtaining pure salt*. The filtrate is collected in an evaporating basin. Evaporate to dryness on a tripod and gauze. When the mixture is almost dry, pieces of hot solution and solid usually jump out of the basin and therefore a small flame must be used. Evaporation on a steam bath avoids this 'spitting' of the hot mixture.

SATURATED AND UNSATURATED SOLUTIONS

Add a little sugar to a beaker half full of water and stir well. An aqueous sugar solution forms. Add a little more sugar. If this dissolves on stirring, the solution is not saturated with sugar because more is able to dissolve; that is, the solution is unsaturated. Continue to add more and more sugar. Finally no more sugar dissolves. The solution is then saturated with sugar.

A saturated solution is one which can dissolve no more of the solute at the same temperature. An unsaturated solution is one which can dissolve more of the solute at the same temperature.

Most saturated solutions of solids become unsaturated when they are warmed and they can then dissolve more of the solid. Hot, unsaturated solutions of solids may become saturated when they cool.

To obtain crystals from solutions

Crystals that can be used include copper sulphate, potassium nitrate, potassium chlorate, sodium carbonate and chromium potassium sulphate.

(*a*) *Using a cold solution*

1. Add the powdered crystals to a beaker three-quarters full of cold water. Stir well. If all the solid dissolves, add more until some remains undissolved. The solution is then saturated.
2. Stop stirring and let the excess solid settle. Pour off (decant) the clear solution into a second beaker.
3. Leave the cold saturated solution in a warm place for a few days. Water evaporates slowly and some of the solid is deposited as crystals.

(*b*) *Using a hot solution*

1. Make a cold saturated solution of a solute as described above. Now add more solute—this is the mass of solute that will form crystals later in the experiment.
2. Warm the mixture until all the solute dissolves. It may be necessary to boil the water and even to add more water.

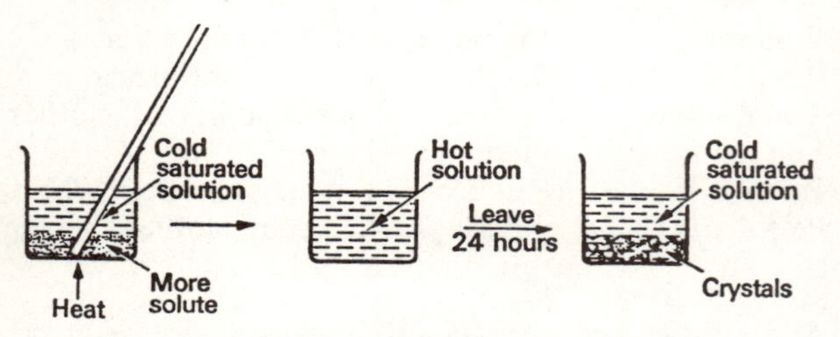

Fig. 1.10 To obtain crystals of a solute

3. The hot solution is saturated or almost saturated. Cover the beaker loosely with paper to keep out dust. Let the solution cool slowly and then stand for at least twenty-four hours. Crystals are deposited. Crystals which form slowly are usually larger and a better shape than those which are formed quickly.

To measure the solubility of a solid in water at room temperature

1. Prepare a saturated solution of the solid by adding the solid to a beaker of cold water, warming a little above room temperature, and stirring well. Some solid must remain undissolved, so add more solid if it all dissolves. Allow the solution to cool to room temperature, or cool it in a dish of water.
2. Weigh an evaporating basin.
3. Add the saturated solution to the basin, and make sure that no solid is added.
4. Weigh the basin and solution.
5. Evaporate the solution to dryness by using a small flame or a steam bath. If a large flame is used, some solution or solid may jump out of the basin.
6. Allow the dish and solid to cool and then weigh.

$$\text{Mass of basin} = x \text{ gram}$$
$$\text{Mass of basin} + \text{solution} = y \text{ gram}$$
$$\text{Mass of basin} + \text{solid (solute)} = z \text{ gram}$$
$$\therefore \text{ the mass of water} = (y - z) \text{ gram,}$$
$$\text{and the mass of solute} = (z - x) \text{ gram.}$$
$$(y - z) \text{ g of water dissolve } (z - x) \text{ g of solute,}$$
$$\therefore \text{ 100 g of water dissolve } \frac{100(z - x)}{(y - z)} \text{ g of solute.}$$

This is the solubility of the solute, because solubility

is the mass of solute which saturates 100 g of solvent at the particular temperature.

If the solubility is measured at other temperatures, the solution must be saturated at each particular temperature. For example, if the solubility at 50 °C is required, dissolve the solute in water at about 55 °C until the solution is saturated. Then allow the solution to cool to exactly 50 °C and pour it into the weighed basin. The rest of the method is exactly the same.

SOLUBILITY CURVES

These curves are obtained by plotting the solubility of solutes against the temperature. The solubilities of most solutes increase with temperature. Some curves are shown in the diagram of Fig. 1.11. Note that the solubility of sodium chloride varies very little with temperature because the curve is almost a horizontal line, and that the solubility of potassium nitrate rises rapidly with temperature because the curve slopes sharply upwards.

The composition by mass of a saturated solution at any temperature can be read off a solubility curve. For example, saturated sodium nitrate solution at 70 °C contains about 140 g of nitrate in 100 g of water, i.e. in 240 g of solution. At 30 °C, 100 g of water dissolve about 90 g of sodium nitrate, forming 190 g of solution.

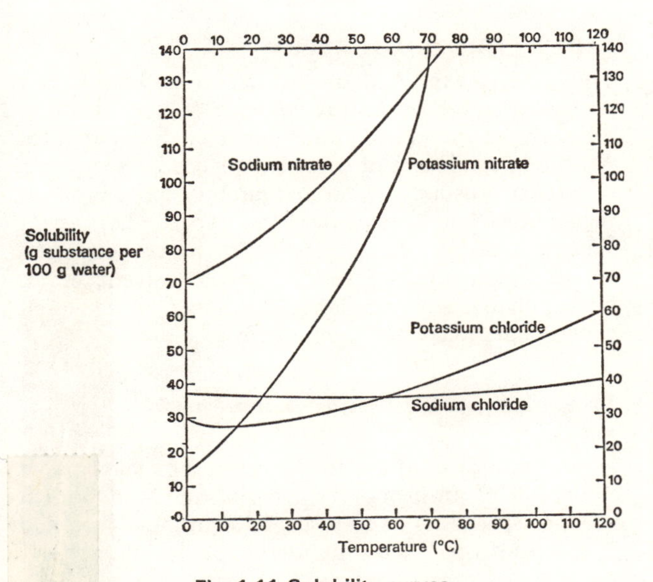

Fig. 1.11 Solubility curves

It follows that if 240 g of saturated sodium nitrate solution is cooled from 70 °C to 30 °C, the mass of nitrate which separates is (140 − 90) g = 50 g. Therefore, solubility curves can be used to calculate the mass of a solute which crystallizes from a given saturated solution when it is cooled.

DISTILLATION

To obtain pure water from sea-water or ink

1. Add sea-water or ink to a distillation flask. Add pieces of broken porcelain which ensure steady boiling during the experiment and prevent 'bumping' of the boiling liquid.
2. Connect a water-cooled condenser (Liebig condenser) to the flask. Pass a steady stream of cold water from a tap up the condenser because the water jacket does not remain full if the water passes down.

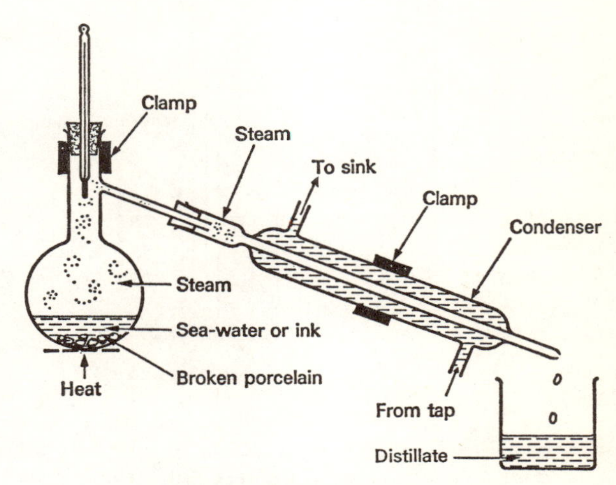

Fig. 1.12 Distillation apparatus

3. Boil the liquid in the flask. Collect the distillate in a beaker or flask. Read the temperature from a thermometer which has its bulb near the outlet of the distillation flask. This is the temperature of the vapour as it passes from the flask into the condenser.
4. The distillate is distilled water. Taste it. Mix the distilled water with the residue in the flask to show that sea-water or ink is formed again.

This experiment can be used to determine the boiling point of a liquid. A few cm^3 of the liquid must be available and a small distillation flask is used.

To study the distillation of crude oil (petroleum)

1. Add 2 to 4 cm^3 of crude oil to a test-tube. Push a loose plug of glass wool into the oil to help the oil to boil without bumping. Arrange the apparatus as in the diagram of Fig. 1.13 on p. 6.
2. Heat the oil gently. Collect the distillate in four parts (called fractions), e.g. up to about 70 °C, 70–120 °C, 120–170 °C, and 170–220 °C.
The actual temperatures are not important.
3. Place glass wool on four separate watch-glasses. Add a little of one of the fractions to a watch-glass and light it. Repeat with each fraction in turn and note how readily it burns and what kind of flame it forms.

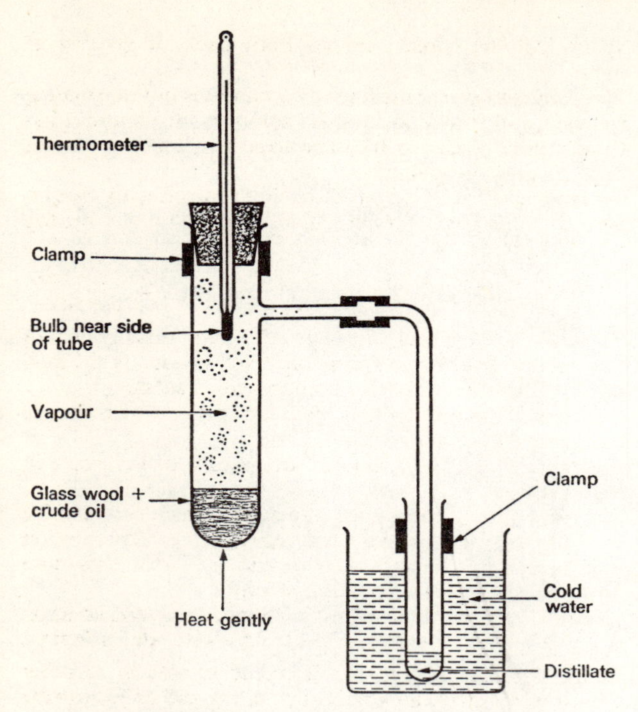

Fig. 1.13 Fractional distillation of crude oil

4. Mix a little of the four fractions in one test-tube and observe if the mixture resembles the original crude oil.

The experiment shows that crude oil is a mixture of several substances. The mixture actually contains petrol, kerosine, fuel oil, diesel oil and other useful materials; each of these is itself a mixture of many substances.

Distillation is the process of boiling a liquid and then condensing the vapour. The liquid that collects from the vapour is the distillate. Fractional distillation is the process of separating two or more liquids by distillation; the distillate is collected in fractions. These fractions are parts of the whole which boil at different temperatures.

To obtain ethanol by fractional distillation of ethanol and water

1. Mix ethanol (or methylated spirit) with two to three times its volume of water.
2. Arrange the apparatus as in the left of Fig. 1.14. The fractionating column is packed with pieces of glass, such as balls, rods or tubing. The diagram in Fig. 1.14 shows two other kinds of column on the right. They all have a large inner surface so that vapour passing up the column meets and mixes thoroughly with liquid falling down.
3. Boil the liquid steadily. Collect the fraction that boils over below about 90 °C. This liquid contains about 50 per cent ethanol.
4. Empty the distillation flask of the remaining liquid and place the fraction of distillate in the flask. Distil once again, collecting the fraction which passes over between 78 °C (the boiling point or b.p. of ethanol) and about 82 °C. This distillate contains about 95 per cent ethanol. It is not possible to separate all the water from ethanol by fractional distillation.

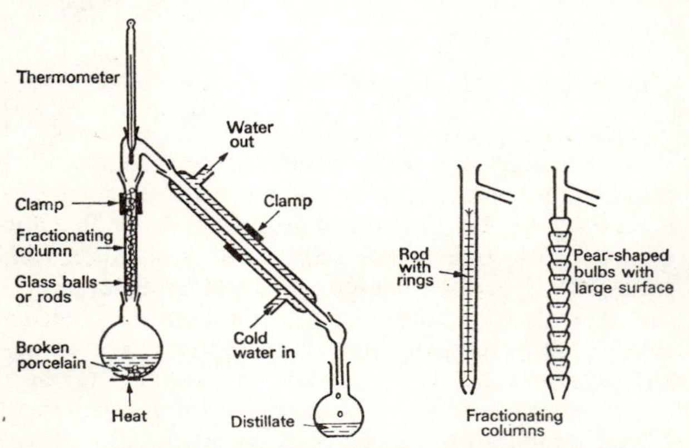

Fig. 1.14 Fractional distillation apparatus

Methanol (b.p. 66 °C) can be completely separated from a methanol-water mixture by fractional distillation.

PAPER CHROMATOGRAPHY

This term means 'colour-writing on paper'. It is a process used for separating coloured solutes by allowing them to move over filter paper at different rates. A little of a mixture of solutes is placed on the paper. A solvent, usually water, ethanol or propanone is then allowed to spread over the paper. The moving solvent dissolves the solutes which move with it. A solute which is more soluble in the solvent moves farther than one which is not so soluble. Some simple experiments make the method clear.

To separate various solutes by chromatography

Ink

1. Add one drop of ink to the centre of a filter paper on top of an evaporating basin, and wait until it stops spreading. Add a second drop, and then add a third drop. Wait a minute or so.

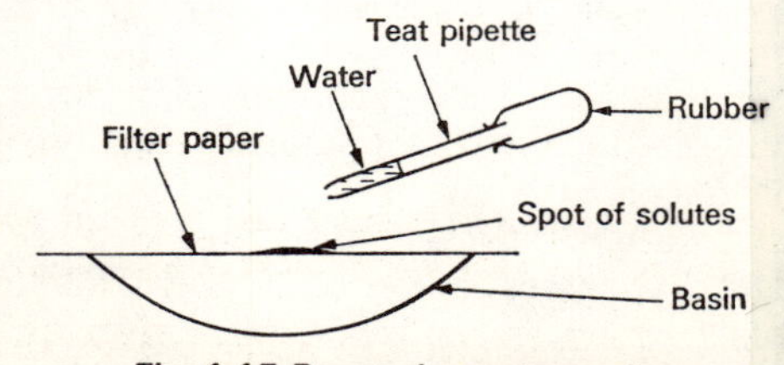

Fig. 1.15 Paper chromatography

2. Use a glass tube or teat pipette to add one drop of water to the centre of the ink spot. This water, which is the moving solvent, spreads across the paper by capillary attraction. This is a force which causes many liquids with the right physical properties to be drawn into the fine pores of a material. Glossy paper cannot therefore be used. When the water stops spreading, add a second drop. Continue until a ring of coloured solute almost reaches the edge of the paper.

The dyes, which are solutes in the ink, usually form well-separated rings. The test shows that ink is not a pure substance.

Grass

3. Cut grass or spinach into tiny pieces. Place the pieces in a mortar and crush with about 5 cm^3 of ethanol or propanone (acetone). A green solution forms. Decant the solution into a test-tube, leaving the solid material behind in the mortar.
4. Use filter paper to separate the two solutes into two coloured bands in the same way as the previous experiment with ink. Use drops of ethanol or propanone as the moving solvent. The green band is chlorophyll and the outer orange band is xanthophyll. Sometimes carotene forms an inner band, especially if benzene or methylbenzene is the solvent.

Screened methyl orange may also be used in this experiment. It contains two dyes in a mixture of ethanol and water. Separate the dyes as above using water as a moving solvent.

A filter paper with bands of separated solutes is called a *paper chromatogram.*

Chromatography has been used to detect the use of a different ink on forged documents, to identify a poison in blood, and to identify in a few minutes the amino-acids in a complex mixture. Mixtures of gases can be analysed easily by using special methods of chromatography. It is no longer necessary for the solutes to be coloured—the name chromatography is therefore not a good one. When two colourless solutes have been separated, their colours can be changed by treatment with suitable chemicals or by looking at them under ultra-violet or other light. Instead of paper, thin layers of material such as silica gel, aluminium oxide, or cellulose which are mounted on glass are often used because paper tends to fall apart when treated with strong chemicals.

SUMMARY

Flame: space in which gases burn and give out heat and light. Inside the flame there are unburnt gases; this space is cool.
Parts of Bunsen burner: chimney, collar, gas jet, and base.
Luminous flame: unsteady and larger; four burning zones; yellow colour; cooler; cannot burn back. It gives most light.
Non-luminous flame: most used because it is hotter, smaller, steadier, and forms no soot. Its colour is bluish and it has three zones of burning. It can be noisy and sometimes burns back. It gives little light.
Burning back: caused by too much air mixed with the gas. Result: flame burns noisily at the gas jet and not at the top of the chimney.

States of matter: solid, liquid, and gas; defined by their ability to change in shape and in size when compressed.

Solution: homogeneous mixture (the same throughout) of two or more substances, one of which is usually a liquid *solvent*; the others dissolve and are called *solutes.*
Suspension: liquid contains visible particles which can be mechanically separated.
Solutes: can be solids, liquids, or gases which dissolve in liquid solvent.
Solubility: the greatest mass of a solute which can dissolve in 100 g of solvent at any one temperature. The solution which contains this mass of solute and can dissolve no more at the same temperature is *saturated.*
Crystals: can be obtained by cooling a saturated solution or by evaporating excess liquid from a saturated solution.

Distillation: separation of a liquid from solutes or other liquids of higher boiling point by boiling and collecting the vapours as liquid *distillate* by cooling (condensing).
Fractional distillation: collection of distillate in *fractions* to obtain better separation of two or more liquids.

Chromatography: separation of solutes by their different rates of movement in a porous medium caused by a moving solvent.

QUESTIONS

1. Draw diagrams of the Bunsen flame: (*a*) with the air hole closed; (*b*) with the air hole open. How can the flame be made to strike back, and why does this happen? (C.)
2. Describe how you would show that the inside of a non-luminous Bunsen flame is not very hot. Suggest a method of finding out whether or not the inside of a candle flame contains unburnt wax.
3. Explain why the luminous Bunsen flame is rarely used in chemistry laboratories. Mention four differences between this flame and the non-luminous Bunsen flame.
4. Draw a labelled diagram of the apparatus you would use to obtain a few cm^3 of dissolved gas from a sample of tap-water. Describe the method of the experiment and state what you would expect to observe.
5. Describe carefully how you would prepare a saturated solution of sodium chloride at room temperature. How would you determine the solubility of the sodium chloride?
6. Explain how you would determine whether an aqueous solution of potassium nitrate was: (*a*) unsaturated, or (*b*) saturated, or (*c*) supersaturated. Outline briefly the experiments that must be done to obtain the solubility curve of a solute in water between 20 °C and 80 °C.

7. How would you obtain a reasonably pure sample of ethanol from wine or any other dilute solution of ethanol? (C.)
8. Describe two experiments you have done or observed in which solutes were separated by paper chromatography.
9. A reference book states that the solubility of copper sulphate in water at 15 °C is 19·0 g. What is meant by this statement? The solubility at 75 °C is 55 g. What mass of crystals should be deposited if 77·5 g of copper sulphate solution, saturated at 75 °C, is allowed to cool to 15 °C?
10. The following figures are for the solubility of potassium nitrate in water at various temperatures. Plot the solubility curve, with the temperature marked horizontally and the solubility vertically.

Temperature (°C)	20	30	40	60	70
Solubility (g/100 g)	33	50	65	105	138

11. Use the figures in the above question to calculate the mass of potassium nitrate in 100 g of saturated aqueous solution at 30 °C. What mass of crystals form when this solution is cooled to 20 °C?
12. How would you prove the truth of the statement that blackboard chalk is almost insoluble in water?

100 g of water at 15 °C dissolve at saturation 37 g of sodium chloride and 25 g of potassium nitrate; at 70 °C the corresponding masses are 38 g and 140 g per 100 g of water respectively.

100 g of a mixture of the above two salts, in equal proportions by mass, are shaken with 100 g of water at 70 °C until equilibrium is reached. The solution is filtered hot. The hot filtrate is slowly cooled at 15 °C and again filtered. The final filtrate is evaporated to dryness. What will be the masses and compositions of the three residues, assuming that no solution is left in either of the filter papers? (C.)

2 Action of Heat on Materials

WHAT HAPPENS WHEN MATERIALS ARE HEATED?

Candle-wax, kerosine, petrol, wood and coal are substances which change completely when they have been heated. They burn readily and seem to disappear. The metal filament in an electric lamp or in a torch bulb changes when it is heated, but when it is cool again no permanent change can be noticed. The iron or aluminium pots and pans used for cooking food also do not change after heating, or else they change very slowly.

Substances can be divided into two different classes by the effect of heat on them:

1. Substances which change on heating and which change back to the original substances on cooling.
2. Substances which change on heating and which do *not* change back to the original substances on cooling.

The first change is temporary and the second one is permanent. Although some substances cannot be made hot enough in the school laboratory to change at all, we can heat certain substances and observe carefully all the changes.

To study the effect of heat on substances

Heat a little of each material, gently at first, under a Bunsen flame and then allow to cool. Then heat the material strongly. At all times observe what happens. Note down which changes are temporary and which changes are permanent.

Sulphur Powder some roll sulphur in a mortar and half fill a test-tube with the powder. Use a small Bunsen flame about 3 cm high to melt the sulphur and then cool. Heat again strongly and test for any gas evolved with a lighted splint.

Iodine Add one crystal to a dry test-tube. Hold the tube horizontal and heat the crystal so that the top of the tube remains cool. Do not smell the poisonous vapour.

Ammonium chloride Heat the crushed powder as for iodine.

Naphthalene Crush a moth-ball to powder and repeat as for iodine.

Ammonium dichromate Heat one large crystal on a metal tray; alternatively powder the crystal and heat it on a flat asbestos paper. Remove the flame as soon as a change is observed.

Copper carbonate Heat as for sulphur or on asbestos paper.

Potassium manganate(VII) Heat as for sulphur; test any gas formed with a glowing splint.

Sodium chloride and *potassium chloride* Heat strongly.

METALS

Platinum wire Hold in tongs in a hot flame and allow to cool.

Nichrome wire Test similarly. Nichrome is used for electrical heating elements and contains nickel, chromium and iron.

Magnesium ribbon Hold in tongs with one end in a flame over a sheet of asbestos to protect the bench. Do not look directly at the burning metal as its light can damage your eyes.

Zinc dust Heat on asbestos paper. Break up the crust that forms, so as to expose the metal to the air.

Tin, lead, copper Heat each metal in a piece of porcelain. If it melts, stir with a steel needle.

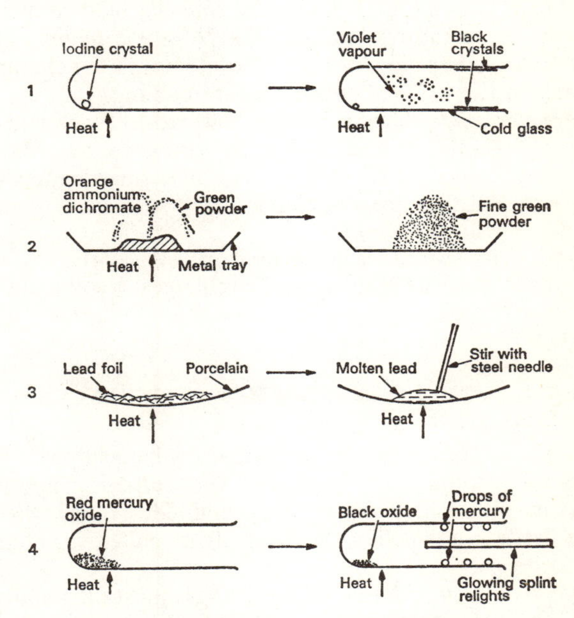

Fig. 2.1 Action of heat on some substances

OXIDES

Mercury (II) oxide, HgO Heat gently in a test-tube. Quickly pour the warm oxide on to paper and watch it cool. Replace in the test-tube and heat strongly as for iodine with the tube horizontal. Test for any gas formed with a glowing splint. Do not smell the poisonous mercury vapour.

Zinc(II) oxide Heat in a test-tube or on asbestos paper.

Lead (II) oxide, PbO Heat on asbestos paper. Lead compounds react with glass and spoil the test-tube.

Red lead oxide Heat in a test-tube and test for any gas formed with a glowing splint. The oxide is Pb_3O_4.

TEMPORARY CHANGES OBSERVED WHEN SUBSTANCES ARE HEATED

Sulphur The yellow solid melts to form an amber liquid which then turns brown and finally black. The liquid changes back to yellow solid when it cools.

Iodine does not melt. The black shiny crystal changes to a violet vapour. This changes back to tiny crystals when it reaches the cold glass of the test-tube.

Ammonium chloride also does not melt, but forms a white vapour which changes back direct to a solid when it cools.

Naphthalene melts and the liquid forms a vapour when strongly heated, but the vapour changes back directly to solid naphthalene when it is cooled slowly.

The change of a solid directly into a vapour or of a vapour directly into a solid is called *sublimation.* No intermediate liquid state is formed. Iodine, ammonium chloride and naphthalene sublime. The solid which condenses from the vapour is called a *sublimate.*

Sodium chloride and *potassium chloride* melt at very high temperatures (about 800 °C). *Platinum* and *nichrome* wires glow red hot or white hot. They change back to the original substances on cooling.

Mercury oxide changes colour from red to black when heated gently. White *zinc oxide* turns yellow when heated. Yellow *lead (II) oxide*, PbO, turns reddish-brown. All three oxides change back to their original colours when cool.

No new substance is formed in any of the above changes which are called *physical* changes. No chemical change occurs.

PERMANENT CHANGES WHEN SUBSTANCES ARE HEATED

Sulphur The black liquid formed when sulphur is heated strongly finally boils. The sulphur vapour burns with a blue flame when lighted. A gas with an irritating smell is evolved. It is called sulphur dioxide.

Ammonium dichromate The orange substance splits up to form colourless gases (these are steam and nitrogen) and a solid residue of a green powder that is chromium oxide.

Copper carbonate The green substance forms a black solid that is copper oxide and gives off a colourless gas which is carbon dioxide.

Potassium manganate(VII) forms a gas that re-lights a glowing splint.

Magnesium burns with a brilliant white light and forms a white powder called magnesium oxide.

Zinc, tin and *lead* melt when heated and slowly change to form coloured oxides.

Copper forms black copper oxide. If the copper is thick, the inside does not change because air cannot reach it.

Mercury (II) oxide splits up when heated strongly. It forms tiny drops of silvery mercury on the cold glass of the test-tube; a gas is formed which re-lights a glowing splint.

Red lead oxide splits up and forms the same gas which re-lights a glowing splint. It also forms a solid residue of lead (II) oxide which is also called litharge.

One or more substances are formed in each of the above changes which are permanent. They are called *chemical* changes. The original substance is not re-formed when the new substances cool.

Ammonium dichromate, copper carbonate, potassium manganate(VII), red mercury oxide and red lead oxide are all made up of parts which split up to form two or more substances. These five substances are all examples of *compounds* which change chemically by *decomposition.* Because heat is used for the change, the process is called *thermal decomposition.*

All the other substances are examples of *elements* which cannot decompose chemically into anything simpler. The change is a chemical *reaction.* In the above reactions the element combines with oxygen gas from the air. Elements combine together to form a *compound.* Oxides are compounds in which an element is combined with oxygen.

To study the action of heat on some crystals

Copper(II) sulphate crystals

1. One-quarter fill a hard-glass test-tube with dry, powdered copper sulphate crystals. Arrange the apparatus as shown in Fig. 2.2. Any vapours formed pass into the cooled test-tube and condense to a liquid.

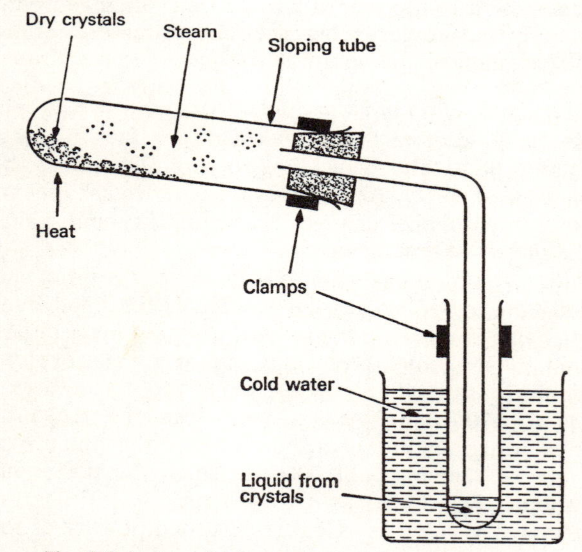

Fig. 2.2 Action of heat on copper sulphate crystals

2. Heat the blue powder until no more change is observed. Use a moderate flame as a very hot flame can split up the solid product and form a black powder. The heated tube slopes downwards slightly so that any liquid which condenses in it cannot run onto and crack the hot glass.
3. Remove the delivery tube out of the liquid in the cooled test-tube before removing the flame, otherwise some liquid may be sucked back into the hot tube as it cools and so break it.

4. Let the solid residue cool. Hold the tube in the palm of the hand and add a few drops of the liquid that has been obtained. Is the original substance re-formed?
5. Repeat the experiment with each of the following crystals:

    Cobalt chloride—pink colour
    Sodium carbonate—clear colour
    Sodium sulphate—clear colour
    Sodium chloride—white colour
    Potassium chloride—white colour

CRYSTALS WITH WATER AND CRYSTALS WITHOUT WATER

A crystal is a solid with a regular geometrical shape. Some, called *hydrates*, contain water of crystallization or hydration. These crystals lose their shape when they are heated and change to a powder as the water comes off. Some crystals, such as sodium and potassium chlorides, do not contain water and are *anhydrous* which means 'without water'. The powders formed when hydrates are heated are also anhydrous.

Some substances have no regular shape. They are called non-crystalline or *amorphous*. The anhydrous substances left when hydrates are heated are amorphous. Charcoal, lime, soot, powdered glass and clay are other examples of amorphous substances.

The chemical and common names of some hydrates are given here. Pupils who have already studied chemical formulae may find these a help also. The dot in each chemical formula of a hydrate means that the water is loosely combined and comes off readily on warming.

Chemical name	*Common name*	*Formula*
Copper(II) sulphate	Blue vitriol	$CuSO_4.5H_2O$
Iron(II) sulphate	Green vitriol	$FeSO_4.7H_2O$
Zinc(II) sulphate	White vitriol	$ZnSO_4.7H_2O$
Sodium carbonate	Washing soda	$Na_2CO_3.10H_2O$
Sodium sulphate	Glauber's salt	$Na_2SO_4.10H_2O$
Magnesium sulphate	Epsom salt	$MgSO_4.7H_2O$

CHEMICAL CHANGES IN WORDS

We can write equations in words to represent some of the changes already observed:

Sulphur + oxygen = sulphur dioxide (colourless)

Copper carbonate (green) = carbon dioxide + copper oxide (black)

Magnesium + oxygen = magnesium oxide (white)

Red mercury oxide ⇌ oxygen + mercury (silvery)

Red lead oxide ⇌ oxygen + lead oxide (yellow when cold)

Copper sulphate-5-water (blue) ⇌ water + copper sulphate (white)

Cobalt chloride-6-water (pink) ⇌ water + cobalt chloride (blue)

Sodium carbonate-10-water (clear) ⇌ water + sodium carbonate (white)

Sodium sulphate-10-water (clear) ⇌ water + sodium sulphate (white)

Six of these changes can be reversed. For example, the original hydrates can be re-formed by adding water to the anhydrous solid. These changes are reversible and the arrows ⇌ are used to show this.

No test was done in the above experiments with hydrates to prove that the colourless liquid formed is water. This is done by finding the boiling point of the liquid which, if it is water, boils at 100 °C. See p. 25 in chapter 5.

THE MASS OF AIR

Our next experiments find out if air has mass and determine any changes in mass when substances are heated. If a substance loses weight, something must pass from it into the air. If a substance gains weight, something must pass to it from the air, or from the flame or the container.

To find if air has mass

1. Choose a large plastic container which is fitted with a tap. The container should be at least 25 cm in length, width and breadth. Force air into the container by means of a foot pump.
2. Weigh the container filled with air under pressure.
3. Fill a large flask, at least 1000 cm³, with water and invert it on a stand in a trough of water. Fit rubber tubing to the container.

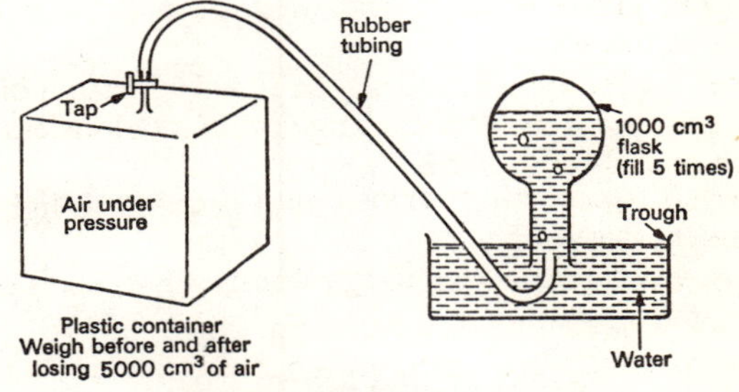

Fig. 2.3 Finding if air has mass

4. Open the tap slightly so that the air passes from the container into the flask. When the flask is just full, close the tap, fill the flask with water once again, and then fill it with air from the container. Repeat this several times until at least 5000 cm³ of air has been let out of the container.
5. Weigh the container again. Any loss is the weight of the air let out of the container.

Mass of container + air at start = p g
Mass of container + air at finish = q g
Volume of air let out of container = r dm^3
Mass of air let out of container = $(p - q)$ g

$$\therefore \text{ the mass of 1000 cm}^3 \text{ of air} = \frac{(p-q)}{r} \text{ g}$$

To find if there is a change in mass when substances are heated in air

Magnesium

1. Place asbestos paper on the bottom of a crucible to prevent the hot magnesium reacting with the porcelain. Weigh the crucible with its lid and the asbestos.
2. Add magnesium powder or clean magnesium ribbon to the crucible. Weigh the complete crucible and its contents.
3. Place the crucible on a pipe-clay triangle. Adjust the lid so that you can just see inside the crucible. Heat strongly until the magnesium starts to burn. Use tongs to lift the lid about 1 cm every ten seconds so that air can enter but very little white fumes of magnesium oxide can escape.

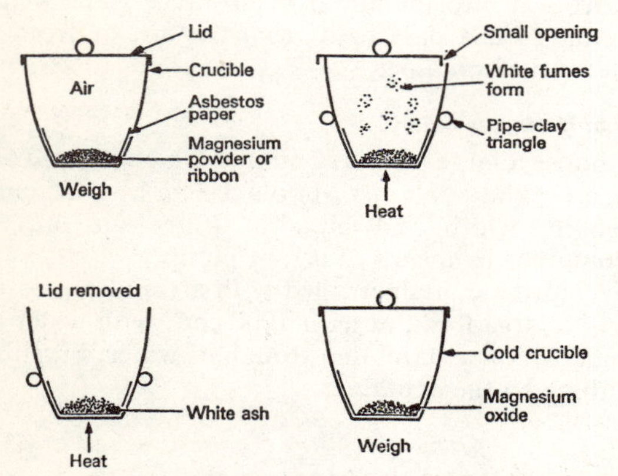

Fig. 2.4 Is there a change in mass when magnesium burns?

4. When the reaction seems complete, remove the lid and heat strongly.
5. Allow to cool. Weigh the complete crucible and contents again.

Mass of crucible, lid and asbestos paper = k g
Mass of crucible, lid, asbestos and magnesium = l g
Mass of crucible, lid, asbestos and oxide = m g
$\therefore$ increase in mass of magnesium = $(m - l)$ g
Original mass of magnesium = $(l - k)$ g

Repeat the experiment with copper, dry sand, sodium chloride, red lead oxide, potassium manganate(VII), and copper sulphate crystals. Which substances gain in mass, which lose, and which do not change?

Potassium manganate(VII) and red lead oxide are two substances which lose mass. The next experiment shows what happens to the substance lost.

To collect any gas formed when a solid is heated

1. Add potassium manganate(VII) (or red lead oxide) to a test-tube. Cover it with a loose plug of glass wool, which prevents tiny pieces of solid passing out of the tube during heating.
2. Arrange the apparatus as shown in Fig. 2.5. The piston should be left fully in the syringe.

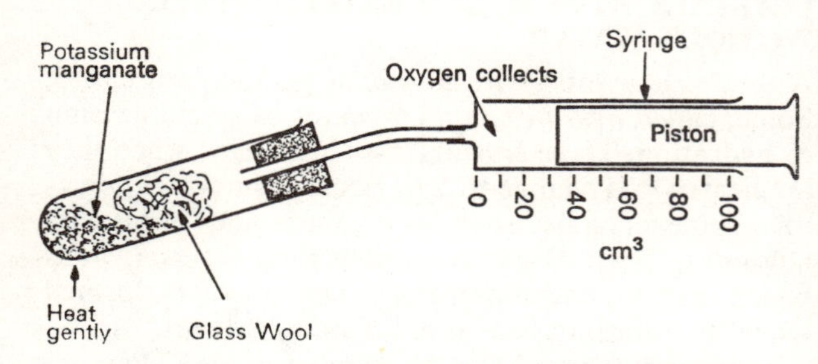

Fig. 2.5 Collecting the gas from heated potassium manganate(VII) (or red lead oxide)

3. Heat the solid. Any gas formed forces the piston out and collects in the syringe. Test the gas with a glowing splint.

To find if heated copper uses up air

1. Add copper that has been freshly made from black copper oxide (because this copper is the most reactive) to a test-tube. Connect the test-tube to a gas syringe containing about 50 cm^3 of air. Note the exact volume.
2. Heat the copper vigorously for some time.
3. Allow the apparatus to cool. Note the volume of gas left in the syringe. The decrease is the volume of air used. If necessary, weigh the test-tube and copper both before and after heating. The change in mass is the mass of gas used up.
4. Disconnect the syringe and expel the gas slowly over a glowing splint. Is the gas different from air?

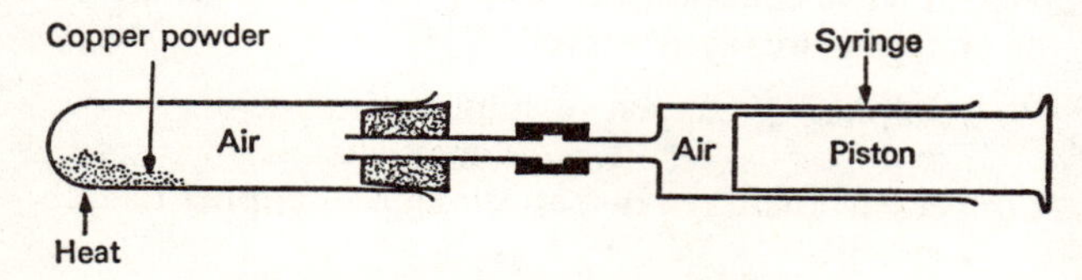

Fig. 2.6 Does heated copper use up air?

Volume of air in syringe at start = a cm^3
Volume of gas in syringe at finish = b cm^3
Volume of air used up = $(a - b)$ cm^3

If the method of weighing is used, then:

Mass of test-tube + copper at start = c g
Mass of test-tube + product at finish = d g
Mass of air used up = $(d - c)$ g

REACTIONS OF ELEMENTS WITH EACH OTHER

Many elements are able to combine with each other to form a compound. The formation of a compound from its simpler parts is called *synthesis*. For example, magnesium oxide is synthesized by burning magnesium metal in oxygen from the air; copper oxide is made by synthesis when copper is heated in air.

The element sulphur provides further examples of synthesis because it is able to combine with many metals. A compound that contains sulphur and one other element is called a sulphide; this word means 'combined with sulphur'.

In the next experiments we shall learn how to make a sulphide, how its two parts are different from each other and how a sulphide compound is different from a physical mixture of the two parts of the compound.

To study the reaction between sulphur and a metal

Zinc

1. Mix zinc dust with an equal volume of powdered roll sulphur in a pestle and mortar.
2. Fold a sheet of asbestos paper into a cone and half-fill it with the mixture.

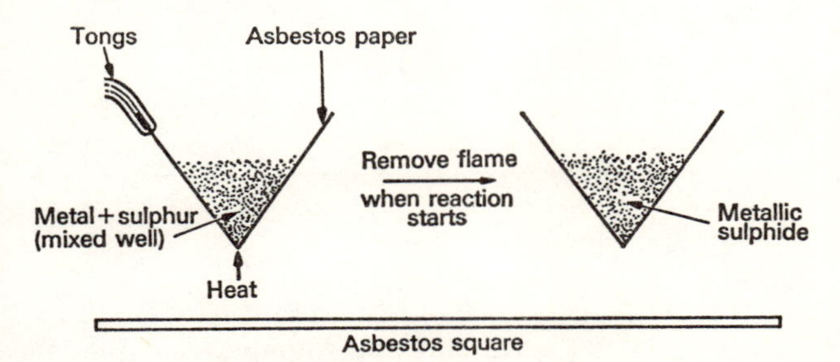

Fig. 2.7 The reaction between a metal and sulphur

3. Use tongs to hold the asbestos. Heat the mixture until reaction begins and then remove the flame at once. Hold the mixture over an asbestos square to protect the bench.
4. Let the product cool and powder it in a mortar.

Other Metals

Copper powder or non-greasy iron filings can be used equally easily in the laboratory, but the reaction with zinc is the most vigorous.

Zinc + sulphur = zinc(II) sulphide (white)
Copper + sulphur = copper(II) sulphide (black)
Iron + sulphur = iron(II) sulphide (black)

Note Iron filings are usually greasy. The grease prevents the metal reacting easily with chemicals like sulphur, air or moisture. The grease should be removed by leaving the filings in methylated spirit for twenty-four hours and then allowing to dry in air.

To study the differences between sulphur and some metals

1. Take twelve test-tubes. To three of the tubes add powdered roll sulphur to a depth of 0·5 cm. Do the same with non-greasy iron filings, copper powder and zinc dust.
2. Half fill one tube of each substance with water. Shake well and allow to settle.
3. Add dilute acid (hydrochloric or sulphuric) to one tube of each substance; add a few drops of concentrated acid if no reaction occurs. If a gas forms, cover the mouth of the tube with paper until the tube is full of gas. Test the gas with a flame.
4. Half fill one tube of each substance with carbon disulphide. Shake and allow to settle.
   Note Carbon disulphide vapour is both poisonous and burns very easily just like petrol. Use it with care and make sure that there are no flames alight on the bench.
5. Observe the following differences between the metals and sulphur:

	Metal	*Sulphur*
Water	Metal sinks.	Most sulphur sinks, but some floats.
Dilute acid	Reaction occurs; a gas is evolved which burns with a 'pop'.	No reaction.
Carbon disulphide	Does not dissolve.	Dissolves readily to a yellowish solution.

To find any differences between a compound and a mixture of sulphur and a metal, e.g. zinc, copper, or iron

1. Look at them both carefully through a lens.
2. Add the sulphide and the mixture to separate test-tubes containing water. Shake and allow to stand.

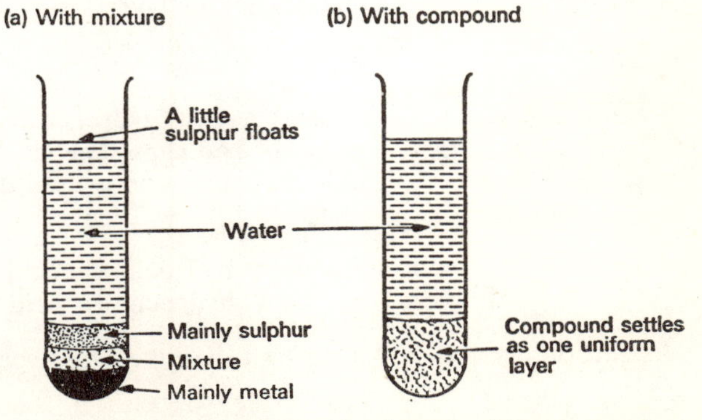

Fig. 2.8 Action of water on a mixture and a compound

3. Add dilute acid (hydrochloric or sulphuric) to the sulphide; add a few drops of concentrated acid if little or no reaction occurs. Cover the mouth of the test-tube with paper until the tube fills with gas. Test the gas with a flame or a burning splint. Repeat the test with the mixture.

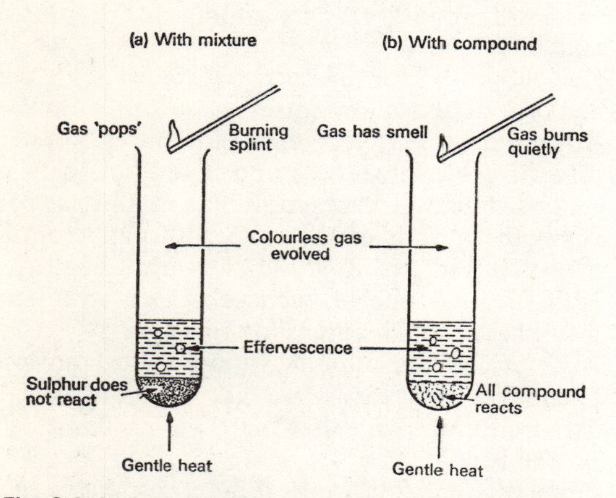

Fig. 2.9 Action of acids on a mixture and a compound

4. Cover one end of a bar magnet with paper to prevent substances sticking to it. Place this end of the magnet first in the mixture and then in the compound. Observe any difference. This test is only for iron and its compounds because zinc and copper are not magnetic.

5. Add carbon disulphide (take care with this poisonous inflammable liquid) to some of the compound in a dry test-tube. Shake well and then filter through a *dry* filter paper; carbon disulphide does

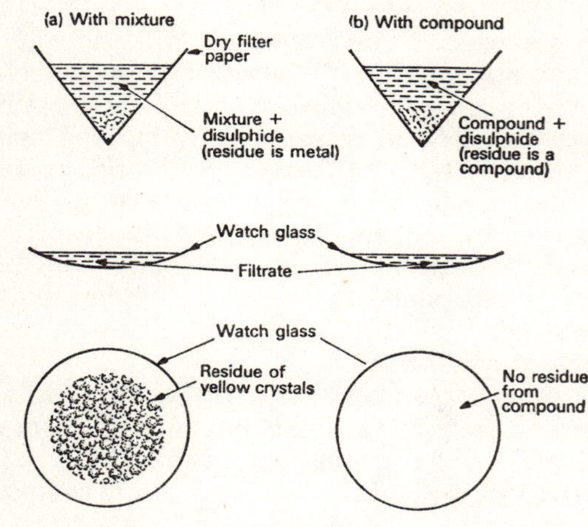

Fig. 2.10 Action of carbon disulphide on a mixture and a compound

not pass through wet paper because it does not mix with water. Collect the filtrate on a watch-glass and let it evaporate in the open. Repeat the test with the mixture.

Differences between a compound and a mixture

	Compound (zinc sulphide)	*Mixture* (zinc and sulphur)
1. Appearance	Dirty white. It is the same throughout (homogeneous) and no zinc or sulphur is visible.	Bluish and yellowish, i.e. the colours of both zinc and sulphur are present. Both substances are visible through a lens.
2. Water	Settles as one layer.	The denser zinc tends to settle first before the less dense sulphur, but separation is only partial.
3. Dilute acid	Forms a gas (hydrogen sulphide) with a smell of bad eggs. The gas burns quietly. All the compound is used up if excess acid is used.	Forms a gas (hydrogen) which burns with a 'pop'. All the zinc is used up if excess acid is used, but the sulphur does not react.

zinc sulphide + hydrochloric acid = hydrogen sulphide + zinc chloride
zinc + hydrochloric acid = hydrogen + zinc chloride

4. Carbon disulphide	Does not dissolve.	The sulphur dissolves. If excess disulphide is used, all the sulphur dissolves. The zinc does not dissolve.

ELEMENTS, COMPOUNDS AND MIXTURES

An element is any substance which cannot be split up by chemical means into simpler substances. The name was first suggested by Boyle in 1661. We now know that almost 90 elements exist in the world, and possibly in the universe. Other elements have been made artificially in science laboratories and a total of 105 has been recognized.

The most important element is oxygen, for half of the earth's crust and atmosphere consists of this element, either as the gas or as compounds containing it. Another quarter of the earth's crust is the element silicon which is present in compounds in clay, sand and many rocks.

Two or more elements joined together chemically form a *compound.* Two or more substances which are mixed but not joined together chemically form a *mixture.* The substances in a mixture may be elements or compounds.

Every substance we meet in everyday life is really a mixture because it is very difficult to obtain a pure element or compound except under special conditions in laboratories.

SUMMARY

Physical change: a change in which no new chemical substance is formed.
Examples of physical changes: melting, freezing, boiling, condensing, mixing, dissolving, changing of colour, subliming.
Sublimation: change of solid directly into a gas on heating, or change of gas directly into a solid on cooling. Sublimate is solid formed from gas.

Chemical change: a change in which one or more new chemical substances are formed by a reaction.
Examples of chemical changes: burning, rusting, exploding, and all chemical reactions.
Element: substance which cannot be split into simpler substances by chemical means.
Compound: substance which consists of two or more elements chemically combined; for example, iron sulphide.
Mixture: consists of two or more substances which are mixed together physically but not chemically combined; for example, iron and sulphur.
Decomposition: the splitting up of a compound to form two or more substances. Thermal decomposition is by means of heat.
Synthesis: the making of a compound directly from its elements by chemical reaction; e.g. *oxides* from oxygen and a metal, *sulphides* from sulphur and a metal.

Crystal: solid that has a regular geometrical shape; usually formed from a solution or the liquid state.
Amorphous solid: one with no definite shape; non-crystalline; e.g. soot, lime, clay, *anhydrous* copper sulphate.
Water of crystallization or water of hydration: the definite amount of water which combines with some compounds when they form crystals from aqueous solutions.
Anhydrous substance: one without water of crystallization.
Hydrate: compound that contains water of crystallization.

QUESTIONS

1. Describe what happens when each of the following substances are heated gently in a test-tube: sulphur, mercury oxide, sodium chloride.
2. What further changes occur when the three substances named in question 1 are heated separately in a hot flame?
3. Mercury is an element and mercury oxide is a compound. What is meant by this statement? Describe briefly how you would obtain a specimen of mercury if you were given about 1 g of mercury oxide.
4. Describe all the changes which can be observed when (*a*) copper sulphate crystals and (*b*) cobalt chloride crystals are heated in a test-tube.
5. Water is added drop by drop to white copper sulphate powder in a beaker. Describe and explain briefly what happens.
6. What do you understand by the following terms? In each case, give one example.

   Sublimation Decomposition Synthesis
7. Suggest a simple experiment by which you could show that less than half of sodium carbonate crystals consists of sodium carbonate. Show clearly what weighings you would make.
8. Give three characteristics of chemical change. Illustrate each of the points you mention by reference to the reaction between powdered sulphur and iron filings. (C.)
9. Define an element. The elements in a mixture of elements can be separated from one another by physical methods but the elements in a compound can only be separated from one another by chemical means and even then the elements may not be actually isolated. Illustrate this statement by reference to a mixture of iron and sulphur and the compound iron sulphide, FeS. (C.)
10. You are given a mixture of clean iron filings and sulphur powder and also some iron sulphide powder, FeS. Describe in outline experiments you would perform (using water, dilute acid and carbon disulphide) which would enable you to distinguish clearly between the mixture and the compound. Mention carefully the results of each experiment.

3 Action of Electricity on Materials

Our experiments in chapter 2 showed that heat energy from a Bunsen burner helps to produce physical and chemical changes. Therefore it is possible that electrical energy from a battery or mains supply may do the same. Electricity passes readily through copper wire which is used to bring the electrical supply to houses, and it also passes through the metal case of an electric torch from the battery through the lamp filament, the switch and the case, and back to the bottom of the battery.

The metals are able to conduct electricity. Rubber or plastic is used to cover copper wire because these substances are non-conductors of electricity. Air is also a non-conductor, otherwise batteries would never remain charged because electricity would pass from them through the air. However, lightning passes through the atmosphere; therefore air can conduct electricity under special conditions.

Our first experiment is to find which substances conduct electricity when joined to a battery.

To find which substances conduct electricity

1. Connect two carbon or steel rods to a 6-volt battery (or a 6-volt d.c. supply) through a switch and an ammeter or an electric lamp which will show when an electric current is flowing.

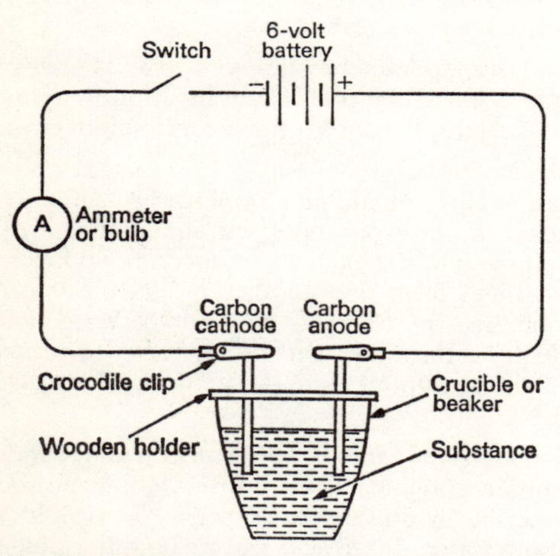

Fig. 3.1 Which substances conduct electricity?

2. Add distilled water to a crucible, a beaker or a test-tube and let the rods dip into it. Press the switch down. Does the ammeter show a reading or does the electric lamp light?
3. Repeat the test with various liquids in place of the distilled water. Wash and clean the rods after use in any liquid, and also clean the crucible. Suitable liquids are:
   (*a*) ethanol or petrol,
   (*b*) dilute sulphuric acid or sodium chloride solution,
   (*c*) mercury.
4. Repeat the test with various solids in powder form or in small pieces. Suitable solids are:
   (*a*) sulphur, sugar, or polythene,
   (*b*) iron, lead, or aluminium,
   (*c*) sodium chloride, lead(II) bromide and iodide, or potassium iodide.
5. If any of these solids is a non-conductor, add distilled water to it and observe if the solution or mixture is a conductor.
6. Repeat the experiment with molten lead bromide (or lead iodide), molten potassium iodide, and molten sulphur. Refer to pp. 17 and 18.

Use your observations to classify substances into the groups shown below.

(*a*) those which conduct electricity but do not decompose,
(*b*) those which conduct electricity and seem to change,
(*c*) those which only conduct when molten, and
(*d*) those which conduct when in solution.

ELECTROLYTES AND NON-ELECTROLYTES

Ethanol, petrol, sugar, sulphur, and polythene do not conduct electricity. Sodium chloride, lead bromide, lead iodide and potassium iodide do not conduct electricity when solid; they are all conductors when in aqueous solution and when molten. The solution seems to decompose because gas is evolved at the carbon rods. Mercury, iron, and other metals are conductors, but they do not decompose.

Substances which conduct electricity when they are molten or in solution and are decomposed by it are called *electrolytes*. Substances which do not conduct electricity when molten or in solution are non-electrolytes. Substances which do conduct electricity, but are *not* decomposed by it, for example mercury, are also non-electrolytes. Water is a non-electrolyte under normal conditions, although it can conduct a little if a very high voltage is used.

The process of decomposing an electrolyte by an electric current is called *electrolysis*. The apparatus used includes a beaker or crucible and electrode rods and is called a voltameter. The carbon or metal rods are *electrodes*. Carbon and platinum are the best electrodes because they do not react readily with electrolytes or with the substances formed by electrolysis. The positive electrode is the anode, and it is connected to the positive terminal of the battery. The negative electrode is called the cathode.

To study the electrolysis of acidified water

1. Arrange the apparatus as in Fig. 3.2. The carbon electrodes pass through the stopper of a wide vessel. The water is acidified with dilute sulphuric acid. The rheostat is used to alter the current and should be first set so that the current is the smallest possible. A platinum anode gives the best results.

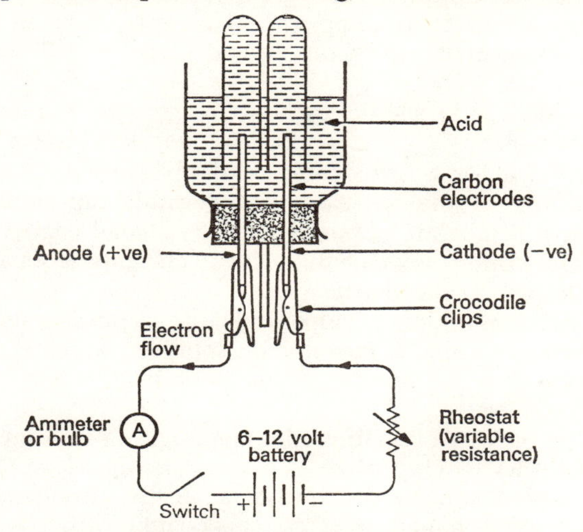

Fig. 3.2 Electrolysis of dilute sulphuric acid

2. Switch on the current. Adjust the rheostat until a steady stream of gases forms at the electrodes. Note the quantity of gases which collects in the test-tubes.
3. Test each gas with a burning splint and with a glowing splint.
4. Repeat the experiment using a copper anode. Observe any difference.

Note Solutions of dilute sodium chloride or dilute sodium hydroxide may equally well be used as electrolytes in this experiment.

To study the electrolysis of copper(II) sulphate solution

1. Set up the same electrical circuit as before, shown in Fig 3.3, and connect the circuit to a copper

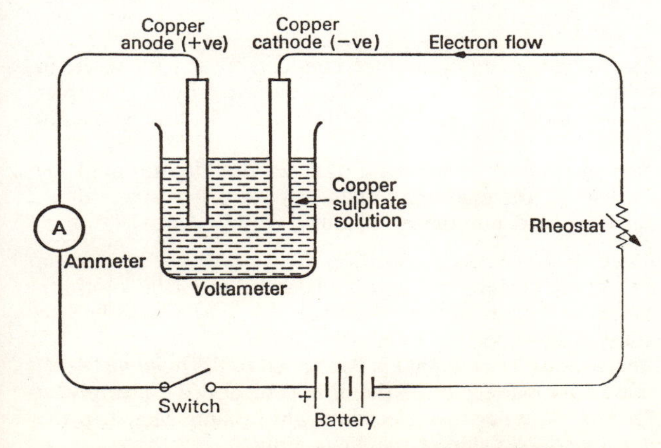

Fig. 3.3 Electrolysis of copper sulphate solution

voltameter. Use copper sulphate solution as electrolyte and copper electrodes in place of the carbon electrodes of the last experiment.
2. Clean the electrodes and note their appearance. Electrolyse the copper sulphate and again examine the electrodes.
3. If required, weigh each electrode both before electrolysis and after electrolysis. After electrolysis, clean the electrodes before weighing by dipping in distilled water and then in ethanol which quickly evaporates and leaves them dry.
4. Repeat the electrolysis with a platinum anode, and collect any gas formed, as shown in Fig. 3.4. Test the gas with a glowing splint.

Use the same voltameter (platinum anode) to electrolyse copper chloride solution. Test any gas formed by its smell, colour, and action on damp litmus paper.

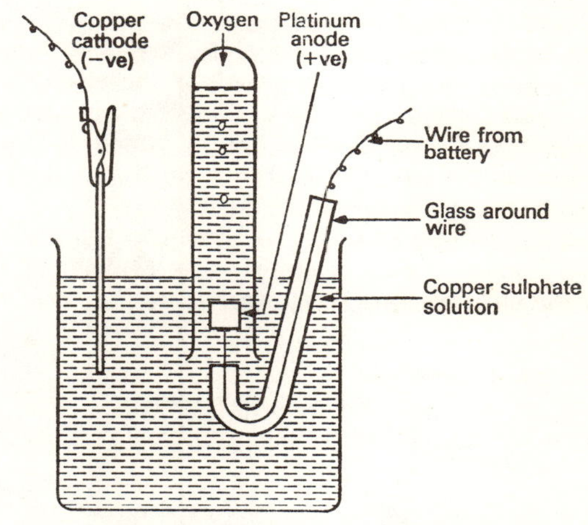

Fig. 3.4 Electrolysis of copper sulphate solution (platinum anode)

To study the electrolysis of molten compounds

1. Add crystals of lead(II) bromide to a depth of about 3 cm to a hard-glass test-tube. Heat them until they just melt. Electrolyse the molten compound, using carbon rods as electrodes. Refer to Fig. 3.5.
2. Note the colour of any gas formed and smell it carefully. Observe what is formed at the cathode.
3. Repeat with lead iodide crystals.
4. Repeat with potassium iodide crystals. The change at the cathode is not visible. Potassium bromide can also be used.

RESULTS OF ELECTROLYSIS

Water acidified with dilute sulphuric acid Hydrogen (H_2) forms at the cathode and oxygen (O_2) at the anode. The volume of hydrogen is twice that of the oxygen when the electrolyte is saturated with both gases. Other quantitative experiments show that the acid is not decomposed. The only chemical change is the decomposition of water which may be represented by the equation:

$$2H_2O = 2H_2 + O_2$$

If a copper anode is used, no oxygen is evolved. Instead, some of the copper dissolves from the anode and forms a blue solution. This shows that the material of the electrodes can influence the results of electrolysis.

Dilute sodium chloride and sodium hydroxide The results are the same as those with dilute sulphuric acid. Their electrolysis can also be called the 'electrolysis of water'.

Copper(II) sulphate solution Copper is transferred from anode to cathode, and no gas is evolved. The mass of copper deposited on the cathode equals the mass removed from the anode, and the concentration of the electrolyte does not change.

If a platinum anode is used, oxygen is evolved at the anode and copper is deposited at the cathode. This copper comes from the electrolyte, and the copper sulphate is gradually decomposed. If the electrolysis is done long enough, the blue colour of the electrolyte disappears and a colourless liquid, which is dilute sulphuric acid, is left. The acid then forms oxygen and hydrogen.

Copper(II) chloride solution A greenish gas with an irritating smell, which is chlorine, is evolved at the anode; it bleaches litmus. Copper is deposited at the cathode.

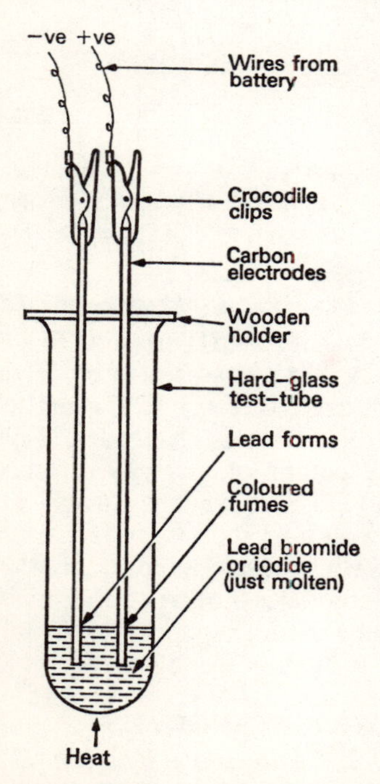

Fig. 3.5 Electrolysis of fused lead bromide or iodide

Molten lead(II) bromide or iodide At the anode, a brown gas with an irritating smell (bromine) or a violet vapour (iodine) is formed. Lead is formed at the cathode.

Molten potassium iodide Violet iodine vapour forms at the anode. Potassium forms at the cathode but it cannot be seen.

To plate various articles with copper

1. Remove impurities and grease from the surface of the article by rubbing with steel wool, dipping in sodium hydroxide solution, and washing well with distilled water.
2. Make the article the cathode in a copper sulphate voltameter, with a copper anode. Pass a small current for a few minutes and revolve the article slowly so that all sides face towards the anode. Best results are obtained with very small currents. The copper deposited by large currents is easily rubbed off the article.

Nickel ammonium sulphate is the electrolyte used in nickel plating. Zinc sulphate solution, which contains a little sulphuric acid and boric acid, is used in zinc plating.

The process of coating one metal with a thin layer of another by electrolysis is called *electroplating*. The metal that is coated is always made the cathode. In this way iron and steel can be coated with silver, chromium, nickel, tin or zinc; this improves the appearance of the article and prevents rusting.

Note: a voltameter is sometimes called a coulometer.

SUMMARY

Electrolyte: compound that conducts electricity when in solution or molten, *and* that is decomposed by it. Examples include most salts such as sodium chloride, and acids and alkalis.

Non-electrolyte: compound that is *not* decomposed by electricity; for example, ethanol, petrol, kerosine, sugar, sulphur, wax, polythene, mercury, naphthalene.

Electrolysis: the decomposition of an electrolyte by passing an electric current through it when molten or in solution.

Electrode: plate or rod that dips into the electrolyte to allow electric current to enter or leave.

Anode: the electrode that is connected to the positive (+ve) side of the battery or of the direct current (d.c.) supply.

Cathode: the negative electrode which is connected to the negative (−ve) side of the d.c. supply.

Results of electrolysis

(*a*) ***Platinum or carbon electrodes.*** Electrode does not dissolve.

	Electrolyte	*At cathode*	*At anode*
1. Dilute solutions in water:	Sulphuric acid, Sodium chloride, Sodium hydroxide	Hydrogen (2 volumes)	Oxygen (1 volume)
	Copper sulphate	Copper	Oxygen
	Copper chloride	Copper	Chlorine
2. Molten salts:	Lead bromide	Lead	Bromine (brown gas)
	Lead iodide	Lead	Iodine (violet vapour)
	Potassium iodide	Potassium	Iodine (violet vapour)

(*b*) *Copper electrodes.* Anode dissolves.

Electrolyte	*At cathode*	*At anode*
Dilute sulphuric acid (turns blue)	Hydrogen	No oxygen
Copper sulphate solution	Copper deposited	Copper dissolves

QUESTIONS

1. Explain what is meant by the following terms: (*a*) electrolyte; (*b*) non-electrolyte; (*c*) electrolysis. Describe how you would obtain a test-tube full of oxygen by electrolysis of a suitable solution.
2. Name two liquids or solutions which are electrolytes and two which are non-electrolytes. What is the essential difference between an electrolyte and a metallic conductor? What do you deduce from the fact that solid sodium chloride does not conduct electricity but aqueous sodium chloride is an electrolyte?
3. How would you obtain a small specimen of lead by electrolysis of a lead compound? Sketch the voltameter you would use and indicate the nature and polarity of the electrodes.
4. A direct current of electricity is passed, using platinum electrodes, through a dilute solution of copper sulphate until the solution becomes colourless, and for some time thereafter. Describe all that happens during the electrolysis. Briefly describe how a metal is copper-plated. (C.)

4 Reactions with Oxygen of the Air

When magnesium, copper and many other metals are heated sufficiently to burn in air, they increase in mass. Part, but not all, of the air is used up when these metals are heated in it. When magnesium burns, it forms a white powder which is magnesium oxide. Copper forms a black solid oxide.

Our first experiments will examine the products formed by other substances which react with air.

To study the products formed by a burning candle

1. Place a candle on a gas-jar cover. Light the candle and lower a dry gas-jar over it. What happens?
2. Add a little white copper sulphate powder, or blue cobalt chloride, to the inside of the jar to show if the liquid there contains water. If the powder changes colour, water is present.
3. Light the candle again and lower a second gas-jar over it. Remove the jar, add lime water and shake to test for carbon dioxide. If this gas is present, the lime water turns milky.

To study the reaction between air and aluminium

1. Half fill a porcelain basin with mercury.
2. Clean aluminium foil by rubbing it with sandpaper to remove any film of oxide on it. Use cotton wool to rub the clean foil with mercury for some time. The mercury forms an amalgam with the aluminium on the surface.
3. Hold the metal in your hand for one or two minutes and note what happens.

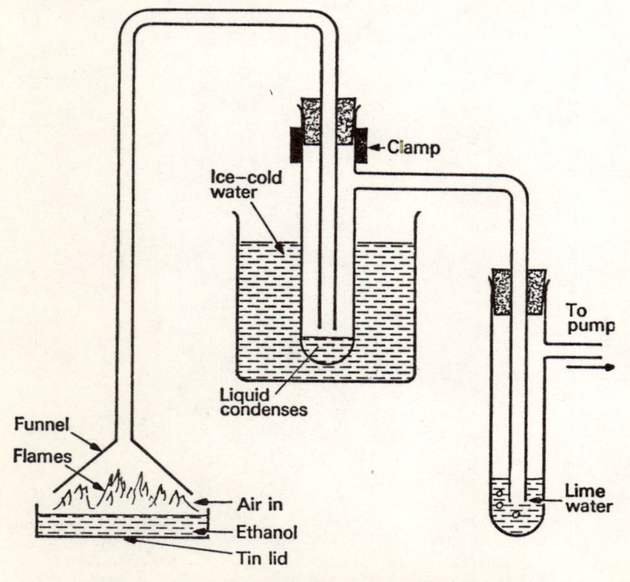

Fig. 4.1 Products formed by burning ethanol

To study the products formed by burning ethanol

1. Add a few cm^3 of ethanol or other liquid that burns easily to a tin lid or clock-glass. Arrange the rest of the apparatus as shown in Fig. 4.1. A pump sucks the products of combustion (burning) through a cooled tube and then through a tube containing lime water.
2. If a liquid condenses in the cooled tube, test it with white copper sulphate or blue cobalt chloride.

To study the products of respiration

1. Breathe on a sheet of glass and add white copper sulphate powder to the mist that forms.
2. Breathe into a gas-jar, add lime water at once and shake. Add lime water to a jar of air and shake. Observe any difference between the two jars. Shaking the jar of air with lime water is a control experiment which allows a comparison between it and the other jar.

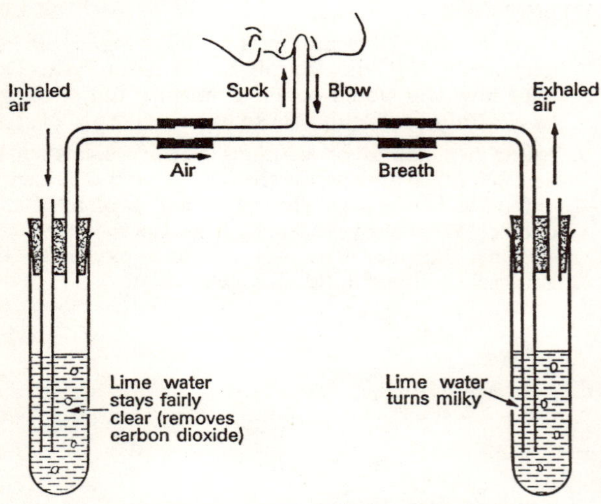

Fig. 4.2 Exhaled air contains more carbon dioxide than ordinary air

The diagram in Fig. 4.2 shows an apparatus used to show that exhaled air contains more carbon dioxide than ordinary air. The lime water on the left removes carbon dioxide from the air before it enters your lungs. Observe how the delivery tubes are arranged so that air enters through one boiling tube and breath passes out through the other.

To find if air is used up when a candle burns

1. Stick a candle to a shelf in a trough of water. Light it.
2. Lower a dry gas-jar quickly over the burning candle. Observe how the water level changes.
3. Repeat the experiment by burning ethanol, petrol or paper on a small tin lid floating on the water or placed on top of the shelf.

The part of the air which is used up is called active air and the part which is not used up is inactive air.

Active air is oxygen; inactive air is nitrogen with about 1 per cent of other gases.

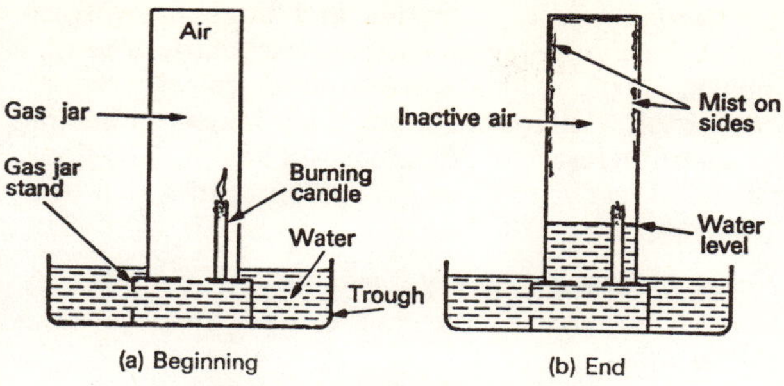

Fig. 4.3 A burning candle uses up air

To find if there is a change in mass when iron rusts

1. Add non-greasy iron filings to a clock-glass. Weigh the glass and filings.
2. Fill the glass carefully with distilled water and leave for several days. Add more water from time to time if it evaporates. The iron rusts (rusting is a slow combustion).
3. Leave the clock-glass and rust in bright sunshine in order to dry them, or dry in a moderate oven at about 100 °C. Weigh the glass and rust. Is there an increase or decrease in mass?

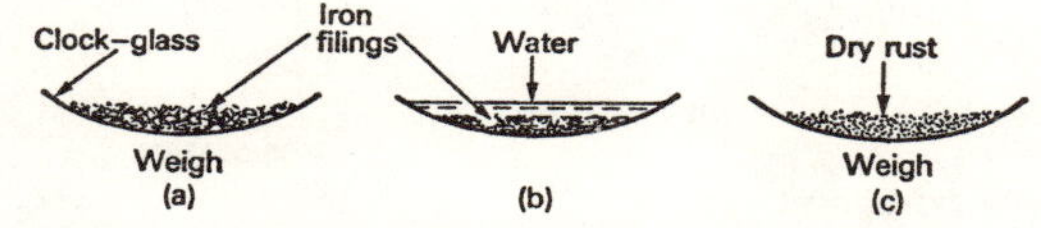

Fig. 4.4 Is there a change in mass when iron rusts?

To measure the fraction of air used up when iron rusts

1. Drop non-greasy iron filings to the bottom of a graduated tube. Add a few cm³ of water. Use a long rod to push a loose plug of glass wool near the filings so that they are held in position when the tube is inverted.

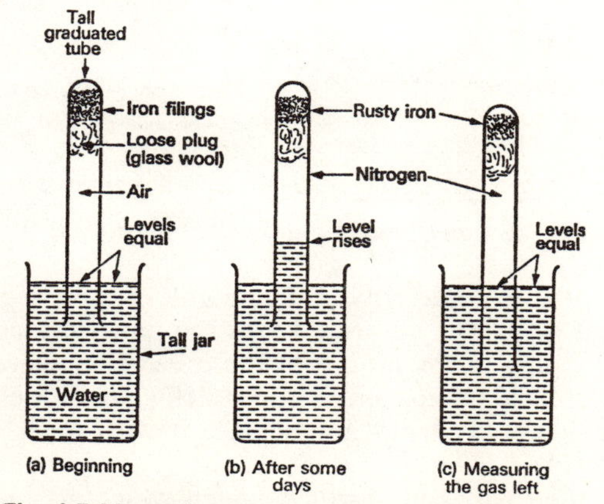

Fig. 4.5 Measuring the air used up when iron rusts

2. Invert the tube in a tall jar almost full of water. Make the water levels inside and outside equal so that the air in the tube is at atmospheric pressure. Note the volume of this air.
3. Leave the apparatus for several days until the iron rusts and the water level in the tube changes no more.
4. Make the water levels equal once again by lowering the graduated tube. Note the volume of inactive air and then test it with a burning splint.

$$\text{Volume of air before rusting} = V\ \text{cm}^3$$
$$\text{Volume of gas after rusting} = v\ \text{cm}^3$$
$$\text{Volume of air used up} = V - v\ \text{cm}^3$$

The fraction of the air used up is $\dfrac{V - v}{V}$.

To measure the fraction of air used up by phosphorus

White phosphorus catches fire so easily that it is stored under water. Always hold it with tongs and cut it under water. It is a soft wax-like solid and is sometimes called yellow phosphorus. Red phosphorus is much less reactive.

1. Cut, under water, a piece of white phosphorus about the size of a small pea. Use tongs to place it at the bottom of a graduated tube and repeat the method of the previous experiment. The phosphorus smoulders (burns slowly) in air and the reaction is finished in one or two hours.
2. Observe the fumes which are formed and what happens to them. After noting the volume of gas before and after the experiment, test the gas left with a burning splint and add litmus paper to the water.

To find if there is a change in mass when a candle burns

1. The U-tube contains lumps of soda-lime which absorb water and carbon dioxide. Place a loose plug of glass wool at each end of the U-tube to hold the lumps in position.

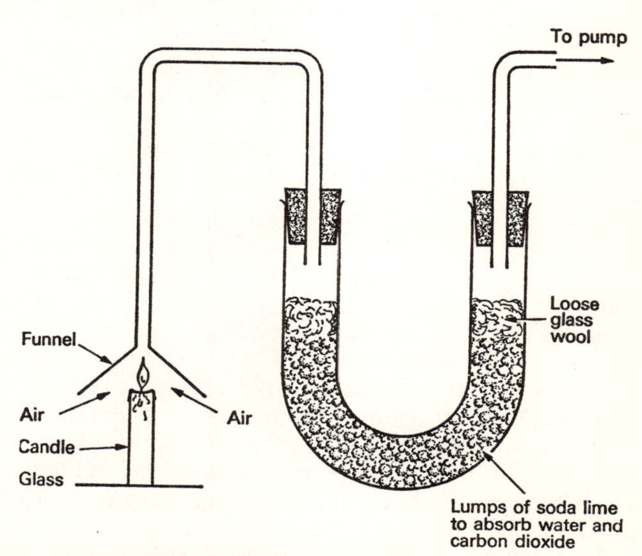

Fig. 4.6 Is there a change in mass when a candle burns?

2. Weigh the whole apparatus shown in the diagram.
3. Light the candle. Use a pump to suck the gases formed during combustion through the U-tube at a steady rate.
4. After about five minutes, turn off the pump and allow the apparatus to cool. Weigh the whole apparatus again.

   Some candle wax has been used up, and water and carbon dioxide have been formed.

   Mass of apparatus at start = ?

   Mass of apparatus at finish = ?
5. *Control experiment* Now suck air through the apparatus at about the same rate as before and for the same time. Weigh the whole apparatus again. This experiment shows if the change in mass is caused by the burning candle or by the air passing through the apparatus. For example, if the mass of the apparatus increases by 0·3 g when the candle is burning and by 0·1 g when the candle is not burning, the change due to the burning candle is 0·3 − 0·1 = 0·2 g.

COMBUSTION, RESPIRATION, AND RUSTING

Only about one-fifth of the air by volume is used up when copper is heated in air, when iron rusts or when phosphorus smoulders in air. Therefore one-fifth of the air is active air (oxygen) and four-fifths is inactive air (nitrogen and other gases).

Combustion in air is a chemical change during which elements or compounds combine with oxygen to form oxides, and heat is also produced. Rusting is a slow combustion during which the iron combines with oxygen to form an oxide; heat is produced but it is not easy to demonstrate that this is so. Respiration is also a slow combustion. Oxygen combines with red haemoglobin of the blood to form oxyhaemoglobin, which reacts with digested food in the body to form water, carbon dioxide and energy.

Combustion, rusting and respiration all:

(*a*) produce oxides, which are heavier than the original substances,

(*b*) use up about one-fifth of the air, and

(*c*) produce heat or energy.

Candle The wax burns to form carbon dioxide and water. Sometimes a little soot (carbon) forms if there is not enough air. The candle uses up part of the air only (the active part called oxygen). The flame usually goes out before all the oxygen has been used up.

Aluminium Normally this does not react with air because a film of oxide on its surface prevents further reaction. Mercury removes this oxide. The aluminium then reacts quickly with oxygen of the air. Soft, feathery aluminium oxide grows on its surface and the metal becomes too hot to hold owing to the great heat evolved.

Ethanol, petrol, kerosine, and benzene All burn in a plentiful supply of air to form carbon dioxide and water only. Soot is formed if there is not enough air for complete combustion.

Respiration Carbon dioxide and water are formed. Exhaled air contains much more carbon dioxide than inhaled air (actually about 100 times more).

Phosphorus White phosphorus smoulders in air and forms white fumes of phosphorus(V) oxide. The fumes react with water to form phosphoric acid which turns litmus red.

To investigate the rusting of iron

1. Add non-greasy iron nails or iron filings to three test-tubes.
2. To tube (*a*), which is the control tube, add distilled water and place a loose plug of glass wool in its mouth to keep out dust.

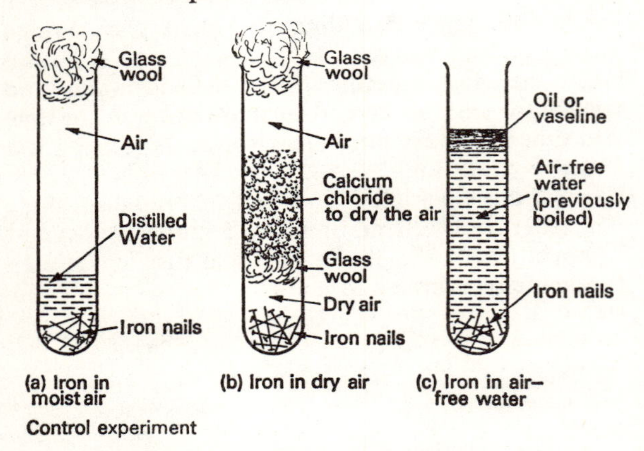

Fig. 4.7 Investigating the rusting of iron

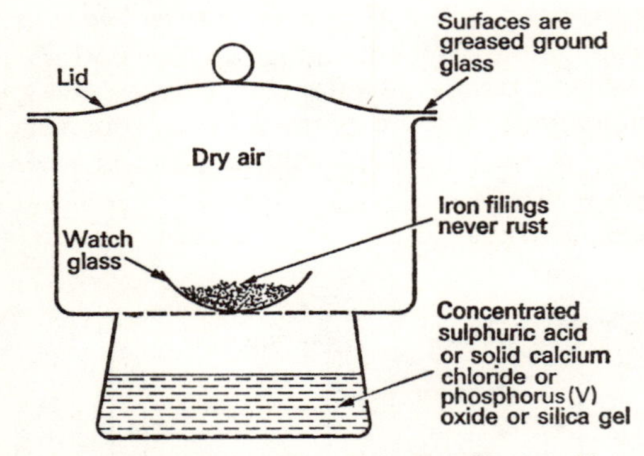

Fig. 4.8 Iron stored in a desiccator (in dry air)

3. To tube (*b*), add solid calcium chloride on a glass wool plug. The chloride keeps the air dry. Place a loose plug in the mouth of the tube. This prevents the chloride going moist too quickly. If it does go damp, knock it out and replace it with fresh, dry chloride.
4. To tube (*c*), add distilled water which has been boiled for a few minutes to drive out any dissolved air. Cover the water with a layer of oil or vaseline to prevent air dissolving. The vaseline melts on the hot water and then solidifies as the water cools.

5. Leave the three tubes for several days or even weeks. Observe which iron rusts.

Iron can be stored in a desiccator for years and it does not rust.

PREVENTING THE RUSTING OF IRON

Rusting of iron and steel can be prevented or made slow only by keeping air and water away from it. To do this, the metal is covered with oil, grease, paint, tar, other metals or plastics.

Galvanized iron is iron coated with zinc; tin-plate is iron or steel coated with tin. Tin-plate rusts quickly if the coating of tin is broken in any part. Galvanized iron does not rust easily even if the zinc coat is broken; experiments show that the zinc is corroded before the iron. Galvanized iron cannot be used to hold food as zinc is poisonous.

SUMMARY

Combustion in air: a chemical change in which substances combine with oxygen to form oxides and to produce energy.
Examples of combustion: burning of candle-wax, kerosine, petrol, benzene, which form carbon dioxide and water. Soot (carbon) is formed if there is not enough air.
Respiration produces carbon dioxide and water.
Aluminium forms aluminium oxide.
Phosphorus forms white phosphorus(V) oxide which reacts with water to form phosphoric acid.
Iron slowly forms iron rust which is a hydrated oxide of iron. Both water and air (or oxygen) must be present.

Products of combustion weigh more than the original substance. The extra mass is the mass of oxygen used up. Only one-fifth of the volume of the air present during combustion is used up.

Prevention of rusting: cover iron or steel with oil, grease paint, tar, plastics or other metals.
Galvanized iron: iron coated with zinc.
Tin-plate: iron coated with tin.

QUESTIONS

1. Describe how you would determine, to within about 1 per cent, the percentage by volume of oxygen in a sample of ordinary air. Draw a diagram to show how you would take your measurements of volume. Explain why this experiment cannot be done satisfactorily by burning a candle in a measured volume of air. (C.)
2. Describe the methods and results of two experiments which show that air is a mixture. How would you show experimentally that water and oxygen are necessary for the rusting of iron?
3. Mention four different methods by which the rusting of iron can be prevented. What is the essential feature of all the methods? Your answer should include references to galvanized iron and tin-plate.
4. A large flask is completely filled with water and fitted with a rubber bung and delivery-tube which is also completely filled with the same water. The end of the delivery-tube is placed under water in a small trough and a graduated tube, also full of water, is placed over this end. The water in the flask is heated until it boils, and the whole apparatus is allowed to cool to the original temperature. Some gas is found to have collected in the graduated tube. The volume of gas is 24 cm^3. A piece of white phosphorus on a long wire is inserted into the graduated tube and left until no further change is apparent. The volume of gas left is found to be 16 cm^3. (*a*) What do you think was the purpose of the experiment? (*b*) Why was the delivery-tube also filled with water? (*c*) Of what did the 24 cm^3 of gas consist? (*d*) What was the purpose of the white phosphorus? (*e*) From the figures given calculate the percentage of one of the constituents of the gas in the tube. (*f*) How does your result in (*e*) compare with that which you would expect to get with air? (L.)

5 Water and Hydrogen

To find what happens when certain substances are left in air

1. On separate clock-glasses place specimens of the following solids:

Anhydrous copper sulphate	Sodium sulphate (hydrated)
Sodium hydroxide	Potassium hydroxide
Calcium chloride	Iron(III) chloride
Sodium nitrate	Sodium carbonate (hydrated)
Calcium oxide (lime)	

2. Add ethanol to a beaker and concentrated sulphuric acid to another beaker to a depth of about 2 cm. Put a label on the outside of the beakers to mark the surfaces of the liquids.
3. Leave all the specimens in the open air. Observe any visible changes after (*a*) five to ten minutes, (*b*) a few hours, and (*c*) a few days.
4. Place crystals of blue copper sulphate on a clock-glass and leave them inside a desiccator for some time. Observe any visible changes.

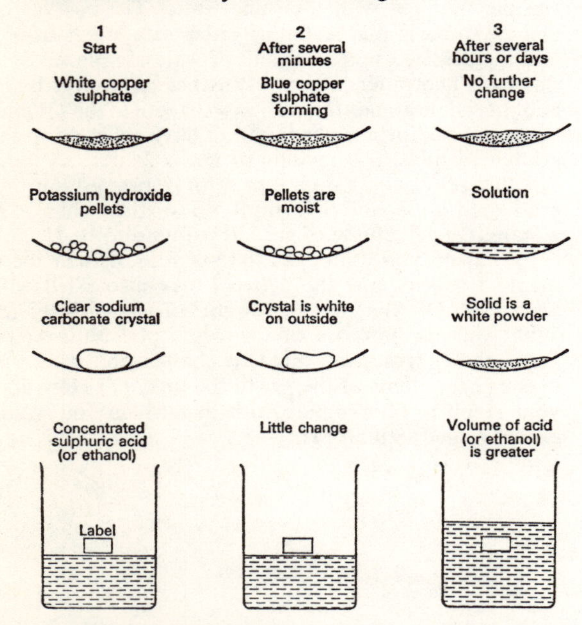

Fig. 5.1 Changes when substances are left in air

WATER IN THE ATMOSPHERE

Ordinary common salt usually becomes damp and sticky when left in the air, particularly on humid days. It is an example of a *hygroscopic* substance—one which absorbs water from the air. Pure sodium chloride, however, is not hygroscopic. Some substances absorb so much water that they dissolve in it; they are *deliquescent*, which means 'turn to a liquid', and the process is called deliquescence.

Clearly all deliquescent compounds must be hygroscopic. Not all hygroscopic compounds are deliquescent, because some absorb only a little water and do not absorb enough to form a solution.

The hydrates of sodium carbonate and sodium sulphate give up some or all of their water of crystallization to the atmosphere, particularly on dry days or when they are placed in the dry air of a desiccator. The clear hydrated crystals become covered with a white powder and gradually change completely to a white, amorphous powder. The giving up of water of crystallization by a crystal to the atmosphere is called *efflorescence*, and the substances are efflorescent.

Deliquescent	*Hygroscopic*
Sodium hydroxide	All deliquescent substances
Potassium hydroxide	Impure common salt
Calcium chloride	Copper oxide
Iron(III) chloride	Calcium oxide
Phosphorus(V) oxide	Concentrated sulphuric acid
Sodium nitrate	Ethanol

Hydrated sodium carbonate (washing soda) and hydrated sodium sulphate are examples of efflorescent substances. However, they do not effloresce if the atmosphere is humid. Copper sulphate crystals are not usually efflorescent, but they effloresce if the atmosphere is unusually dry and also if they are kept in a desiccator.

DRYING AGENTS

Hygroscopic and deliquescent substances are used to dry gases and other compounds. A damp gas is dried by bubbling it through a bottle containing concentrated sulphuric acid or by passing it through a U-tube packed loosely with calcium chloride, phosphorus(V) oxide or silica gel.

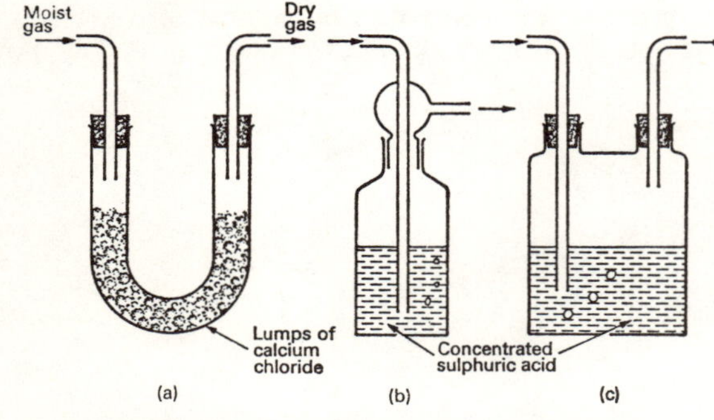

Fig. 5.2 Three different methods of drying gases

Some liquids, e.g. petrol or benzene, can be dried by leaving solid calcium chloride in them for a day or two. The chloride absorbs the water. Solids can be dried, or be kept dry, by keeping them in a *desiccator* at the bottom of which is solid calcium chloride, concentrated sulphuric acid or silica gel. Silica gel is frequently used to keep scientific equipment such as microscopes, cameras, and lenses dry in humid climates. Silica gel has no chemical action on most substances. After it has become damp through constant use, it is easily dried by moderate heat in an oven and is then ready for further use.

To collect and test the liquid formed by burning gas

1. Pass a fuel gas, such as town gas, coal gas, petrol gas, or hydrogen from a cylinder, through two U-tubes containing calcium chloride or silica gel in order to dry it.

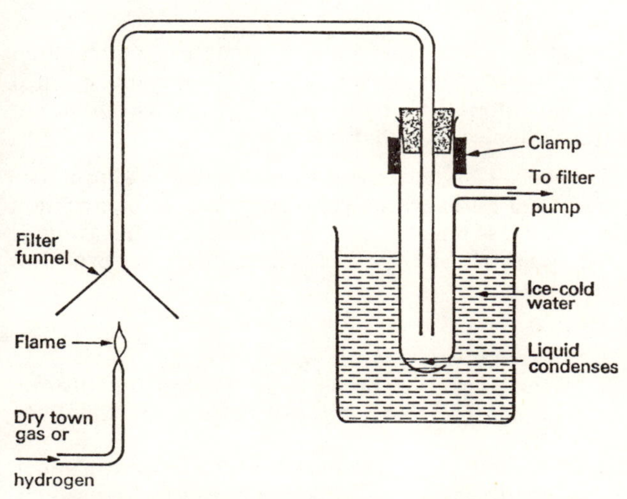

Fig. 5.3 Collecting the liquid formed by burning gas

2. Light the gas and obtain a flame about 3 cm high. Let it burn under a funnel and suck the hot gases through a test-tube cooled in a beaker of cold water. If possible, collect about 2 cm^3 of any liquid formed.

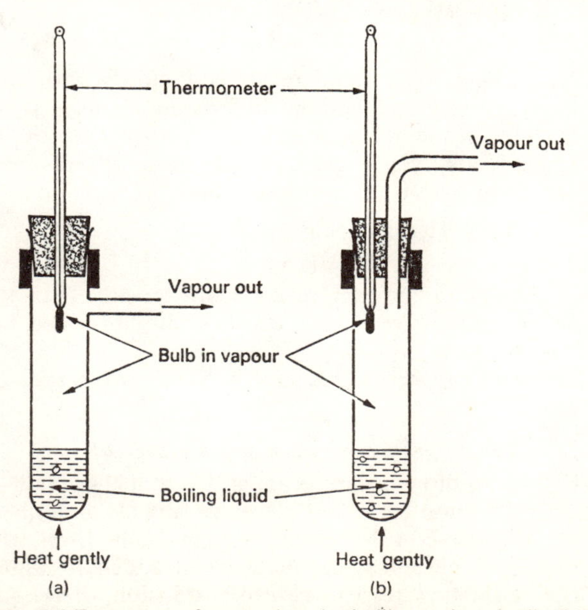

Fig. 5.4 Two ways of measuring the boiling point of a liquid

Testing the liquid

(*a*) Add two drops of the liquid to white, anhydrous copper sulphate.

(*b*) Add two drops of the liquid to blue, anhydrous cobalt chloride or to blue cobalt chloride paper.

(*c*) Find the boiling point of the remaining liquid, if there is enough, using the apparatus shown in the diagram of Fig. 5.4.

Note It is very dangerous to prepare hydrogen from chemicals and to use it in this experiment. If a mixture of hydrogen and air is lit, the whole apparatus can be blown apart, and there is great danger from flying glass. The experiment can be done by a teacher, but all precautions should be taken and a safety screen is essential. Hydrogen from a cylinder is much safer.

IS WATER AN OXIDE?

What conclusion can be drawn from each of the three tests done on the liquid? Does the liquid contain water, and possibly other products? Is the liquid only water and nothing else?

We have already learned that most substances form oxides when they burn in air. Water is probably an oxide of hydrogen, or a hydride of oxygen, which is the same. We know (p. 39) that magnesium removes oxygen from the oxides of copper, lead, and carbon. Therefore the metal might remove oxygen from water or steam. Since sodium and calcium have a greater affinity for oxygen than magnesium has, they too might remove oxygen from water. Lithium and potassium are two reactive metals that resemble sodium, and their actions on water can also be investigated.

To study the reactions between water and metals

Sodium and potassium can cause serious injuries if they are used carelessly. Always use *small* pieces and have a safety screen to protect the face and eyes in case of explosions. Experiments with these two metals should be done *only* by the teacher.

Sodium

1. Use dry tongs to take sodium out of a bottle in which it is stored under kerosine or naphtha. Place the sodium on a filter paper to dry. Cut a few pieces, each about the size of a rice grain, with a dry knife. Remove any corrosion from the outside of the sodium.
2. Add the pieces of sodium, one at a time, to cold water in a basin, beaker, or trough. Use a safety screen in case any of the metal flies out of the water.
3. When the reaction is finished, rub a little of the final solution between two fingers and then wash them.

4. Dip litmus paper or Universal Indicator paper into the solution.
5. Evaporate some of the solution to dryness in an evaporating basin.
6. Now float a filter paper on some fresh cold water in a trough. Add a small piece of sodium to the moist paper which stops the metal moving about. Note what happens.

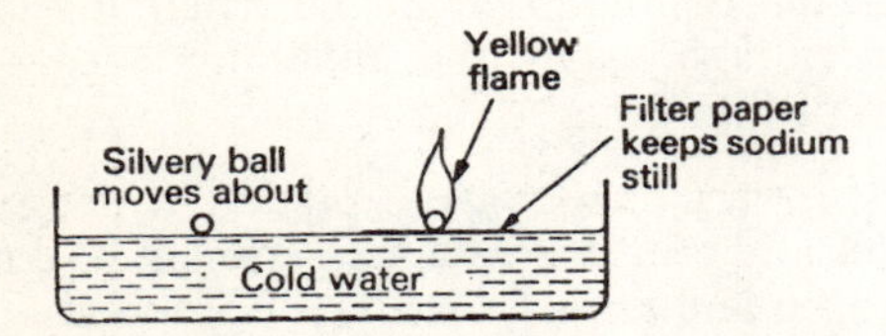

Fig. 5.5 The reaction between sodium and cold water

Lithium and potassium Repeat the above test with these metals. The reaction with lithium is moderate and fairly safe. The reaction with potassium can be very dangerous and a safety screen is essential.
Calcium Drop calcium turnings (small chips of the metal) into a beaker of water and also into a test-tube of water. Test any gas formed with a burning splint.
Magnesium Use sandpaper to clean magnesium ribbon. Drop the metal into cold water in a beaker. Observe the changes, if any, after a few minutes and after a day or two.

Add clean magnesium ribbon to hot water and observe any changes.
Zinc and copper Repeat the test with these metals.

REACTIONS BETWEEN WATER AND METALS

The reaction with potassium is most vigorous, and it can be violent and explosive. Sodium also reacts vigorously. Lithium and calcium react steadily. Magnesium reacts very slowly with cold water, and careful observation is essential to observe any changes; it reacts more quickly with hot water, but the reaction is still slow.

Observations you should have made are as follows.

Sodium
1. The metal floats on water.
2. It melts to a silvery ball which makes a hissing sound with the water. The melting point (m.p.) is 97 °C.
3. The ball moves rapidly over the water surface because it is pushed by steam and by hydrogen which is formed. Sometimes it sticks to the glass of the container and catches fire; its flame is yellow. The flame is caused by hydrogen burning, and its colour is due to the sodium in it.
4. The silvery ball becomes smaller. Finally a transparent ball of sodium hydroxide is left and this breaks up into pieces, making a cracking noise.
5. The solution feels soapy or slippery. It is alkaline. A white solid, sodium hydroxide, is left when it is evaporated.

$$2Na + 2H_2O = 2NaOH + H_2$$
$$\text{or } 2Na + 2H^+ = 2Na^+ + H_2$$

Potassium The metal reacts in the same way as sodium, but the reaction is more vigorous. The hydrogen formed always burns, and the colour of the flame is lilac. Potassium hydroxide, sometimes called caustic potash, is formed in the solution.

$$2K + 2H_2O = 2KOH + H_2$$
$$\text{or } 2K + 2H^+ = 2K^+ + H_2$$

In the above reactions, a sodium or potassium atom supplies an electron to a hydrogen ion, formed by the water. Two hydrogen atoms then combine to form a molecule of hydrogen gas.

Lithium This reacts less violently than sodium. The hydrogen burns with a crimson flame. As the reaction quickens, so much steam is evolved that the flame is extinguished. Lithium hydroxide, LiOH, is formed in solution.

Calcium
1. The metal sinks in the water.
2. A steady effervescence occurs. The gas is hydrogen.
3. The water becomes warm if there is not too much of it.
4. The solution is alkaline. A little calcium in much water forms a clear solution because the product, calcium hydroxide, is slightly soluble. The hydroxide is formed as a white precipitate when sufficient calcium is added to a small volume of water, e.g. water in a test-tube.

$$Ca + 2H_2O = Ca(OH)_2 + H_2$$
$$\text{or } Ca + 2H^+ = Ca^{2+} + H_2$$

Magnesium The metal reacts very slowly with cold water. White magnesium hydroxide forms on the outside of the metal and stops further reaction or makes it slow. Hydrogen is formed at the rate of about one test-tube of gas in several days.

$$Mg + 2H_2O = Mg(OH)_2 + H_2$$
$$\text{or } Mg + 2H^+ = Mg^{2+} + H_2$$

Magnesium powder reacts more quickly than the ribbon, and the reaction is quicker with hot water, but it is still slow.
Zinc and copper These metals have no reaction on water.

To collect the gas formed when metals react with water
Figure 5.6 shows a simple apparatus that is suitable.

Cut sodium into tiny pieces as already described. Wrap the pieces loosely in a cage made from wire gauze, which sinks the metal when added to water. Use a safety screen in case of explosion. *Never* use potassium in this method.

Calcium turnings or magnesium powder must first be wrapped in filter paper to stop them spreading too much, before adding to water. Magnesium ribbon can be rolled into a ball and dropped into the water. Lithium can also be used in this apparatus.

Test the gas formed by using a burning splint.

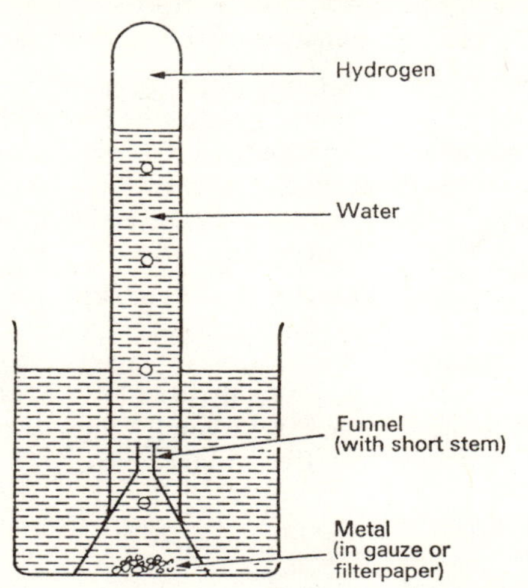

Fig. 5.6 Collecting hydrogen from a metal and water

To study the action of steam on heated magnesium

1. Push a loose plug of wet glass wool or asbestos wool to the bottom of a boiling-tube.
2. Clean magnesium ribbon with sandpaper and push the metal half-way down the tube. Clamp the tube horizontally so that it is easy to heat the metal.

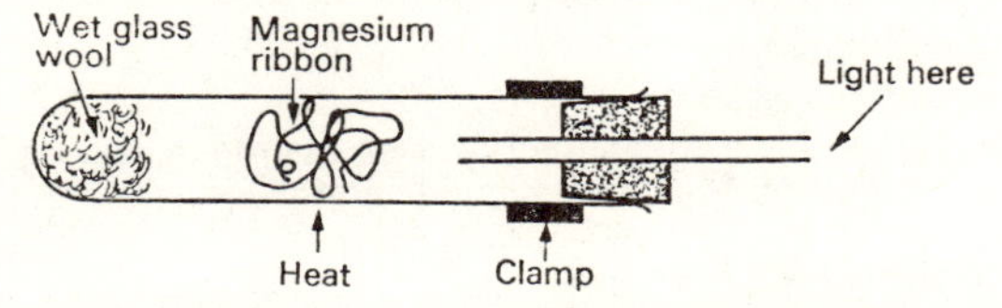

Fig. 5.7 Action of steam on heated magnesium

3. Heat the magnesium strongly. There is no need to heat the water in the wool which is heated sufficiently by conduction.
4. Observe all the changes. The gas formed usually catches fire; if not, light it with the flame. The boiling-tube always cracks or breaks, and the glass turns black. White magnesium oxide is formed because the temperature is so high that the hydroxide cannot exist.

$$Mg + H_2O = H_2 + MgO\text{, white}$$

If great care is used, it is possible in this way to collect a gas-jar of hydrogen over water. A wide delivery tube must be used otherwise the gas, which is formed so quickly, blows out the rubber stopper of the boiling-tube. At the end of the reaction, remove the stopper at once to prevent water from the trough being sucked back into the hot boiling-tube and causing a minor explosion as it turns to steam. Pupils should *not* try this experiment.

To study the action of steam on heated iron or zinc

1. Arrange the apparatus as shown, with iron filings or zinc powder in the middle of the boiling-tube.

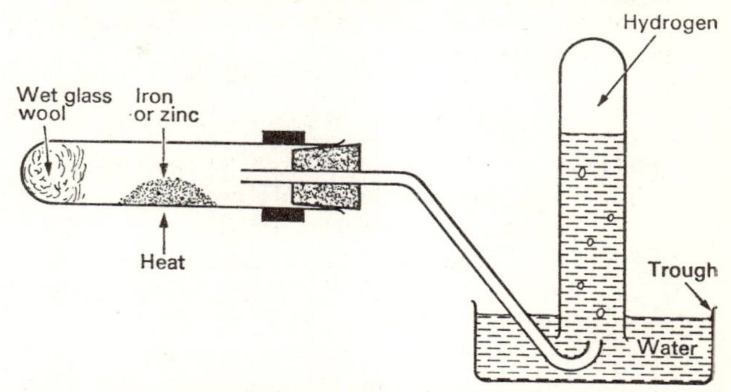

Fig. 5.8 Action of steam on heated iron or zinc

2. Heat the metal. The water on the glass wool is heated by conduction and boils steadily. Collect the gas formed. Remove the end of the delivery tube from the water, to avoid 'sucking back', before removing the flame at the end of the experiment. Note the colour of the solid residue.

$$3Fe + 4H_2O = 4H_2 + Fe_3O_4$$
$$Zn + H_2O = H_2 + ZnO$$

To study the reactions between metals and dilute acids

1. Add dilute hydrochloric acid to magnesium ribbon in a boiling-tube. Cover the mouth of the tube with paper so that the gas that forms gradually fills the tube and does not escape almost as fast as it is produced.
2. Hold the tube with its mouth near a flame, and then remove the paper. Note what happens.
3. Repeat the test with dilute sulphuric acid, and then with dilute nitric acid.
4. Repeat the whole test with zinc, iron, and copper. Warm gently if the reaction is slow with the cold acid, but do not boil.
5. *Magnesium and very dilute nitric acid* Mix dilute nitric acid with twice its volume of water. Use this dilute solution in the apparatus of Fig. 5.6 to react either with magnesium powder wrapped in paper or with magnesium ribbon. Collect and test any gas formed.

To study the reactions between metals and alkalis

1. Add sodium hydroxide solution to aluminium powder. If a gas is formed, test it with a flame.
2. Repeat the test with potassium hydroxide solution.
3. Repeat the test with zinc and each of the two alkalis. If reaction is slow, add one or two pellets of the solid alkali in order to increase the concentration of the solution, and warm gently but do not boil.

DISPLACEMENT OF HYDROGEN FROM ACIDS AND ALKALIS

Potassium, sodium, lithium, and calcium displace hydrogen from acids, but their reactions are so violent that they should never be demonstrated.

Magnesium, zinc, and iron react with dilute hydrochloric and sulphuric acids to form hydrogen:

$$Mg + 2HCl = H_2 + MgCl_2$$
$$\text{or } Mg + 2H^+ = Mg^{2+} + H_2$$

The reactions are the same as those with water, and they differ only because the acids contain a much greater concentration of hydrogen ions.

Lead and copper do not form hydrogen with dilute acids, but lead forms the gas with hot, concentrated hydrochloric acid.

Nitric acid forms hydrogen only with magnesium and only when it is very dilute. This acid forms oxides of nitrogen, NO_2 and NO, when it reacts with other metals (see p. 128).

$$Mg + 2HNO_3 = H_2 + Mg(NO_3)_2$$

Aluminium and zinc react with solutions of sodium hydroxide and potassium hydroxide to form hydrogen:

$$2Al + 2NaOH + 6H_2O = 3H_2 + 2NaAl(OH)_4, \text{ sodium aluminate}$$
$$Zn + 2NaOH + 2H_2O = H_2 + Na_2Zn(OH)_4, \text{ sodium zincate}$$

To prepare gas-jars of hydrogen

1. Dilute some concentrated hydrochloric acid by adding it to an equal volume of water. Alternatively dilute some concentrated sulphuric acid by adding it to three times its volume of cold water. Stir well and then allow to cool. Use either of these diluted acids in the preparation.
2. Place some granulated zinc in a flask and arrange the apparatus as shown on the left of the diagram below.

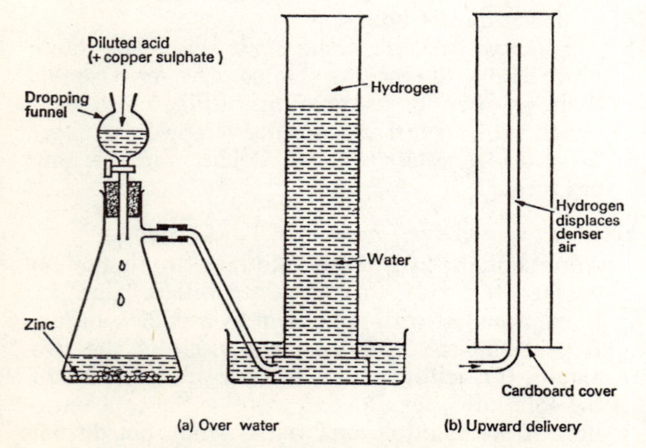

Fig. 5.9 Preparation and collection of hydrogen

3. Add the acid from a dropping funnel to the zinc. If the reaction is slow, add a few cm^3 of copper sulphate solution.

$$Zn + 2HCl = ZnCl_2 + H_2$$
$$\text{or } Zn + 2H^+ = Zn^{2+} + H_2$$
$$Zn + H_2SO_4 = ZnSO_4 + H_2$$
$$\text{or } Zn + 2H^+ = Zn^{2+} + H_2$$

4. Allow the air in the apparatus to escape and then collect the hydrogen in gas-jars or boiling-tubes.

The hydrogen is collected over water, because it is only slightly soluble. It can alternatively be collected by upward delivery, shown on the right of Fig. 5.9, because it is much lighter than air.

Kipp's apparatus This can be used to supply hydrogen when required. It is useful in laboratories which do not have hydrogen in cylinders. Zinc is placed in the middle bulb (Fig. 5.10) and dilute hydrochloric acid in the lower bulb and in the reservoir. When the tap is opened, acid rises into the middle bulb.

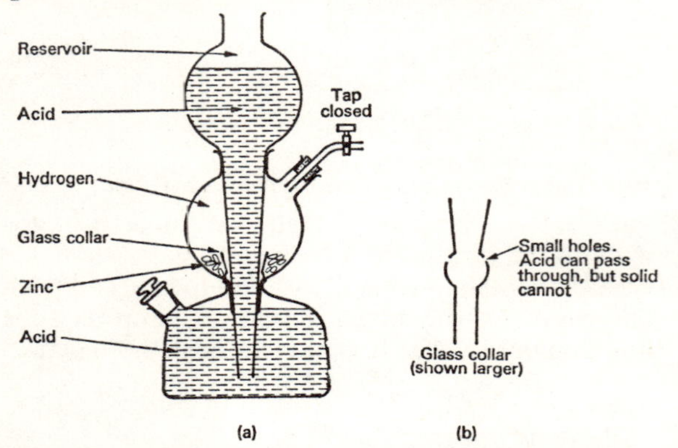

Fig. 5.10 Kipp's apparatus (used for hydrogen, carbon dioxide, hydrogen sulphide and chlorine)

Experiments with hydrogen

1. Note the colour and smell of the gas. Any smell is due to traces of impurities.
2. Test the solubility of the gas by inverting a tube or jar of hydrogen in cold water.
3. Put moist litmus paper and Universal Indicator paper in the gas.
4. Compare the density of the gas with that of air:
   (*a*) Take two jars of hydrogen, each with a cover. Hold one jar mouth downwards and the second mouth upwards. Remove the covers at the same time. After ten to fifteen seconds let another person test the gas in the jars with a burning splint. Which gas burns with a 'pop'?
   (*b*) Hold the delivery tube of the apparatus used to prepare hydrogen in a detergent solution in order to form bubbles. How do the bubbles move in air?
   (*c*) Seal a large polythene bag which measures at least 50 cm × 50 cm with adhesive tape, leaving a small hole in one corner. Alternatively use a

large rubber balloon. Fill it with dry hydrogen, or coal gas which contains 50 per cent hydrogen, and then release it. What happens?

5. Test how the gas burns.
   (*a*) Hold a dry jar of hydrogen mouth downwards. Remove the cover and insert a long burning splint into the gas. What happens?

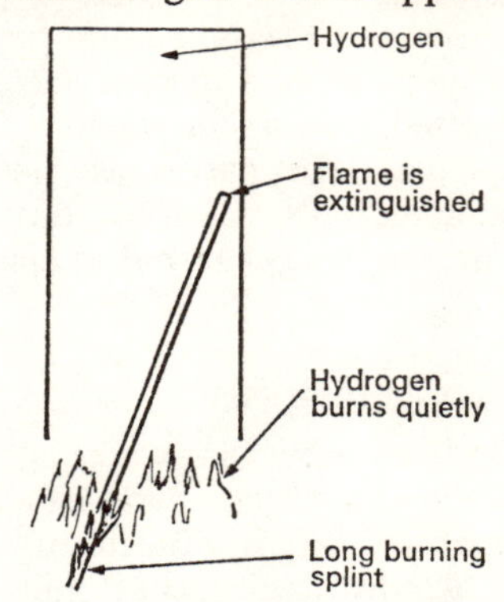

Fig. 5.11 Combustion of hydrogen

   (*b*) Put a flame to the mouth of a jar containing a mixture of hydrogen and air. What happens?
   (*c*) Fill a jar with a mixture of hydrogen and oxygen. Remove the cover and put a flame to the mixture. Be prepared for an explosion.

To study the action of hydrogen on heated oxides

This experiment can be dangerous if a hydrogen-air mixture forms in the apparatus because it burns explosively. It is safer to use coal gas or town gas.

1. Place samples of the oxides of copper, lead, iron, and zinc in separate porcelain boats.
2. Place all the boats in a combustion tube. The tube should slope downwards so that, if any liquid is formed, it remains on the cold glass and does not fall on and crack the hot glass.
3. Pass coal gas or town gas (or dry hydrogen from a cylinder) through the tube to drive out the air. Heat the oxides and note all the visible changes.

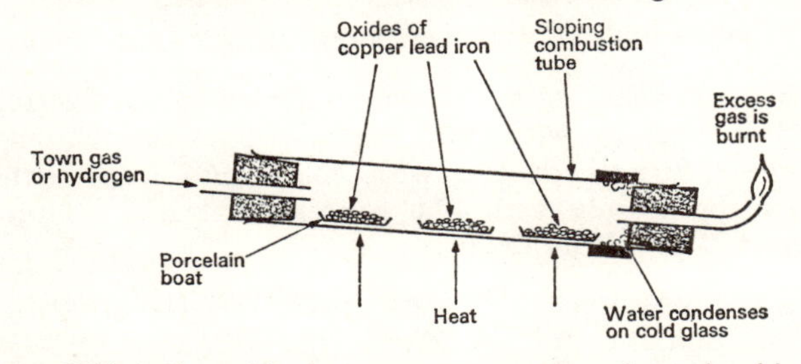

Fig. 5.12 Action of hydrogen or town gas on heated oxides

Simpler apparatus Fig. 5.13 shows a simpler method of doing the same experiment, using a boiling-tube. The oxide is placed in a tube directly, and no porcelain boat is required. Lead oxide ruins glass, and therefore it should be placed either in a porcelain boat or, better still, on a piece of asbestos paper inside the tube.

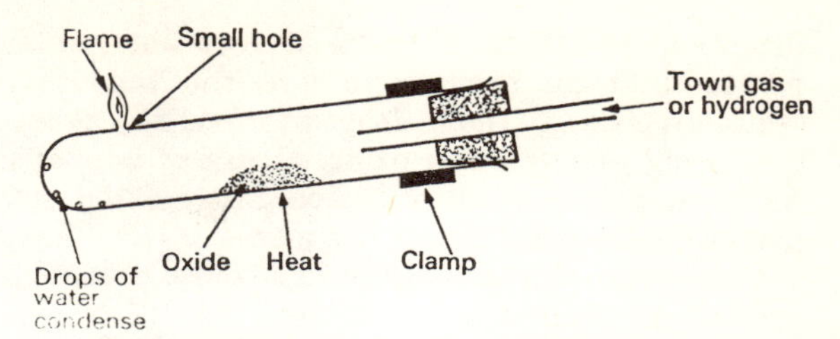

Fig. 5.13 Action of hydrogen or town gas on a heated oxide (simple apparatus)

REDUCING ACTION OF HYDROGEN ON OXIDES

Black copper oxide glows red-hot and changes to reddish-brown copper; yellow lead oxide changes to silvery balls of molten lead; black iron oxide changes to the grey metal; zinc oxide is not changed by the hydrogen. Drops of water are formed on the cold part of the combustion tube.

Oxide	+ hydrogen	= water +	metal
CuO + (black)	H_2	$= H_2O$ +	Cu (reddish-brown)
PbO + (yellow)	H_2	$= H_2O$ +	Pb (silvery ball)
Fe_3O_4 + (black)	$4H_2$	$= 4H_2O$ +	$3Fe$ (grey)
ZnO + (white)	H_2	= no reaction	

The oxides lose oxygen and the hydrogen gains oxygen. The oxides are reduced to metals and the hydrogen is oxidized to water. The oxides are called *oxidizing agents* because they supply oxygen.

The hydrogen is a *reducing agent* because it removes oxygen. Hydrogen does not reduce oxides of metals which are in the top part of the reactivity series, such as zinc, aluminium, and magnesium because their affinity for oxygen is greater than that of hydrogen.

The reactivity series is on p. 30.

REVERSIBLE REACTIONS

Hydrogen reacts with hot iron oxide to form iron and water, but steam reacts with hot iron to form iron oxide and hydrogen. Clearly this reaction can go in either direction. The reaction is reversible and, as we have seen in chapter 2, the sign $\rightleftharpoons$ is used in the equation to show this.

$$3Fe + 4H_2O \rightleftharpoons Fe_3O_4 + 4H_2$$

Equations for other reversible reactions already met are:

$$CuSO_4 + 5H_2O \rightleftharpoons CuSO_4.5H_2O$$
$$CoCl_2 + 6H_2O \rightleftharpoons CoCl_2.6H_2O$$
$$2H_2 + O_2 \rightleftharpoons 2H_2O$$

The physical conditions at the time, such as temperature, pressure, and concentration of reagents, determine in which direction a reversible reaction proceeds. Steam reacts with hot iron because the steam drives away the hydrogen formed and prevents it reacting with the iron oxide. Hydrated copper sulphate decomposes when heated because the steam passes away into the atmosphere; water changes white copper sulphate back to its hydrate because it is present in sufficient concentration.

MANUFACTURE OF HYDROGEN

1. *From natural gas* This gas contains methane, CH_4. It is partially oxidized by steam at about 900 °C in the presence of a nickel catalyst†. Refer to page 162.

$$CH_4 + H_2O = CO + 3H_2 - \text{heat}$$

The carbon monoxide is removed as in the Bosch process (below). This is the best method in countries which have ample supplies of cheap natural gas.

2. *From water* (Bosch process) Water gas is made by passing steam through white-hot coke (p. 118). The gas is a mixture of hydrogen and carbon monoxide. More steam is added to the mixture, and the gases are passed over an iron catalyst at about 500 °C:

$$CO + H_2O = CO_2 + H_2$$

The carbon dioxide (CO_2) is removed by washing with water or better still with potassium carbonate solution under pressure:

$$K_2CO_3 + CO_2 + H_2O = 2KHCO_3$$

3. *Electrolysis* Hydrogen is a by-product from the manufacture of sodium hydroxide by electrolysis of brine; see p. 65.

USES OF HYDROGEN

1. *Fuels* Town gas, coal gas, and water gas contain about 50 per cent of hydrogen by volume. Liquid hydrogen has been used as a fuel in certain types of rockets.
2. *Manufacture of ammonia* Hydrogen and nitrogen react to form ammonia when iron is present as a catalyst; see p. 125.
3. *Manufacture of hydrochloric acid* Hydrogen chloride is formed by burning hydrogen in chlorine. It dissolves in water to form the acid.
4. *Hardening* (*hydrogenation*) *of oils* Vegetable and animal oils, such as olive oil, whale oil, palm oil, and groundnut oil, combine with hydrogen under pressure and in the presence of a nickel catalyst.

$$\text{Oil} + \text{hydrogen} = \text{fat}$$

The fats are used for making margarine, candles, soap, and for cooking.

5. *Petrol* Hydrogen under great pressure reacts with powdered coal (or with a liquid called creosote which is obtained from coal) to form a liquid which contains petrol. The process is called the hydrogenation of coal. The process is only used in countries which have no petroleum but plenty of cheap coal.
6. *Oxyhydrogen flame* A mixture of hydrogen and oxygen can burn and produce a temperature of over 2000 °C. The flame is used to melt and to cut metals, and it can be used even under water.
7. *Filling balloons* The gas is used to fill weather balloons and children's balloons, but not balloons which carry passengers because it is too inflammable.

WATER IS A COMPOUND

The following facts show that water is not a mixture.

Separation The hydrogen and oxygen in water cannot be separated by physical means, such as cooling, boiling, or distillation, but only by chemical means.

Properties The properties of water are very different from those of the two elements in it. Water is a liquid; hydrogen and oxygen are gases that are difficult to liquefy. Water does not burn; hydrogen is explosive and oxygen relights a glowing splint, but water is used to extinguish fires.

Energy Heat, and sometimes light and sound, are given out when water is made from its elements.

Composition Hydrogen and oxygen combine in definite proportions by mass (1·008 : 8) to form water. If water was a mixture, the elements could be present in any proportions.

REACTIVITY SERIES FOR METALS

The reactivity of various metals with oxygen was studied in chapter 2 on page 9. We have seen in this chapter (pp. 26–7) that the order of reactivity of metals with water is:

potassium sodium calcium magnesium iron

More metals can be included in the list by considering their reactions with dilute acids. Copper and mercury never form hydrogen with acids, and lead forms hydrogen only with hot concentrated hydrochloric acid. The full series is:

1. Potassium	5. Aluminium	9. (Hydrogen)
2. Sodium	6. Zinc	10. Copper
3. Calcium	7. Iron	11. Mercury
4. Magnesium	8. Lead	12. Silver

You can remember the order by using the 'words'

PoSo CaMAl ZILHyCoMS

in which Po reminds you of potassium, So of sodium, Z of zinc, Co of copper, and so on.

† The meaning of the word *catalyst* is explained on p. 37 in chapter 7.

SUMMARY

Hygroscopic substance: absorbs water from the atmosphere. Examples: calcium oxide, concentrated sulphuric acid, ethanol.
Deliquescent substance: absorbs so much water from the atmosphere that it forms a solution. Examples: calcium chloride, sodium hydroxide, sodium nitrate, phosphorus(V) oxide.
Efflorescent substance: hydrated salt that loses water of crystallization to the atmosphere. Examples: sodium carbonate, sodium sulphate and, if the air is very dry, blue copper sulphate.
Drying agents: used to dry gases, liquids or solids. Examples: calcium chloride, concentrated sulphuric acid, phosphorus(V) oxide, silica gel.

Water is an oxide of hydrogen because it is formed by burning hydrogen or other gases that contain hydrogen and because active metals can displace hydrogen from it leaving the metal oxide or hydroxide. Water is not a mixture.
Action of metals: potassium, sodium, lithium, calcium, magnesium all displace hydrogen from water; potassium is the most reactive and magnesium the least. Alkaline hydroxides are formed. Magnesium, zinc, and iron react with steam to form oxides.

Hydrogen: lightest gas known, colourless, odourless, very inflammable, explodes with oxygen.
Dilute acids: release hydrogen with magnesium, zinc, and iron. Lead and copper do not react. Dilute nitric acid forms oxides of nitrogen, except with magnesium if very dilute.
Alkalis: aluminium and zinc release hydrogen from sodium hydroxide and potassium hydroxide.
Preparation: from zinc and dilute hydrochloric or sulphuric acid. Collect over water or by upward delivery. Kipp's apparatus can be used.
Chemical properties: removes oxygen from active oxides of metals; for example, iron oxide or copper oxide.
Manufacture: from methane (in natural gas) and steam with nickel; from water gas (which contains carbon monoxide and hydrogen) by the Bosch process using iron; by electrolysis of brine (salt solution).
Uses: in gaseous fuels; in manufacture of ammonia, hydrochloric acid, fats, petrol; for very hot oxy-hydrogen flame and for filling balloons.

Reactivity series for metals: order of chemical activity of metallic elements.
Reversible reaction: proceeds either way depending on physical conditions. Examples include formation of hydrates, reaction between steam and iron.
Reducing agent: removes oxygen from compound. For example, hydrogen.
Oxidizing agent: adds oxygen to an element or compound. For example, metal oxides.

QUESTIONS

1. Describe and explain what happens when a named efflorescent substance is exposed to air. If copper sulphate crystals were introduced into the 'vacuum' above the mercury column in a simple barometer, what might be observed?
2. What is the effect of leaving solid sodium hydroxide exposed to air? When 30 g of washing soda crystals are left in dry air for some time a loss in mass of about 17 g occurs; how can you account for this?
3. Give two chemical tests by which you could determine if a liquid contained water, and suggest two further tests that you could use to show that the liquid was pure water. Give three reasons why water is considered to be a compound of oxygen and hydrogen. (C.)
4. Describe and explain what happens when : (i) metallic calcium is dropped into water; (ii) magnesium ribbon is heated in steam. State two respects in which the action of iron on steam differs from the reaction between magnesium and steam. (C.)
5. Describe the preparation and collection of hydrogen in the laboratory from a suitable acid. How would you use this hydrogen to make and collect a few drops of water? (C.)
6. Describe how you would show that hydrogen is (*a*) a reducing agent, and (*b*) lighter than air.
7. Explain what is meant by the statement that the reaction between steam and hot iron is a reversible reaction. Give three other examples of reversible reactions. How would you show experimentally that one of these three reactions is reversible?
8. Hydrogen gas may be prepared in the laboratory by the action of a metal on an acid. Draw a diagram of an apparatus suitable for this preparation. (*a*) Which of the following metals would be most suitable? Copper, zinc, magnesium or sodium. Give a reason for rejecting each of the metals you do not select. (*b*) Which of the following acids would you choose? Dilute sulphuric, concentrated sulphuric, dilute nitric, concentrated nitric. Say why you would not use each of the acids you reject. (*c*) It is necessary to run the apparatus for some time before attempting to collect the gas. Why? How do you test when to start collecting? (*d*) How would you modify your apparatus in order to collect dry hydrogen? (L.)

6 Acidic and Alkaline Solutions

INDICATORS

Litmus is a compound that can be extracted from plants called lichens. Its colour is either red, blue or purple. Litmus paper is paper that has been soaked in litmus solution and then dried.

Dip litmus paper into some of the following liquids or aqueous solutions of the solids.

water	soap	stomach powder
common salt	detergent	bicarbonate of soda
vinegar	lemonade	sodium hydroxide
sugar	soda water	hydrochloric acid
washing soda	toothpaste	sulphuric acid
lemon juice	tea	nitric acid
lime juice		

Some solutions turn litmus red; these are called acidic and usually they have an acid or sour taste. Other solutions turn litmus blue; these are called alkaline. Some liquids and solutions have no action on litmus. A substance which changes its colour, depending on whether it is in an acidic or an alkaline solution, is an indicator.

Two other indicators that are often used in chemical laboratories are methyl orange and phenolphthalein, but one of the best is a mixture of substances that is called Universal Indicator or U.I.

Use U.I. to test the acidity or the alkalinity of some of the substances listed above.

Because Universal Indicator is a mixture of several substances, it has a range of seven colours:

acid red orange yellow green blue indigo violet *alkaline*

U.I. shows that hydrochloric acid is more acidic than vinegar or lemon juice, and that sodium hydroxide is more alkaline than washing soda or bicarbonate of soda. The colour of U.I. indicates the degree of acidity or alkalinity. Instead of referring to the colours, it is more convenient to use a scale of numbers, as follows:

red	orange	yellow	green	blue	indigo	violet
0–4	5	6	7	8	9	10–14

The numbers are called pH numbers and the scale is the pH scale. pH is an abbreviation for 'the strength of hydrogen ions'. The reason for the name of this scale and the particular numbers need not concern us at this stage.

Pure water has a pH of 7 and is neutral. Solutions with a pH of less than 7 are acidic. A solution with a pH of 4 is more acidic than one with a pH of 6. Solutions with a pH of more than 7 are alkaline. If the pH of a solution is 10, it is more alkaline than one with a pH of 8.

To prepare and test some natural indicators

1. Collect coloured parts, such as flowers or petals and leaves, from fruits, plants, and vegetables.
2. Add each separately to a mortar and crush with a little warm methylated spirit and water. This usually dissolves the colouring material. Sometimes the crushed mixture has to be warmed gently for ten to twenty minutes before the colouring material dissolves. Filter the mixture.
3. Add each coloured solution to an acidic solution and to an alkaline solution. Observe whether the colour changes when added to the acidic or to the alkaline solution.

To study the properties of dilute acids

Dilute hydrochloric or sulphuric acid can be used in these tests. However, tartaric acid is better because it forms precipitates in certain tests and makes the reactions clearer.

1. Prepare tartaric acid solution by adding the powder to a depth of 2 cm in a test-tube, filling with water and shaking well.
2. Put one drop of the acid solution on the tip of the tongue—take care.
3. Add a little acid to four test-tubes and test with a few drops of litmus, U.I., methyl orange, or phenolphthalein in each tube.
4. Add magnesium ribbon to the acid solution in a boiling-tube. Cover the mouth of the boiling-tube until it fills with gas, and then uncover it near a flame. What is the gas?
5. Add a little calcium carbonate to the acid. Test any gas with lime water. See Fig. 6.1a. The gas may be withdrawn from the test-tube with a teat pipette and then forced out through the lime water. See Fig. 6.1b.

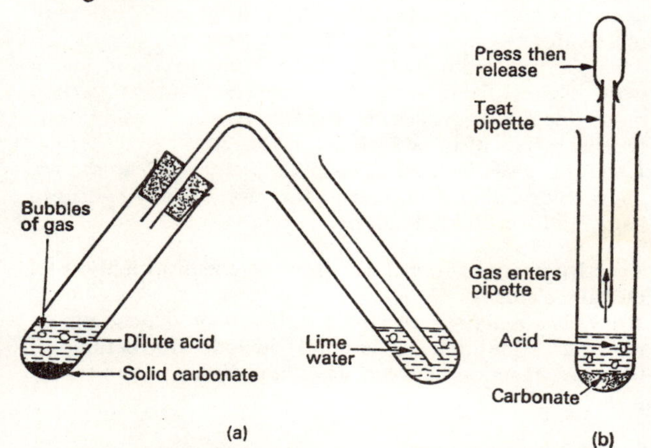

Fig. 6.1 Action of dilute acid on a carbonate

Repeat this test with sodium carbonate and sodium hydrogencarbonate. Repeat with sodium sulphite and sodium hydrogensulphite, and warm gently if necessary. Smell any gas formed, but do not test with lime water.

6. Add black copper oxide to warm dilute sulphuric acid or hydrochloric acid.
   Repeat this test with blue copper hydroxide and with white zinc oxide.
7. Add tartaric acid slowly to ammonia solution until the product just turns litmus red. Then add a few more drops of acid. Let the product stand for a time and observe any change.

ACIDS AND ALKALIS IN COMMON USE

Acids that are commonly used in everyday life are:

Citric acid, in lemons, limes and other citrus fruits

Tartaric acid, in baking powder and effervescent drinks

Ethanoic acid, in vinegar which is a 4 per cent solution of the acid

Carbonic acid, in soda water and other aerated waters.

Alkalis that are commonly used in laboratories are:

Sodium hydroxide or caustic soda, $NaOH$

Potassium hydroxide or caustic potash, KOH

Calcium hydroxide solution or lime water, $Ca(OH)_2$

Magnesium hydroxide suspension or milk of magnesia, $Mg(OH)_2$

Ammonium hydroxide or ammonia solution, $NH_3(aq)$.

Caustic means 'burning'. Concentrated caustic alkalis should *never* be touched with your fingers: they destroy skin and flesh and cause a wound similar to a burn.

PROPERTIES OF DILUTE ACIDS

Taste Sharp and sour.

Indicators Litmus is red; methyl orange, red; phenolphthalein, colourless. U.I. is either red, orange, or yellow, depending on the acid.

Metals Hydrogen is usually formed; for example,

$$Mg + H_2SO_4 = MgSO_4 + H_2$$

$$Fe + 2HCl = FeCl_2 + H_2$$

Carbonates and hydrogencarbonates Carbon dioxide is formed; for example,

$$CaCO_3 + 2HCl = CaCl_2 + CO_2 + H_2O$$

$$NaHCO_3 + HCl = NaCl + CO_2 + H_2O$$

Sulphites and hydrogensulphites Sulphur dioxide is formed; for example,

$$Na_2SO_3 + H_2SO_4 = Na_2SO_4 + SO_2 + H_2O$$

$$NaHSO_3 + HCl = NaCl + SO_2 + H_2O$$

Oxides and hydroxides of metals These are called bases. If the base is soluble, it is called an alkali. All bases react to form water and compounds called salts. For example,

$$CuO + H_2SO_4 = H_2O + CuSO_4 \text{ copper sulphate}$$

$$Zn(OH)_2 + 2HCl = 2H_2O + ZnCl_2 \text{ zinc chloride}$$

$$NH_3(aq) + \text{tartaric acid} = H_2O + \text{ammonium tartrate}$$

To study the properties of dilute alkalis

1. Rub dilute caustic soda between two fingers and then immediately wash them to avoid burning.
2. Note the colour of each of the four indicators litmus, methyl orange, phenolphthalein and Universal Indicator in dilute caustic soda solution.
3. Add dilute caustic soda solution to solid ammonium chloride or ammonium sulphate and warm gently. Smell any gas formed and test the gas with moist litmus paper or Universal Indicator.
4. Add dilute caustic soda solution in turn to solutions of magnesium sulphate, zinc sulphate, iron(II) sulphate, iron(III) chloride, lead nitrate and copper sulphate. If a precipitate is formed, note its colour.
5. Place a few pellets of sodium hydroxide on a watch-glass and leave in air. Note any change after a few minutes, a few hours and a few days. Repeat this test for potassium hydroxide.

PROPERTIES OF DILUTE ALKALIS

Alkalis are bases which are soluble in water. They feel slippery and soapy.

Indicators Litmus is blue; methyl orange, yellow; phenolphthalein, red. U.I. is either blue, indigo or violet, depending on the alkali.

Ammonium salts Ammonia gas (NH_3) is evolved; it has a characteristic pungent smell.

$$NaOH + NH_4Cl = NaCl + NH_3 + H_2O$$

Solutions of salts Coloured hydroxides are precipitated. These precipitates can be used to identify the metal present in some solutions.

$$2KOH + MgCl_2 = 2KCl + Mg(OH)_2,$$
white magnesium hydroxide

$$2NaOH + ZnSO_4 = Na_2SO_4 + Zn(OH)_2,$$
white zinc hydroxide

$$2NaOH + FeSO_4 = Na_2SO_4 + Fe(OH)_2,$$
green iron(II) hydroxide

$$3NaOH + FeCl_3 = 3NaCl + Fe(OH)_3,$$
reddish-brown iron(III) hydroxide

$$2KOH + Pb(NO_3)_2 = 2KNO_3 + Pb(OH)_2,$$
white lead hydroxide

$$2NaOH + CuSO_4 = Na_2SO_4 + Cu(OH)_2,$$
blue copper hydroxide

Deliquescence The two caustic alkalis, $NaOH$ and KOH, are deliquescent; see chapter 5, p. 24.

Metals Only aluminium and zinc are able to release hydrogen from the caustic alkalis; see chapter 5, p. 28.

$$2Al + 2NaOH + 6H_2O = 3H_2 + 2NaAl(OH)_4,$$
sodium aluminate

$$Zn + 2NaOH + 2H_2O = H_2 + Na_2Zn(OH)_4,$$
sodium zincate

REACTION OF DILUTE ACIDS WITH DILUTE ALKALIS

We have seen that the colour of litmus in a dilute acid solution is red and that its colour in a dilute alkaline solution is blue. If alkali is added to an acid solution containing litmus, the colour of the solution changes from red to blue. Clearly a chemical reaction takes place; it is known that water is one of the products that is always formed when an acid reacts with an alkali. For example

$$HCl + NaOH = NaCl + H_2O$$

and in general terms

acid + alkali = salt + water

The reason for this reaction is now known to be because both the acid and the alkali form smaller charged particles in solution called *ions*, while water is not able to split up into parts in the same way so easily. The formation of water in these reactions is in effect the neutralization of oppositely charged ions that are in the acid and in the alkali, namely the *hydrogen* ion (H^+) and the *hydroxide* ion (OH^-).

$$H^+Cl^- + Na^+OH^- = Na^+Cl^- + H_2O$$

or in general terms

$$H^+ + OH^- = H_2O$$

An ion with a positive charge, such as the hydrogen ion (H^+), is called a *cation.*

An ion with a negative charge, such as the hydroxide ion (OH^-), is called an *anion.*

We shall see that many reactions in solution can be explained in this way, and in chapter 13, we shall see how electricity can be used to separate these ions.

SUMMARY

Indicator: a soluble chemical whose colour shows whether a solution is acidic or alkaline.

Examples of indicators

	Acidic	*Alkaline*	*Neutral*
Litmus	red	blue	purple
Methyl orange	red	yellow	orange
Phenolphthalein	colourless	red	pink
Universal Indicator	red or orange or yellow	indigo or violet or blue	green

pH scale: a method of numbering how acid or how alkaline a solution is. The scale ranges from 0 which is totally acid to 14 which is totally alkaline. The colour of Universal Indicator corresponds to these numbers:

acid red	orange	yellow	green	blue	indigo	violet *alkaline*
0–4	5	6	7	8	9	10–14

Water: if it is pure, it is neutral and has pH of 7. Neutral solutions also have a pH of 7.

Properties of acids: sour taste, affect indicators.
With many metals: release hydrogen.
With carbonates and hydrogencarbonates: release carbon dioxide.
With sulphites and hydrogensulphites: release sulphur dioxide.
With oxides of metals and hydroxides: form salts and water.

Properties of alkalis: feel soapy, affect indicators; caustic alkalis are deliquescent.
With ammonium salts: release ammonia gas.
With some metal salts: give coloured precipitates.
With zinc and aluminium: release hydrogen gas.

Ions: electrically charged particles of chemicals in solution.
Anion: negatively charged part of ionized chemical, e.g. hydroxide ion, OH^-.
Cation: positively charged part of ionized chemical, e.g. hydrogen ion, H^+.
Acid-alkali reaction: formation of water from hydrogen and hydroxide ions in solution

acid + alkali = salt + water

QUESTIONS

1. Write equations for the action of sodium on water and the action of magnesium on dilute hydrochloric acid. Suggest why these two reactions could be considered as fundamentally similar.
2. When Universal Indicator is added to dilute sulphuric acid and to vinegar (another dilute acid) the colours obtained are not the same. Explain why this is so, and in your answer refer to pH values.
3. Give short accounts of the action of dilute sulphuric acid on iron, copper, and magnesium. Use your answer to place these elements in an order of activity.
4. Tests carried out on a white solid resulted in the following observations being made. (*a*) When several pieces of the solid were dissolved in water contained in a small beaker, the beaker became warm. The solution had a soapy feel and turned red litmus paper blue. (*b*) When a small piece of the solid was placed on a watch-glass and left exposed to the air for some time, a clear liquid formed which turned red litmus paper blue. Some time later, the liquid disappeared leaving behind a white crystalline solid. (*c*) Some of the solution from (*a*) was added to ammonium chloride in a test-tube and heated. A gas was given off that had a pungent smell. (*d*) When a few drops of the solution from (*a*) were added to a solution of zinc chloride, a white precipitate was formed. Identify the white solid and explain each of the observations made. (C.)
5. Account for the formation of (*a*) an inflammable gas when zinc dust is treated with caustic soda solution;

(*b*) a pungent smelling gas when ammonium chloride is treated with caustic soda solution; (*c*) a crystalline precipitate when carbon dioxide is passed into a cold, fairly concentrated solution of caustic soda; (*d*) a reddish-brown precipitate when caustic soda solution is added to iron(III) chloride solution. (L.)

6. Suggest two ways by which you could distinguish between aqueous solutions of ammonia and caustic soda. Using aqueous sodium hydroxide as the only reagent, how would you identify three white solids known to be magnesium sulphate, ammonium sulphate, and zinc sulphate?

7 The Atmosphere and Oxygen

GASES OF THE ATMOSPHERE

The atmosphere is a mixture of many gases which surrounds the earth's crust and it is about two hundred kilometres thick. It is estimated to contain 1 200 000 000 000 000 000 kilograms or $1{\cdot}2 \times 10^{18}$ kg of oxygen and just less than 4×10^{18} kg of nitrogen. The atmosphere also contains six thousand million, million kilograms of argon which was once called by a most unsuitable name—a rare gas. The average composition of dry air is:

	Percentage by volume	*Percentage by mass*
Nitrogen	78	75·5
Oxygen	21	23·2
Argon	0·93	1·3
Carbon dioxide	0·03	0·05

It also contains small amounts of other 'rare' gases, such as helium and neon, some gases produced by combustion of fuels including sulphur dioxide, hydrogen sulphide and carbon monoxide, together with dust, bacteria, soot, and other matter. Ordinary air always contains water vapour, usually between one and four per cent.

Oxygen is the most important gas because it is vital to all living things. Nitrogen dilutes the oxygen and therefore reduces the rates of respiration, combustion, rusting, corrosion of metals and other processes that use up oxygen.

Noble gases For many years these were called 'rare' gases, and then they were called 'inert' gases because it was believed that they did not form compounds. Recently, some compounds have been prepared and so these gases have been renamed noble gases. Five of them, helium, neon, argon, krypton, and xenon, exist in the atmosphere, but the sixth one, radon, does not.

In 1785 Cavendish found that 'azote', which was the name for nitrogen from the air, combined with oxygen when sparked. A small residue, about one per cent, remained no matter how long the sparking was continued. The residue was the noble gases, although Cavendish did not realize this. Over a century later in 1892, Rayleigh found that a large globe could hold 2·310 g of 'atmospheric' nitrogen, but only 2·299 g of 'chemical' nitrogen; that is, the nitrogen obtained from the atmosphere was denser than nitrogen prepared by chemical methods. This showed that the atmospheric nitrogen contained other gases. Later he isolated the first noble gas, argon.

In 1868, the existence of a new element called helium was detected in the sun by examination of lines in the solar spectrum. The gas was not detected on earth until 1895. Later, neon, krypton, and xenon were obtained by careful evaporation of a large amount of liquid air. The sixth gas, radon, is formed during the radioactive decomposition of radium.

DISCOVERY OF OXYGEN

Oxygen was first isolated by Scheele in 1773, and he was the first to recognize that combustion and respiration depended on this gas. In 1774, Priestley discovered quite independently how to prepare oxygen by heating red mercury oxide.

At about the same time Lavoisier established the correct theory of combustion. He heated mercury at its boiling point in a fixed volume of air for about twelve days. The fraction of air used up was about one-fifth by volume and a red powder was formed on the mercury. Lavoisier collected and heated the red powder and obtained a gas which re-lit a glowing splint. The volume of this gas equalled the volume of air used up by the hot mercury. Lavoisier called this gas oxygen, which means acid-producer, because he incorrectly believed that all acids contain this gas. Let us now repeat Lavoisier's experiment.

To study the reaction between air and boiling mercury

1. Cover the bottom of a flask with mercury. Place a funnel in the mouth of the flask to help retain the poisonous mercury vapour during the experiment.
2. Heat the mercury in a fume cupboard until it is almost, but not quite, boiling. Continue the heating for three to seven days.

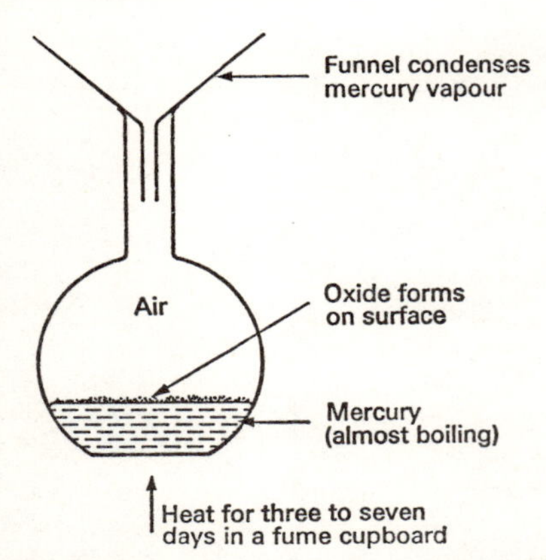

Fig. 7.1 Reaction between air and boiling mercury

3. Allow the contents of the flask to cool. Filter the product through a filter paper with a tiny pin-hole in it. The mercury passes through but the mercury oxide does not. A grey mixture of oxide mixed with some mercury remains on the paper.
4. Heat the grey mixture in a horizontal test-tube until the mercury distils off and leaves the oxide.

To study the action on hydrogen peroxide of manganese(IV) oxide

1. One-third fill a boiling-tube with hydrogen peroxide.
2. Weigh out about 1 g of dry manganese(IV) oxide, a black powder. Note the exact weight. Add all the powder very slowly to the peroxide and note any reaction. Test the gas with a glowing splint.
3. When the reaction is complete, filter the mixture. Wash the tube several times with distilled water to ensure that every particle of the manganese oxide is on the filter paper.
4. Dry the paper and oxide in sunshine or in a moderate oven.
5. Find the mass of the oxide by weighing and allowing for the weight of the filter paper. What conclusion do you draw?

To study the action of copper oxide on potassium chlorate(V)

1. One-quarter fill a test-tube with potassium chlorate, a white powder.
2. Weigh out about 1 g of dry copper(II) oxide, a black powder. Note the exact weight.
3. Heat the chlorate until it just melts at 370 °C. Observe if it seems to decompose.
4. Add the copper oxide to the molten chlorate and note any reaction. Test any gas with a glowing splint.

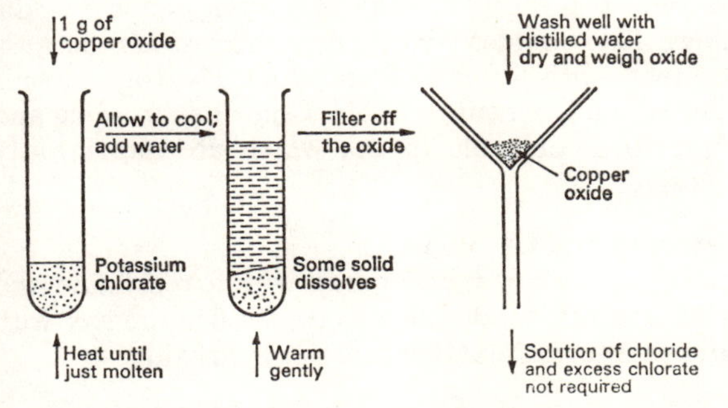

Fig. 7.2 The action of copper(II) oxide on potassium chlorate

5. *Finding the mass of copper oxide left* Allow the mixture to cool, add cold water and warm gently to dissolve the soluble portion of the mixture completely. Pour the solution and the remaining solid into a filter paper held in a funnel. Wash out the test-tube with distilled water into the filter paper and wash the filter paper. The insoluble copper oxide is held on the filter paper; when it has been washed with distilled water, dry and weigh how much oxide is left—allow for the weight of filter paper.

Manganese(IV) oxide, zinc oxide or silica can be used in place of copper oxide.

CATALYSIS

Certain chemicals called *catalysts* are able to change the speed of a reaction without themselves being used up in the process. There are many examples of this action that are used both in the laboratory and in industrial processes. The reaction may be either made slower or made quicker, but of course the most useful catalysts make the reaction quicker. The method of action of a catalyst is thought to be due to the formation of intermediate chemicals between the reagents that either block the reaction or help it along.

In the last two experiments, we saw that manganese (IV) oxide helped the release of oxygen from hydrogen peroxide, and that copper(II) oxide helped the release of oxygen from potassium chlorate. Zinc oxide, manganese(IV) oxide or silica also act as catalysts to potassium chlorate, so that the reaction proceeds at a faster rate and at a lower temperature.

Catalysts do not have to be oxides; many metals are also able to speed up a reaction by acting as catalysts. Can you describe the three reactions in chapter 5 that used iron or nickel as catalysts?

To prepare oxygen

From red mercury oxide

1. Place red mercury(II) oxide in a boiling-tube. Fit a stopper and delivery tube to the mouth of the boiling-tube as shown in Fig. 7.3a and clamp it so that the mouth is slightly lower than the base. The delivery tube leads to a gas-jar containing water over a trough for collecting the gas.

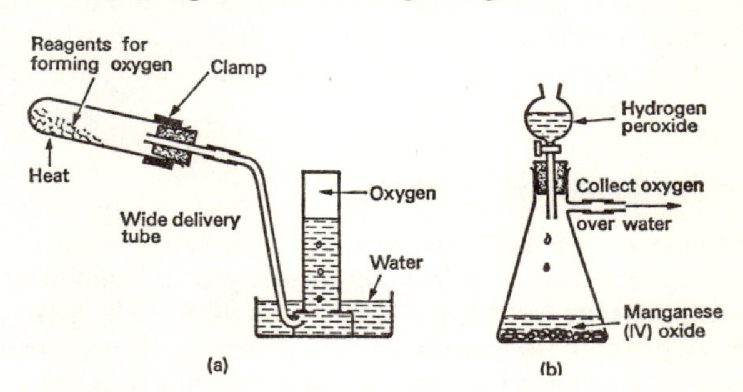

Fig. 7.3 Preparation of oxygen

2. Heat the oxide. When the air has been displaced from the apparatus, collect the oxygen in the gas-jar. Oxygen cannot be collected simply by displacing the air from a gas-jar because its density is about the same as that of air. The gas may be dried with concentrated sulphuric acid or with calcium chloride.
3. Note how silvery drops of mercury form on the cold parts of the tube.

$$2HgO = 2Hg + O_2$$

Note Red lead oxide, Pb_3O_4, may be used in place of mercury oxide.

From potassium chlorate(V)

1. Mix potassium chlorate with about one-tenth of its mass of copper oxide to help the release of oxygen.
2. Place the mixture in the apparatus shown in Fig. 7.3a and heat. The boiling-tube should be about one-third full if several gas-jars of oxygen are required, but leave plenty of space over the mixture to allow the oxygen to escape easily.

$$2KClO_3 = 2KCl + 3O_2$$

Note Manganese(IV) oxide can be used in place of copper oxide—use about one-quarter of the mass of potassium chlorate. These catalysts allow the decomposition to take place at 200 °C instead of about 600 °C and at a much steadier rate.

From potassium manganate(VII)

Use in place of the chlorate-oxide mixture in the same apparatus.

From hydrogen peroxide

1. Place manganese(IV) oxide at the bottom of a conical flask as shown in Fig. 7.3b on p. 37.
2. Use a dropping funnel to add 20-volume hydrogen peroxide solution as required. Oxygen is evolved without warming.

$$2H_2O_2 = O_2 + 2H_2O$$

This is the most convenient method of preparing oxygen.

Gas jars of oxygen are best obtained by filling them from an oxygen cylinder. Oxygen is manufactured cheaply from liquid air—see p. 40—and is stored under pressure.

PROPERTIES OF OXYGEN

PHYSICAL PROPERTIES

It is a colourless, tasteless, odourless gas. It condenses to a pale blue liquid at −183 °C or 90 K. The gas is sparingly soluble in water which dissolves about 4 per cent by volume of oxygen, compared with only 2 per cent by volume of nitrogen. Oxygen is readily absorbed by an alkaline solution of pyrogallol. It has no action on indicators.

CHEMICAL PROPERTIES

Oxygen does not burn. It re-lights a glowing splint; a burning splint and a burning candle both burn very brightly in the gas.

COMBUSTION OF NON-METALS

Sulphur burns with a bright blue flame, forming misty fumes with a choking smell. The product is mainly colourless sulphur dioxide with some white sulphur trioxide. Both gases react readily with water to form sulphurous and sulphuric acids respectively.

$$S + O_2 = SO_2; \quad SO_2 + H_2O = H_2SO_3,$$
sulphurous acid

$$2S + 3O_2 = 2SO_3; \quad SO_3 + H_2O = H_2SO_4,$$
sulphuric acid

Carbon in the form of red-hot charcoal glows brightly and forms colourless carbon dioxide. The gas forms a slightly acidic solution.

$$C + O_2 = CO_2; \quad CO_2 + H_2O = H_2CO_3,$$
carbonic acid

Phosphorus White phosphorus in oxygen bursts into flame on its own (spontaneously). Burning red phosphorus continues to burn in oxygen with a bright yellow flame. Dense white fumes are formed in each case. The fumes are oxides of phosphorus; they settle on the sides of the gas-jar and form acids with water.

$$4P + 5O_2 = P_4O_{10} \text{ or } 2P_2O_5$$
$$P_4O_{10} + \text{water} = \text{phosphoric acid}$$
$$4P + 3O_2 = P_4O_6 \text{ or } 2P_2O_3$$
$$P_4O_6 + \text{water} = \text{phosphorous acid}$$

Two oxides of phosphorus are formed:

P_2O_5 or P_4O_{10} is phosphorus pentoxide, phosphorus(V) oxide or tetraphosphorus decaoxide;

P_2O_3 or P_4O_6 is phosphorus trioxide, phosphorus(III) oxide or tetraphosphorus hexaoxide.

Many non-metals, such as sulphur, carbon, and phosphorus burn vigorously in oxygen to form acidic oxides. These oxides are called *acid anhydrides* because they react with water to form acids. Hydrogen forms water which is a neutral oxide. Chlorine, bromine and iodine form unstable oxides which are dangerously explosive.

COMBUSTION OF METALS

Calcium The hot metal burns with a red flame and forms a white solid, calcium oxide. This reacts with water and forms alkaline calcium hydroxide.

$$2Ca + O_2 = 2CaO; \quad CaO + H_2O = Ca(OH)_2$$

Magnesium The burning metal continues to burn and gives out a brilliant white light. A light ash of oxide is formed which reacts slightly with water to form a weakly alkaline solution of magnesium hydroxide.

$$2Mg + O_2 = 2MgO; \quad MgO + H_2O = Mg(OH)_2$$

Iron Red-hot iron wool or filings burn readily and form a shower of bright sparks. The product is black iron oxide which is insoluble in water and therefore has no action on indicators.

$$3Fe + 2O_2 = Fe_3O_4, \quad \text{triiron tetraoxide}$$

Metals burn in oxygen to form *basic oxides*. If these oxides are soluble in water, they form alkalis.

Mix the carbonic acid formed by burning carbon with the calcium hydroxide formed by burning calcium; try to explain what happens.

Some metals (Pb, Zn and Al) form *amphoteric oxides* which act both as acidic oxides and as basic oxides. Refer to p. 65.

REACTIVITY OF METALS WITH OXYGEN

We have now seen how a number of different metals react with air or with oxygen. From these experiments it is possible to make a list showing the metals in the order of their reactivity with oxygen. The metals at the top of the list react the most easily with oxygen; those at the bottom, the least easily.

1. *Sodium* It burns in air at room temperatures.
2. *Calcium* It reacts less easily at room temperatures, and forms an oxide on the outside surface.
3. *Magnesium* It burns readily at higher temperatures.
4. *Zinc* and *iron* They burn slowly in oxygen only and do not combine in air unless in powder form.
5. *Lead* and *copper* Both form an oxide on the surface which stops them burning, even in oxygen.

We can see that this order is the same as that obtained by studying the actions of metals on water, steam, and acids. Here, sodium and calcium have the greatest affinity for oxygen in this series, and lead and copper have the least. The metals at the top of the series are able to remove oxygen from the oxides of metals at the bottom of the series. Sodium and calcium are so reactive as to be dangerous.

To study the action of magnesium and iron on the oxides of lead and copper

Magnesium

1. Mix magnesium powder with either lead oxide or red copper oxide, which is copper(I) oxide, Cu_2O.
2. Heat the mixture on porcelain or on asbestos paper, and be prepared for a vigorous reaction.

The magnesium removes oxygen from the oxides.

$$Mg + PbO = MgO + Pb;$$
$$Mg + Cu_2O = MgO + 2Cu$$

Iron

1. Mix non-greasy iron filings with copper(II) oxide and heat on asbestos paper. Note if there is any sign of chemical reaction.
2. Repeat the test using iron filings and lead oxide.

The iron removes oxygen from the oxides.

$$2Fe + 3CuO = Fe_2O_3 + 3Cu$$
$$2Fe + 3PbO = Fe_2O_3 + 3Pb$$

In each case we see that the metal that is higher in the reactivity series acts as a reducing agent; that is, it removes oxygen from the oxides. The oxides of the metals lower in the reactivity series act as oxidizing agents by giving oxygen to metals higher in the series.

To study the action of carbon on some oxides

1. Mix powdered charcoal with dry copper(II) oxide. Place the mixture in asbestos paper folded into a cone.
2. Hold the paper with tongs and heat in a strong flame. Note if there is any reaction. Observe the colour of the product.
3. Repeat the test with lead oxide and try to obtain a small ball of lead.

$$C + 2CuO = CO_2 + 2Cu$$
$$C + 2PbO = CO_2 + 2Pb$$

Many metals are extracted from their oxides by the reducing action of carbon on their oxides. The carbon is usually in the form of coke. Magnesium could be used, but coke is far cheaper.

Our next experiment determines whether magnesium or carbon has the greater affinity for oxygen.

To study the action of magnesium on carbon dioxide

1. Fill a gas-jar with carbon dioxide.
2. Tie clean magnesium ribbon to a combustion spoon. Light the metal and lower it into the carbon dioxide.

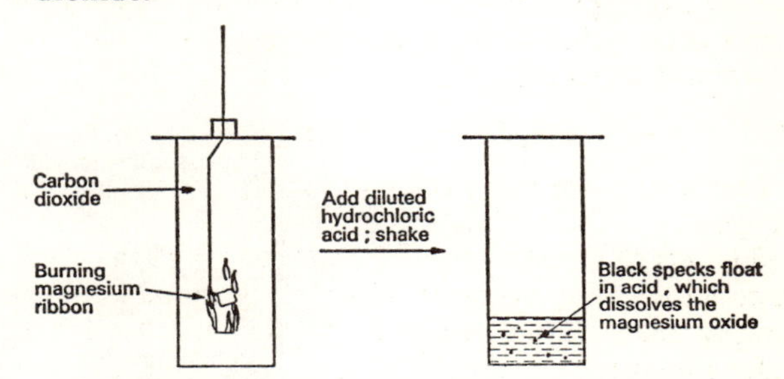

Fig. 7.4 Action of magnesium on carbon dioxide

3. Note the colours of the products. Remove the spoon and unused magnesium.
4. Add dilute hydrochloric acid and a few drops of the concentrated acid. Shake gently until one of the products has dissolved.

$$2Mg + CO_2 = \underset{\text{(black)}}{C} + \underset{\text{(white)}}{2MgO}$$
$$MgO + 2HCl = MgCl_2 + H_2O$$

The acid dissolves the white oxide and leaves black specks of carbon.

INDUSTRIAL OXYGEN

Liquid air is a pale-blue liquid which contains 23 per cent by mass of oxygen. The liquid boils at −193 °C (80 K) and the vapour contains only 13 per cent of oxygen because nitrogen is more volatile and boils at a lower temperature.

The boiling point of nitrogen is 77 K
The boiling point of oxygen is 90 K

Both oxygen and nitrogen are obtained by fractional distillation of liquid air.
Liquefaction of air Air is compressed to about 200 atmospheres; this is a pressure 200 times greater than atmospheric. It is then washed with sodium hydroxide solution to remove carbon dioxide and cooled to about −25 °C to freeze out the water. Carbon dioxide and water would both block the apparatus if they were not removed.

The cooled, compressed air is allowed to expand rapidly. It does work as it expands and therefore loses energy. The loss of energy causes the temperature of the gas to fall.

A common example of this fall in temperature is given when the air in a bicycle tyre is suddenly released: it feels cool. On the other hand, when the tyre is pumped up, the pump begins to feel warm.

The processes of compression, cooling, and then expansion to produce much greater cooling are continued and repeated until the air liquefies.
Fractional distillation of liquid air The liquid air enters a tower which contains many compartments one above the other and each at a slightly higher temperature than the one above it. Boiling liquid air

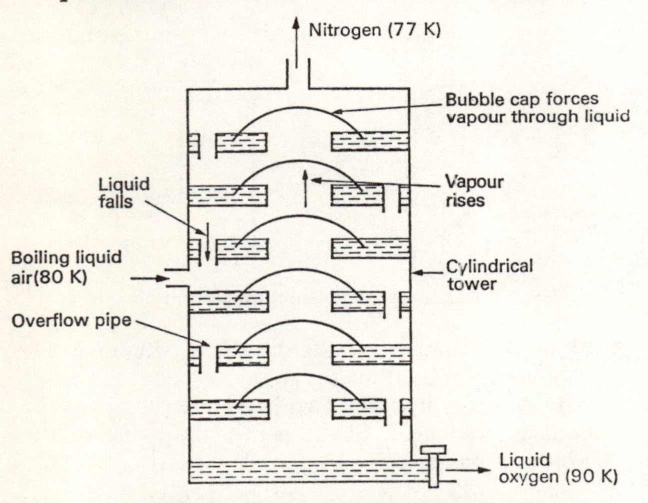

Fig. 7.5 Fractional distillation of liquid air

enters the tower near the centre compartment at a temperature of 80 K. There it separates into vapour which contains 85 per cent nitrogen that passes up the tower, and into liquid which contains about 36 per cent oxygen that passes down the tower. In each compartment of the tower, the mixture of liquid and vapour reaches a new equilibrium that depends upon the temperature. Thus 99 per cent pure oxygen liquid comes out of the bottom of the tower, while nitrogen gas mixed with the noble gases comes out of the top of the tower.

The liquid oxygen is transported to where it is required in large insulated metal tanks, or stored in small cylinders as the compressed gas.

USES OF OXYGEN

Medical Sick or injured persons are sometimes given pure oxygen gas; it is frequently used in operations by dentists and surgeons. Mountain climbers, aircraft pilots and deep-sea divers use the gas for breathing.
Hot flames If oxygen gas is well mixed with a fuel gas before burning, the flame can be used for cutting and joining metals. Gases that are used include bottled gas, which is mainly butane, acetylene, and hydrogen. The oxy-acetylene or the oxy-hydrogen flame gives temperatures over 2000 °C and can even be used under water, for example, to cut the steel of sunken ships. Liquid oxygen is used to burn hydrogen in space rockets.
Steel manufacture Impurities in molten iron, such as carbon, sulphur, and phosphorus can be converted to volatile oxides by blowing air through the mixture. However, the nitrogen in the air makes the steel unsuitable for many purposes and it is now usual to replace the air either with pure oxygen or with a mixture of oxygen and superheated steam. Very large quantities of the gas are needed for this purpose, and chemical plants produce the oxygen very cheaply by the tonne—'tonnage' oxygen.

AIR IS A MIXTURE

Air is not a compound for the following reasons:
Separation Its constituents can be separated by physical means such as cooling to remove the water vapour and fractional distillation to remove the nitrogen. Water dissolves twice as much oxygen as nitrogen from the air; the gas obtained by boiling water to free the dissolved gases contains about 33 per cent oxygen.
Properties These are the average of those of oxygen and nitrogen. Oxygen supports combustion readily, nitrogen does not. Air supports combustion fairly well.
Energy Air can be made by mixing the various gases: oxygen, nitrogen, noble gases, etc. No heat is evolved or absorbed during the mixing, and therefore chemical combination does not take place.
Composition The composition of air varies from time to time and from place to place. It does not correspond to that of a compound with a simple formula. For example, the compound N_4O would contain 20 per cent by volume of oxygen and not 21 per cent, as in air.

SUMMARY

Air: a mixture of gases containing nitrogen (78 per cent), oxygen (21 per cent), noble gases, carbon dioxide and water vapour. Nitrogen dilutes oxygen in the air and slows reactions caused by the oxygen.
Noble gases: helium, neon, argon, krypton, xenon, radon. Originally called 'rare' or 'inert'.

Catalyst: a substance which itself remains unchanged, but which is able to speed up (or to slow down) the rate of a reaction. Examples include manganese(IV) oxide, silica, nickel, platinum, vanadium(V) oxide.

Oxygen: colourless, odourless gas, slightly soluble in water, neutral to indicators.
Laboratory preparation: from hydrogen peroxide and manganese(IV) oxide as catalyst; by heating potassium chlorate (with a catalyst) or potassium manganate(VII); by heating red mercury oxide or red lead oxide.
Chemical properties: re-lights a glowing splint; forms acidic oxides with many non-metals, e.g. sulphur, carbon, and phosphorus; forms basic oxides with metals; forms neutral oxide (water) with hydrogen.
Industrial preparation: from liquid air (after removing carbon dioxide and water) by fractional distillation to give pale-blue liquid at base of tower at −183 °C or 90 K and 200 atmospheres pressure.
Uses: medical, cutting, and joining metal, and in steel manufacture.

Reactivity series: sodium, calcium, magnesium, zinc, iron, lead, copper.
Basic oxide: oxide of metal; if soluble in water, it forms an alkali. Oxides of Pb, Zn, and Al also dissolve in strong alkalis; they are called *amphoteric* oxides.
Acidic oxide: oxide of a non-metal that will form an acid with water, an acid anhydride.
Reducing agent: removes oxygen from a compound, which in turn acts as an *oxidizing agent*. Elements in the reactivity series can reduce oxides of elements lower down the series.

QUESTIONS

1. When 1 g of manganese(IV) oxide is added to hydrogen peroxide solution, effervescence can be observed. How would you collect and identify the gas liberated? Describe briefly what operations you would perform to find out the mass of manganese(IV) oxide left when the reaction was completed.

2. Give three different ways of producing oxygen in the laboratory. For one of these ways, make a diagram of the apparatus you would use for generating and collecting the gas. (C.)

3. Explain, with examples, the meanings of acidic oxide and basic oxide. Describe the preparation of a specimen of an acidic oxide and a basic oxide from two elements. Outline the commercial method by which oxygen and nitrogen are extracted from air.

4. Give short accounts of methods by which iron, copper, and carbon can be obtained from their oxides. Use the results of the experiments to place these elements in an order of reactivity.

5. Describe two chemical tests that would enable you to distinguish between charcoal (carbon), copper(II) oxide, and manganese(IV) oxide. Mention two uses of oxygen which are not connected with respiration. (C.)

6. A sample of the gases dissolved in river water is obtained by boiling the water. How would you determine the percentage by volume of oxygen in the dissolved gases? Explain how and why the oxygen content differs from that of the atmosphere. Of what importance is the dissolved oxygen in river water and in sea water?

8 Atoms and Molecules

DALTON'S ATOMIC THEORY

Over 2000 years ago Greek writers held two conflicting opinions on matter. Some believed that matter is continuous, just as water seems to be continuous in lakes and seas. Others believed that matter consists of particles which cannot be broken down further, just as a sand-hill consists of sand particles that are easy to separate but not easy to break down. These tiny particles of matter were called *atoms*, a word which means 'not cut' in Greek. These ideas were mere opinions and speculations and were not supported by experiments or by facts.

John Dalton, an English schoolmaster, was the first man to state a theory of matter and to support it by experimental evidence. He asserted that:

1. Matter consists of tiny indivisible particles which are called atoms.
2. Atoms cannot be created or destroyed.
3. The atoms of one element are all identical, particularly in mass, and differ from atoms of other elements.
4. Chemical combination occurs only between small numbers of atoms to form 'compound atoms', which are now called molecules.
5. 'Compound atoms' of a compound are identical and differ from 'compound atoms' of other compounds.

Dalton wrote in his *New System of Chemical Philosophy*: 'Whether the ultimate particles of a body, such as water, all are alike, that is of the same figure, weight, etc. is a question of some importance . . . we have no reason to apprehend a diversity in these particulars. . . . If some of the particles of water were heavier than others, if a parcel of the liquid on any occasion were constituted principally of these heavier particles, it must be supposed to affect the specific gravity . . . a circumstance not known. Similar observations may be made on other substances. Therefore we may conclude that the ultimate particles of all homogeneous bodies are perfectly alike in weight, figure, etc. In other words, every particle of water is like every other particle of water: every particle of hydrogen is like every other particle of hydrogen, etc.'

The modern definition of an atom and molecule is as follows:

An *atom* is the smallest uncharged particle of an element that can take part in a chemical change.

A *molecule* is the smallest uncharged particle of an element or compound which can exist on its own.

The word 'uncharged', meaning electrically neutral, appears in the definition because particles with a positive or a negative charge, called ions, can also take part in chemical changes and can sometimes exist on their own.

The number of atoms in one molecule of an element is the atomicity of the element. For example, the atomicities of argon, oxygen, and phosphorus vapour are 1, 2 and 4 respectively, and the formulae of the molecules are Ar, O_2, and P_4.

The movement (diffusion) of solutes and liquids

Solutes Drop a crystal of copper sulphate, ammonium dichromate, or other coloured solute into a tall beaker of water. Leave for several days and observe what happens.

Liquids One-third fill a beaker with water. Leave for about ten minutes until the water stops moving about. Fill a pipette with methylated spirit, hold the tip of the pipette against the inside of the beaker and just above the water surface. Let the spirit run out slowly. It forms a layer over the denser water. Leave for several days and observe what happens.

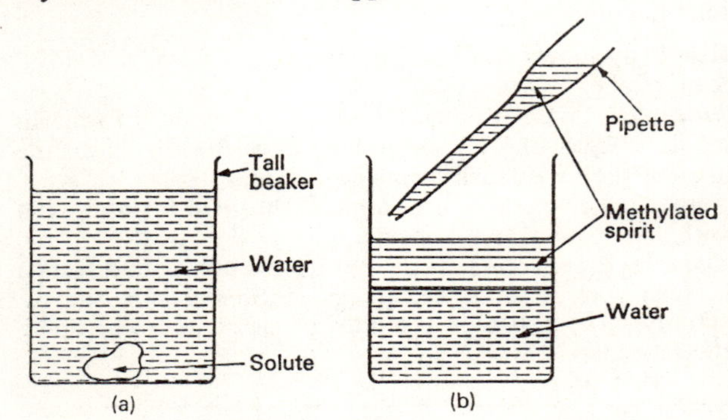

Fig. 8.1 Diffusion of a solute and a liquid

To demonstrate diffusion of gases

1. *Carbon dioxide* Fill a gas-jar with carbon dioxide, which is 1½ times denser than air, and invert it over a jar of air.
2. After ten to twenty seconds, add lime water to the lower jar which originally contained only air, and shake. The test shows that carbon dioxide is a dense gas that quickly falls into a jar of air.
3. Repeat the test with the jar of air over the carbon dioxide. Leave for five to ten minutes, and then test the gas in the upper jar for carbon dioxide.
4. *Hydrogen* Invert a jar of hydrogen over a jar of air, which is 14½ times denser than hydrogen. Leave for five to ten minutes and then test for hydrogen in the lower jar by using a burning splint.
5. *Bromine* This is a corrosive, poisonous liquid which forms a very dense vapour at room temperature. Carefully add a few drops of the liquid to a gas-jar. Invert a jar of air over it and leave for some time. Arrange a large sheet of white paper behind the jars and look through the sides of the jars to see the reddish colour of the bromine vapour.
6. *Nitrogen dioxide* This is a very dense brown gas which may be used in place of bromine. Prepare the

gas by adding a few cm^3 of concentrated nitric acid to copper in a gas-jar.

7. *Ammonia and hydrogen chloride* Take glass tubing about 1 m long and 2 cm diameter and clamp it horizontally in a fume cupboard or in the open air. Hold one piece of glass wool in tongs and dip it into concentrated ammonia solution; place it in one end of the tube. Dip a second piece of glass wool into

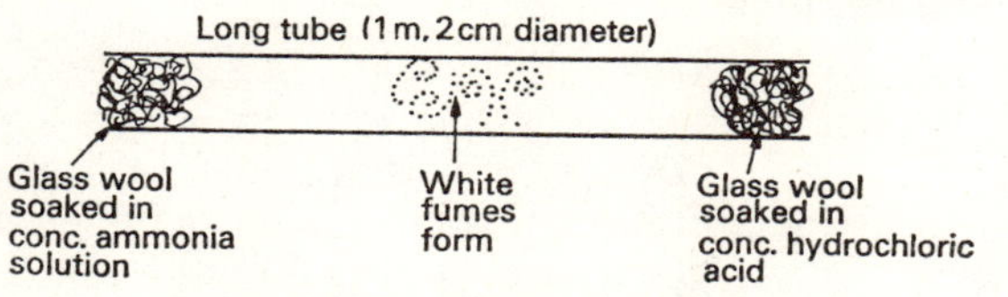

Fig. 8.2 The diffusion of gases

concentrated hydrochloric acid and place it in the other end. Place black paper behind the tube to make any change easily visible, and observe what happens.

8. *Diffusion in a vacuum and in air* The diagram, Fig. 8.3, makes the method and observations clear. Try to explain the results.

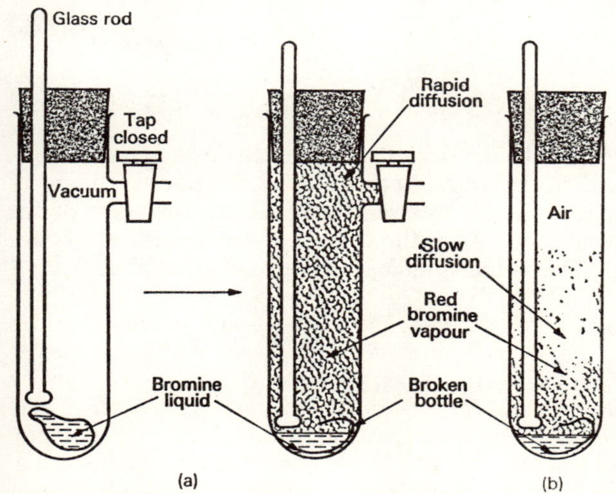

Fig. 8.3 The diffusion of bromine in a vacuum and in air

Diffusion

If a little ammonia solution, petrol or ether is placed on a watch-glass in a closed room, can you smell the substance at the other side of the room? Since air currents do not move the vapour to any great extent, the vapour must move by itself. The motion is called *diffusion.*

We have seen from the experiments in this chapter that solutes diffuse through solvents, liquids diffuse into other liquids, and gases diffuse into gases. We also know that gases diffuse into liquids because water in large reservoirs gradually absorbs air.

Diffusion indicates, but does not prove, that matter is made of particles.

To observe the movement of tiny particles

1. Add a speck of toothpaste or Aquadag, which is a special form of graphite, or one drop of indian ink to about 5 cm^3 of water.
2. Place one drop of the diluted mixture on a microscope slide and look at it under a microscope. The illumination should be from the sides.

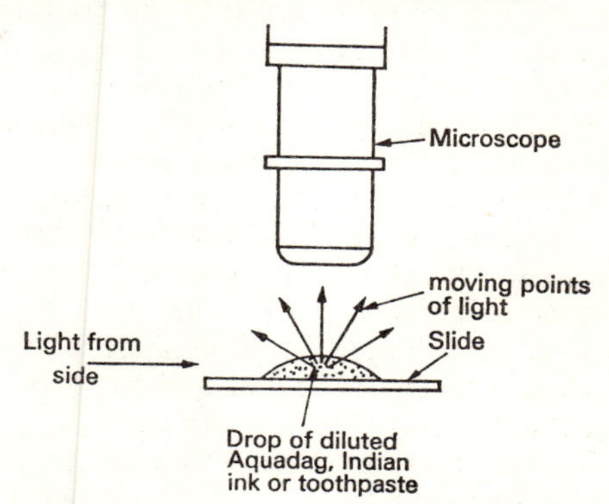

Fig. 8.4 Observing the Brownian movement

Particles can be seen only by light that they scatter. Scattered points of light which move about in a zig-zag manner can be seen.

The continual movement of the particles is the *Brownian movement.* It is explained by the irregular bombardment of a particle by molecules of the water or liquid in which it is suspended. Large particles receive millions of blows each second on all sides which cancel each other out. A smaller particle, however, receives more blows at a particular moment on one side than on the other and therefore it moves. The direction of the force changes all the time and the particle moves irregularly. The movement was first observed by Brown in 1827 when he examined the granules formed by pollen grains when they burst in water.

When sunlight shines into a dark room we see the Brownian movement of suspended dust particles in the air. The Brownian movement is evidence for the existence of tiny particles in a liquid or in a gas and for their perpetual motion.

To show that particles of a solute must be small

1. Dissolve 1 g of potassium manganate(VII) in 1000 cm^3 of water to form a purple solution; 1 cm^3 of solution contains 0·001 g of manganate(VII), and 1 drop contains about 0·000 2 g.
2. Add one drop of solution to 200 cm^3 of water; 1 cm^3 of the new solution contains only 0·000 001 g of manganate(VII). Is the solution still coloured?

The test shows that the particles of manganate(VII) must be very small.

To measure the length of a stearic acid molecule

In this experiment the area of a surface of water that is covered by a known volume of stearic acid is measured. The acid is dissolved either in benzene or in petroleum ether, which quickly evaporates on the surface of the water. The height of the acid molecule can be calculated by assuming that the acid layer is one molecule thick. The experiment can be repeated with palmitic acid or oleic acid.

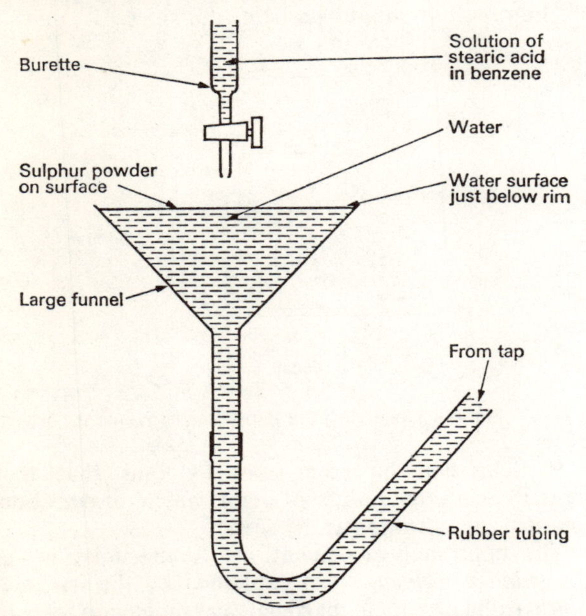

Fig. 8.5 Measuring the length of a molecule of stearic acid

1. Clamp a large funnel over a sink with its rim horizontal. Fill the funnel with running water and let the water run over the rim for a short time. Turn off the tap.
2. Use a small beaker to remove a little water from the funnel so that the water surface is just below the rim and a few drops of acid solution can be added later. Do not touch the water with your fingers as you do this.
3. *Determining the size of one drop* Add benzene to a burette. Note the reading. Let fifty drops run out into a beaker and again note the reading. The difference is the volume of fifty drops. Empty the burette.
4. *Preparing the solution of stearic acid* Make a solution of stearic acid in benzene so that it contains 0·1 g of acid in 1000 cm³ of solvent. Only a few cm³ are required. Add this solution to the burette. The density of stearic acid is known and therefore its volume per 1000 cm³ of benzene solution is known.
5. *Finding the area of the surface film* Clamp the burette with its tip about 1 cm above the water surface.
6. Cover the surface of the water with fine flowers of sulphur from a cloth bag. Add the acid solution one drop at a time, and note how the powder is pushed aside by the solution. At least half of the water surface should be covered by the acid. The solvent quickly evaporates leaving a single layer of acid molecules.
7. Count the number of drops of acid solution that you have added. Measure the diameter of the film of the acid.
8. Turn on the tap to wash away the acid and powder and to clean the rim of the funnel. Repeat the test several times.

SPECIMEN RESULT

Volume of 50 drops of acid solution $= v$ cm³
Density of stearic acid $= 0{\cdot}94$ g/cm³
Concentration of acid solution $= 0{\cdot}1$ g/dm³
Diameter of surface film $= d$ cm
Number of drops added to form the film $= n$

CALCULATION

50 drops of solution have a volume of v cm³
$\therefore$ 1 drop of solution has a volume of $v/50$ cm³
i.e. n drops of solution have a volume of $nv/50$ cm³
1000 cm³ of solution contain 0·1 g of acid

$$\therefore\ nv/50 \text{ cm}^3 \text{ of solution contain } \frac{nv}{50} \times \frac{0{\cdot}1}{1000} \text{ g of acid}$$

$$= \frac{nv}{500\,000} \text{ g of acid}$$

$$= \frac{nv}{500\,000 \times 0{\cdot}94} \text{ cm}^3 \text{ of acid}$$

(since the density of stearic acid is 0·94 g/cm³).

This volume of acid covered a surface of diameter d cm, i.e. an area $\pi d^2/4$ cm². If the thickness of the acid film is t cm, the volume of the film is $\pi d^2 t/4$ cm³.

This is the volume of the acid calculated above,

i.e. $$\frac{\pi d^2 t}{4} = \frac{nv}{500\,000 \times 0{\cdot}94}$$

t can therefore be calculated:

$$t = \frac{4nv}{500\,000 \times 0{\cdot}94 \times \pi d^2}$$

Its value for stearic acid is usually about 2×10^{-7} cm or 2 nanometres.

ABSOLUTE AND RELATIVE MASSES OF ATOMS AND MOLECULES

The absolute mass of all atoms is now known although the methods used to determine them need not concern us. Two values are:

Hydrogen atom: $0{\cdot}167\,35 \times 10^{-26}$ kg
Oxygen atom: $2{\cdot}656\,5 \times 10^{-26}$ kg

The masses of the molecules of these two gases are twice these values because each molecule consists of two atoms. Clearly, the masses are so small that it is not convenient to use the kilogram or the gram for comparing the mass of atoms or molecules.

Originally, the mass of a hydrogen atom, which is the lightest atom, was taken as the standard mass; the masses of other atoms and molecules were compared with it. On this standard the relative atomic mass, sometimes called atomic weight, of oxygen was 15·88, i.e. one atom of oxygen is 15·88 times heavier than one atom of hydrogen.

Later, the mass of an oxygen atom was taken as the standard. The relative atomic mass of oxygen was taken as 16 exactly. On this standard the relative atomic mass or atomic weight of hydrogen was 1·008. The discovery that all atoms of oxygen are not identical and that ordinary oxygen contains three different atoms (explained in chapter 14) meant that the standard mass was merely an average of three different masses.

In 1961, an international standard was adopted. The atomic mass of the commonest carbon atom, called carbon-12 with the symbol ^{12}C, was taken as 12 exactly. This new standard is now used in all science work throughout the world. On the carbon-12 standard, the relative atomic mass of oxygen is 15·999 and of hydrogen is 1·008. The relative molecular masses are twice these values because the molecules of both elements contain two atoms, and the elements are called *diatomic*. Molecules of argon and mercury contain only one atom; the two elements are *monatomic* and their relative molecular masses are the same as their relative atomic masses.

The relative *atomic* mass of an element, or atomic weight

$$= \frac{\text{mass of 1 atom of the element} \times 12}{\text{mass of 1 atom of carbon-12}}$$

The relative *molecular* mass of a substance, or molecular weight

$$= \frac{\text{mass of 1 molecule of the substance} \times 12}{\text{mass of 1 atom of carbon-12}}$$

THE MOLE AND THE AVOGADRO CONSTANT

The basic unit of mass in chemistry is the *mole*; this was formerly called the gram-atom or gram-mole or formula weight.

One mole of a substance is that mass in grams which contains the same number of molecules or atoms as there are atoms in 12 g of carbon-12.

The number of atoms in 12 g of carbon-12 is very large; it has been determined experimentally by a method with which we need not be concerned. This number is called the *Avogadro constant* (symbol L) and its numerical value is $6{\cdot}023 \times 10^{23}$. It follows from the definition of the mole that one mole of any substance contains this number of molecules or atoms.

However, as we shall see, the chemist need not be concerned with the actual number of molecules taking part in a reaction if he refers only to moles, which are the same as either the relative molecular mass of a substance or, if the molecule contains only one atom, as the relative atomic mass. Some examples will make this clear.

Sodium metal is monatomic; it contains only one atom in its molecule. Its relative atomic mass is 23.

$\therefore$ one mole of sodium has a mass of 23 g

Hydrogen gas (H_2) is diatomic; it contains two atoms in its molecule. Its relative molecular mass is the sum of its relative atomic masses.

$$H_2 = 1{\cdot}008 + 1{\cdot}008 = 2{\cdot}016$$

$\therefore$ one mole of hydrogen molecules has a mass of 2·016 g or approximately 2 g, and 1 mole of hydrogen atoms has a mass of 1·008 g.

Similarly, oxygen is diatomic, that is, its molecule contains two atoms of oxygen; 1 mole of oxygen molecules has a mass of $15{\cdot}999 + 15{\cdot}999 = 31{\cdot}998$ g or approximately 32 g, and 1 mole of oxygen atoms is 15·999 g.

Caustic soda has the formula NaOH. One mole of caustic soda therefore has an approximate mass of $23 + 16 + 1 = 40$ g.

A list showing the relative atomic mass of all elements is shown inside the front cover of this book.

We know that 1 mole of all substances contains L molecules or atoms. We therefore know that there are $6{\cdot}023 \times 10^{23}$ molecules of caustic soda in 40 g of caustic soda and $6{\cdot}023 \times 10^{23}$ atoms of oxygen in 15·999 g of oxygen. The concept of mole is also applied to ions, and 1 mole of ions was formerly called one gram-ion. One mole of hydrogen ions is $6{\cdot}023 \times 10^{23}$ ions and it is almost 1·008 g, that is, it is slightly less than 1 mole of hydrogen atoms.

1·00 M SOLUTIONS

A solution which contains 1 mole of a solute is 1000 cm^3 of solution is called a 1·00 M solution, normally shortened to M solution. Thus 1000 cm^3 solution that contains 40 g of caustic soda is M NaOH. Note that this solution, which weighs approximately 1020 g, contains only 980 g of water because the remaining 40 g are caustic soda.

A solution half as concentrated is called 0·50 M; it contains 20 g of caustic soda in 1000 cm^3 of solution and is called M/2 NaOH or 0·5 M NaOH. A solution of caustic soda that is twice as concentrated as M NaOH is called 2 M NaOH.

Full details of the use of these expressions are given in chapter 20.

SUMMARY

Dalton's atomic theory: matter consists of indivisible *atoms*; atoms cannot be created or destroyed; atoms of one element are identical; atoms combine in small, whole numbers.

Atom: smallest, uncharged particle of an element that can take part in a chemical change.
Molecule: smallest, uncharged particle of an element or compound that can exist on its own.
Atomicity: the number of atoms in one molecule of an element. Monatomic elements contain one atom in the molecule; diatomic elements contain two.
Size of atoms and molecules: too small to measure directly.

Proof of existence of atoms and molecules:
Diffusion: mixing of solutes in solution and of gases or liquids amongst each other by unaided movement of their particles.
Brownian movement: irregular, unaided movement of small particles in a gas or a liquid or suspension.
Length of stearic acid molecule: measure area over which acid in benzene solution spreads on water when it pushes sulphur powder to the sides of the water surface.

Mass of atoms, molecules, and ions; measured relative to that of an atom of carbon-12 (^{12}C).
Mole: the mass in grams of a substance which contains as many elementary units (atoms or molecules or ions) as there are carbon atoms in 12 g of carbon-12.
Avogadro constant: $6{\cdot}023 \times 10^{23}$ mol^{-1}.
Relative atomic mass (atomic weight): the mass of an atom of an element relative to that of one twelfth of an atom of carbon-12 ($^{12}C = 12{\cdot}000$).
Relative molecular mass (molecular weight): the mass of a molecule of an element or compound relative to that of one twelfth of an atom of carbon-12.

QUESTIONS

1. Define an atom and a molecule and give examples. Give a brief account of the atomic theory as stated by Dalton.
2. What is the modern standard for relative atomic masses (atomic weights)? Explain why the standard O = 16 fell into disuse.
3. Draw and label an apparatus you would use to show that hydrogen diffuses more rapidly than air. What do you think will happen if a gas-jar of nitrogen is inverted over a similar jar of carbon dioxide and the cover plates removed? Describe a simple test to support your suggestion.
4. A flask is filled with hydrogen at atmospheric pressure. The stopper of the flask does not fit perfectly, and the gap is equivalent to a very small hole. The flask stands in the laboratory for a day or so. State and explain what you think will happen.
5. A horizontal glass tube is 2 m long. One end is closed with glass wool soaked in concentrated ammonia solution and at the same time the other end is closed with glass wool soaked in concentrated hydrochloric acid. Describe what you think will form inside the tube. Explain (*a*) how it forms and (*b*) its position in the tube.
6. 'The study of the Brownian movement has provided the most convincing evidence for the real existence of molecules and for perpetual molecular motion which is postulated by the kinetic theory'. What is the Brownian movement? How would you demonstrate it? Discuss the statement.

9 Chemical Formulae

SYMBOLS OF ELEMENTS

When he was developing his atomic theory, Dalton used symbols for atoms of elements. Here are some of the symbols he used:

⊙ for a hydrogen atom

○ for an oxygen atom

● for a carbon atom

In 1819 Berzelius suggested that the initial letters of elements would be more convenient to use; for example, H for hydrogen, O for oxygen and C for carbon. Finally, a complete system of symbols was drawn up by international agreement, based on the work of Berzelius. Some of the symbols do not agree with the British names for the elements because they are based on the Latin names. The complete list of symbols for all the elements is given inside the front cover of this book. Here are the symbols for some of the elements you will meet in this course.

Aluminium	Al	Hydrogen	H	Oxygen	O
Argon	Ar	Iodine	I	Phosphorus	P
Barium	Ba	Iron	Fe	Potassium	K
Bromine	Br	Lead	Pb	Silver	Ag
Calcium	Ca	Lithium	Li	Sodium	Na
Carbon	C	Magnesium	Mg	Sulphur	S
Chlorine	Cl	Manganese	Mn	Tin	Sn
Copper	Cu	Mercury	Hg	Uranium	U
Fluorine	F	Nitrogen	N	Zinc	Zn

To find the formula of black copper oxide

1. Weigh a porcelain boat.
2. Add dry copper oxide to the boat and weigh again.
3. Pass dry hydrogen from a cylinder or dry town gas over the heated oxide using the apparatus shown in Figs. 5.12 or 5.13 on p. 29.
4. When the reaction is complete, allow to cool and weigh the porcelain boat with the copper that has now been formed.

The following precautions must be taken in this experiment to avoid errors:

(*a*) Copper oxide is hygroscopic. It must be dried by heating to about 400 °C and then allowed to cool in a desiccator.
(*b*) Use analytical grade copper oxide because ordinary copper oxide contains copper as an impurity.
(*c*) If town gas is used, do not heat the oxide too strongly during the experiment, because the gas decomposes and forms some black carbon on the copper.
(*d*) Slope the combustion tube or boiling-tube downwards so that the water formed does not drop onto the hot glass.
(*e*) When reduction of the oxide is complete, turn off the Bunsen flame but allow the gas to pass over the hot copper. If air enters the apparatus, it oxidizes some of the copper back to the oxide.
(*f*) After weighing the boat with the copper at the end of the experiment, replace them in the tube and again heat and pass gas over the copper. Allow to cool and re-weigh. If the mass is less, the reduction was not complete. Continue until two weighings give the same result. This 'heating to constant mass' shows that all the oxide has been reduced to metal.

Specimen results and calculation

1.	Mass of porcelain boat	= ?
2.	Mass of boat + copper oxide	= ?
3.	Mass of boat + copper	= ?
3 − 1.	Mass of copper	= x g
2 − 3.	Mass of oxygen	= y g

y g of oxygen combine with x g of copper.
∴ 16 g of oxygen combine with $16x/y$ g of copper or $16x/63{\cdot}5y$ mole of copper because the relative atomic mass of copper is 63·5.

Thus, 1 mole of oxygen atoms combines with $16x/63{\cdot}5y$ mole of copper atoms. The value of the fraction $16x/63{\cdot}5y$ must be a simple fraction such as $\frac{1}{2}$, 1, $1\frac{1}{2}$ or 2, which corresponds to a formula for copper oxide of CuO_2, CuO, Cu_3O_2 or Cu_2O.

To determine the formula of magnesium oxide

1. Place a little asbestos paper on the inside of a crucible to protect it. Weigh the crucible, lid, and asbestos.
2. Clean some magnesium ribbon with sandpaper and add it to the crucible. Weigh again.
3. Convert the metal to its oxide, and a little nitride, as shown in Fig. 2.4 on p. 12.
4. Allow to cool, add a little water to react with the nitride, and smell the gas formed. Heat the mixture until dry and weigh again.

$$3Mg + N_2 = Mg_3N_2\text{, magnesium nitride}$$

$$Mg_3N_2 + 6H_2O = 3Mg(OH)_2 + 2NH_3$$

$$Mg(OH)_2 = MgO + H_2O$$

Specimen results and calculation

1. Mass of crucible, lid, and asbestos = ?
2. Mass of apparatus + magnesium = ?
3. Mass of apparatus + magnesium oxide = ?

2 − 1. Mass of magnesium = x g

3 − 2. Mass of oxygen = y g

y g of oxygen combine with x g of magnesium.
∴ 16 g of oxygen combine with $16x/y$ g of magnesium, i.e. 1 mole of oxygen atoms combines with $16x/24y$ mole of magnesium atoms, because the relative atomic masses of oxygen and magnesium are 16 and 24 respectively.

The fraction $16x/24y$ (= $2x/3y$) must be a simple fraction such as $\frac{1}{2}$, 1 or $1\frac{1}{2}$, etc., corresponding to a formula of MgO_2, MgO, Mg_3O_2, etc.

To find the formula of water

Before starting this experiment, refer to the precautions in the first experiment in this chapter on p. 47. Only if these precautions are followed will an accurate result be obtained.

1. Dry some copper oxide, CuO, by heating to about 400 °C and then cooling in a desiccator.
2. Put the oxide in a combustion tube and keep it in place with two plugs of glass wool.
3. Weigh the tube, copper oxide, and glass wool.
4. Weigh the whole apparatus shown in Fig. 9.1.
5. Town gas cannot be used in this experiment if it consists mainly of methane. Pass hydrogen from a cylinder through the calcium chloride to dry it and then through the heated oxide until all the oxide is reduced to copper. The concentrated sulphuric acid and calcium chloride on the right absorb all the water formed.
6. Put out the flame and allow the apparatus to cool in a stream of hydrogen.

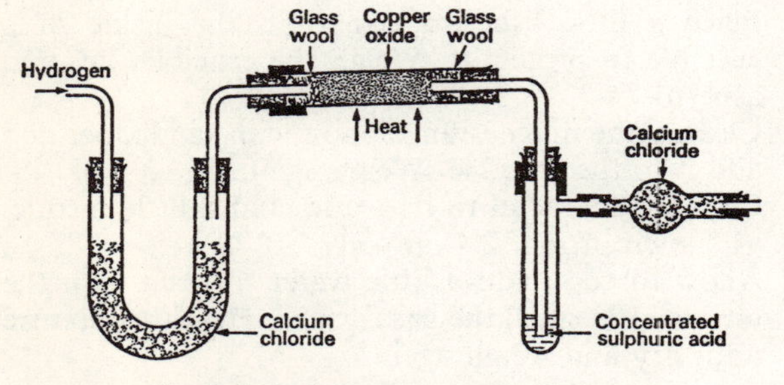

Fig. 9.1 To find the formula of water

7. Weigh the whole apparatus.
8. Reweigh the tube, copper, and glass wool.

Specimen results and calculation

1. Mass of combustion tube + oxide + wool = w g
2. Mass of whole apparatus at start = x g
3. Mass of whole apparatus at end = y g
4. Mass of combustion tube + copper + wool = z g

1 − 4. Mass of oxygen = $(w - z)$ g

3 − 2. Mass of hydrogen = $(y - x)$ g

$(y - x)$ g of hydrogen combine with $(w - z)$ g of oxygen.
∴ 1 g of hydrogen combines with $(w - z)/(y - x)$g of oxygen.

Thus, 1 mole of hydrogen atoms combines with $(w - z)/[16(y - x)]$ mole of oxygen atoms because the relative atomic mass of hydrogen is 1 and that of oxygen is 16. This fraction is either $\frac{1}{2}$, 1, or some other simple value corresponding to a formula for water of H_2O, HO, etc.

THREE CHEMICAL LAWS

The law of conservation of mass This law states that matter cannot be created or destroyed. It means that in a chemical reaction the total mass of the reactants equals the total mass of the products.

The law may be illustrated by two different methods which are shown in Fig. 9.2. In each experiment the whole apparatus is weighed before and after reaction; there should be no change in mass.

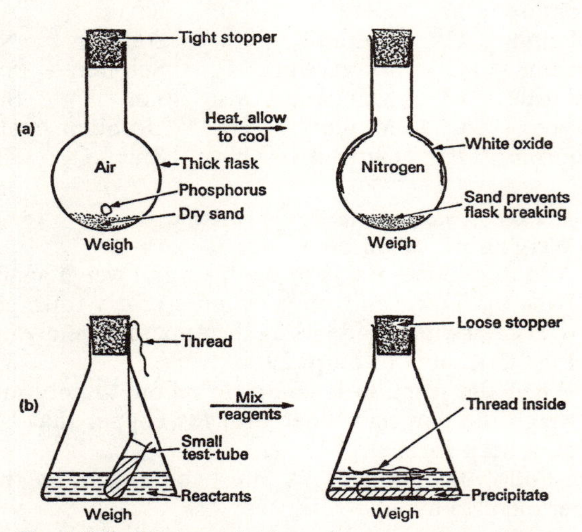

Fig. 9.2 To illustrate the law of conservation of mass

(*a*) White phosphorus is heated on sand in a tightly stoppered flask containing a fixed mass of air.

(*b*) Solutions of two reactants are allowed to combine chemically to form a precipitate by allowing a small tube of one reactant to fall into the other solution contained within a large flask. No gas should be formed by the reaction because it would blow out the stopper or burst the flask. Equations for suitable reactions are:

$$Ba(NO_3)_2 + Na_2SO_4 = 2NaNO_3 + BaSO_4$$
barium sulphate precipitate

$$AgNO_3 + NaCl = NaNO_3 + AgCl$$
silver chloride precipitate

$$Pb(NO_3)_2 + 2NaCl = 2NaNO_3 + PbCl_2$$
lead chloride precipitate

The law of definite proportions This is sometimes known as the law of constant composition. It states that a compound always contains the same elements combined in the same proportion by mass.

To illustrate this law, use the apparatus shown in Fig. 5.12 on p. 29. The three porcelain boats should each contain a sample of copper oxide prepared by three different methods. Weigh each boat empty and also before and after reduction of the oxide to metallic copper—there are nine weighings altogether. Check that the ratio mass of copper: mass of oxygen is the same for each oxide within experimental error.

The law of multiple proportions If elements *A* and *B* combine to form more than one compound, the masses of *A* which combine with a fixed mass of *B* are in a simple ratio.

To illustrate this law, again use the apparatus shown in Fig. 5.12. Use two porcelain boats—one containing black copper oxide and the other red copper oxide. Weigh each boat empty and also before and after reduction of the oxides to metallic copper—six weighings altogether. Check that the mass of copper which combines with 1 g of oxygen in the oxides is a simple ratio. The ratio is 1:2. Three different oxides of lead can be used instead of the copper oxides.

VALENCY

We have seen that hydrogen combines with many different elements. The formulae of some hydrogen compounds formed are:

Hydrogen chloride	HCl
Water	H_2O
Ammonia	NH_3

One atom of an element can combine with one, two, three or more atoms of hydrogen. The number of atoms of hydrogen that combine with one atom of an element is called the *valency* of the element. Sometimes valency is called *combining number*. Some valencies are:

Chlorine	1
Oxygen	2
Nitrogen	3

If an element does not combine with hydrogen, its valency is determined from the number of atoms of an element whose valency is known and with which it does combine. For example:

Element	*Compound*	*Valency*
Sodium	sodium chloride $NaCl$	1
Calcium	calcium oxide CaO	2

Variable valencies Many elements are able to form more than one compound with the same element. In these cases the element has more than one valency which is expressed by a Roman numeral in the name of the chemical. The old method of naming these chemicals is shown against the examples below.

Element	*Compound*	*Valency*
Copper	copper(I) chloride; $CuCl$ (cuprous chloride)	1
	copper(II) chloride; $CuCl_2$ (cupric chloride)	2
	copper(I) oxide; Cu_2O (cuprous oxide)	1
	copper(II) oxide; CuO (cupric oxide)	2
Iron	iron(II) chloride; $FeCl_2$ (ferrous chloride)	2
	iron(III) chloride; $FeCl_3$ (ferric chloride)	3
	iron(II) oxide; FeO (ferrous oxide)	2
	iron(III) oxide; Fe_2O_3 (ferric oxide)	3
Carbon	carbon monoxide; CO	2
	carbon dioxide; CO_2	4

EMPIRICAL FORMULAE

The *molecular formula* of a compound shows the numbers of each kind of atom in one molecule of the compound. The *empirical formula* is the simplest possible formula showing the proportions by mass. 'Empirical' means 'obtained by experiment'. Examples of empirical formulae are given below.

	Empirical formula	*Molecular formula*
Hydrogen peroxide	HO	H_2O_2
Glucose	CH_2O	$C_6H_{12}O_6$
Phosphorus(V) oxide	P_2O_5	P_4O_{10}

Calculation of empirical formula

An oxide of nitrogen contains 30·4 per cent by mass of nitrogen. What is its empirical formula?

	Nitrogen	*Oxygen*
Relative mass in compound =	30·4	69·6
$\frac{\text{Relative mass}}{\text{Atomic mass}}$ =	$\frac{30·4}{14}$	$\frac{69·6}{16}$
i.e. relative number of atoms =	2·17	4·34
(Divide by 2·17) =	1	2

The empirical formula is NO_2.

To determine x in the formula **$CuSO_4 . xH_2O$**

1. Weigh a crucible without its lid.
2. One-third fill the crucible with powdered copper sulphate crystals and weigh again. The difference is the mass of the crystals.
3. Stand the crucible on a pipe-clay triangle which is on a tripod. Heat gently for about three minutes and then heat more strongly until no more steam is given off and the residue is all white copper sulphate. If the heating is too strong, some of the sulphate turns black owing to the formation of copper oxide and sulphur trioxide fumes are given off.

$$CuSO_4 = CuO + SO_3$$

4. Use tongs to place the crucible and sulphate in a desiccator. The copper sulphate cools but does not absorb water from the air.
5. When cold, weigh the crucible and copper sulphate. If necessary, repeat the heating, cooling, and weighing until a constant weight is obtained to prove that the solid does not contain any water of crystallization.

1.	Mass of crucible	=
2.	Mass of crucible + crystals	=
3.	Mass of crucible + anhydrous sulphate	=
2 − 3.	Mass of water	= a g
3 − 1.	Mass of anhydrous sulphate	= b g

If the formula of the crystals is $CuSO_4 . xH_2O$, the molecular mass of the $CuSO_4 = 63{\cdot}5 + 32 + (4 \times 16) = 159{\cdot}5$ g; the mass of the $xH_2O = x(2 + 16) = 18x$ g.

$$\frac{\text{mass of water}}{\text{mass of anhydrous sulphate}} = \frac{18x}{159{\cdot}5} = \frac{a}{b}$$

Since a and b are known from the three weighings, the value of x can be calculated and it must, of course, be a whole number.

SUMMARY

Symbol of an element: one or two letters which usually denotes one atom of the element and sometimes denotes 1 mole of the element. For example, O means one atom of oxygen or 1 mole, 16 g, of oxygen.

Formula: symbols and sometimes numbers which denote one molecule of an element or compound and sometimes denote 1 mole of the substance. For example, H_2O means one molecule of water or 1 mole, 18 g, of water, and Ar means one molecule or 39·9 g of argon (it also means one atom of argon).

To determine some formulae

Copper oxide: reduce the oxide in either hydrogen or town gas that contains hydrogen. Calculate the number of moles of copper atoms which combines with 1 mole of oxygen atoms.

Magnesium oxide: heat the metal in air to form the oxide and a little nitride. Convert the nitride to the oxide by water and heat. Calculate as for copper oxide.

Water: pass dry hydrogen over dry copper oxide heated in a combustion tube. Absorb the water formed in concentrated sulphuric acid and calcium chloride. Calculate the number of moles of hydrogen atoms which combines with 1 mole of oxygen atoms.

Chemical laws

Conservation of mass: matter cannot be created or destroyed in a chemical reaction. To demonstrate, either heat white phosphorus in a closed flask or mix two solutions and weigh before and after they react.

Definite proportions: a compound always contains the same elements combined in the same proportions by mass. Demonstrate by reducing several samples of copper oxide with hydrogen or town gas.

Law of multiple proportions: if elements A and B form more than one compound, the masses of A which combine with a fixed mass of B are in a simple ratio. Demonstrate by reducing the two oxides of copper, or the three oxides of lead, with hydrogen or town gas.

Valency or combining number: the number of hydrogen atoms which combine with one atom of the element. Some elements have variable valencies, e.g. copper, 1 and 2, and iron, 2 and 3.

Molecular formula: shows the numbers of each kind of atom in one molecule.

Empirical formula: the simplest formula. The two formulae for hydrogen peroxide are H_2O_2 and HO.

QUESTIONS

1. Describe briefly an experiment to determine the mass of copper which combines with 1 g of oxygen in black copper oxide.

2. Hydrogen is passed over hot copper oxide in a combustion tube in order to determine the composition by mass of the oxide. (*a*) What is done to ensure that the copper oxide is completely dry before use? (*b*) Why is the copper oxide not heated until all the air has been displaced from the combustion tube? (*c*) Why is the combustion tube tilted slightly instead of being horizontal? (*d*) Why is the copper obtained in the reaction allowed to cool in a stream of hydrogen? (*e*) Name another oxide that may be reduced in this way, and write equations for the two reductions.

3. Two oxides of lead were reduced to the metal by a stream of hydrogen. One oxide contained 90·7 per cent and the other 92·8 per cent of the metal. Calculate for each oxide the mass of lead which combines with 16 g of oxygen; show that the two masses are in a simple ratio. (C.)

4. A flask containing a solution of barium nitrate weighed 140·0 g. A small test-tube containing dilute sulphuric acid weighed 28·2 g. The tube was lowered into the flask, taking care that the two solutions did not mix. The flask was stoppered and weighed. The two chemicals were then mixed and the flask and contents were re-weighed. (*a*) What is the total mass of flask and contents before mixing and after mixing? (*b*) What are the colours of the contents before mixing and after mixing? Write the equation. (*c*) Name and state the chemical law which is illustrated by this experiment. (*d*) Do you consider that this experiment proves the law? Explain your answer.

5. Crystalline barium chloride decomposes when heated according to the equation:

$$BaCl_2 . xH_2O = BaCl_2 + xH_2O$$

Describe, with full experimental details, how you would determine the percentage by mass of water of crystallization in the compound. If, in such an experiment, the percentage is found to be 14·75 per cent, calculate x the number of molecules of water of crystallization in the compound. (Ba = 137; Cl = 35·5; H = 1; O = 16.) (C.)

6. 5·72 g of a hydrate of sodium carbonate formed 2·12 g of anhydrous carbonate. Show that the hydrate is a 10-water hydrate. (Na = 23; C = 12; O = 16; H = 1.)

7. Two anhydrous sulphates of a metal contain respectively 36·8 per cent and 28 per cent of the metal. Calculate its probable relative atomic mass. (S = 32; O = 16.)

8. Calculate the percentage by mass of phosphorus in calcium phosphate, $Ca_3(PO_4)_2$. Find the simplest formula of a compound containing 45·6 per cent tin and 54·4 per cent chlorine. (Ca = 40; P = 31; O = 16; Sn = 119; Cl = 35·5.) (C.)

9. Red lead is an oxide said to have the formula Pb_3O_4. It is easily reduced to lead. Describe how you would attempt to verify the formula. Indicate how you would use your experimental results to arrive at your conclusion. (Pb = 207, O = 16.) (L.)

10 The Periodic Table

CLASSIFICATION OF ELEMENTS

The world seems to be made up of an infinite variety of substances, all of which are different in various ways. If each substance or each group of substances bears no relation to other substances or groups of substances, we could never understand the nature of matter. The Greeks suggested that all matter consisted of four 'elements' only: earth, air, fire, and water, but their idea could not be supported by experiments. Lavoisier and Dalton introduced the new chemistry which led eventually to the first true ideas that the world consists of only a moderate number of different real substances.

In 1789 Lavoisier made a list of thirty-three elements and divided them into five classes:

light and heat which he believed were material substances;
three gases: oxygen, hydrogen, and nitrogen;
seventeen metals which he arranged in alphabetical order;
five earths which are now known to be oxides;
and six other non-metals.

As the number of known elements increased, other chemists divided the metals into several groups; for example, very ductile metals which formed oxides with difficulty, such as gold and silver; ductile metals that did not form acidic oxides, such as iron, copper, and lead; slightly ductile metals, such as zinc and calcium; brittle metals, such as cobalt and nickel, and so on. More groups were added as new metals such as sodium and potassium were discovered which did not fit into any of the stated groups.

In 1803 Dalton published his 'table of the relative weights of the ultimate particles of gaseous and other bodies'; this was the first table of atomic weights. Dalton changed chemists' ideas from a qualitative to a quantitative basis. His work started the Chemical Revolution during the nineteenth century.

DOBEREINER'S TRIADS

By 1829 Dobereiner had shown that there are several groups of three chemically similar elements, which he called triads. He noted that the atomic weight of the middle element of the triad is the arithmetic mean of the other two. Some examples, using modern values, are:

Lithium	7	Chlorine	35·5
Sodium	23	Bromine	79
Potassium	39	Iodine	127

This classification was not satisfactory. Many elements cannot be placed in a triad. Iron, manganese, nickel, cobalt, zinc, and copper were six similar elements, and the four halogens, fluorine, chlorine, bromine, and iodine were difficult to place. Dobereiner actually considered fluorine as the first member of another triad of more active elements.

NEWLANDS' LAW OF OCTAVES

In 1858 Cannizzaro's work, based on the work of Avogadro (see chapter 15), made atomic weights reliable for the first time. In 1864 Newlands arranged elements in ascending order of atomic weight. He wrote: 'the difference between the number of the lowest member of a group and that immediately above it is 7; in other words the eighth element starting from a given one is a kind of repetition of the first, like the eight notes of an octave of music'. He called this relationship the *Law of Octaves*. Later he wrote: 'the numbers of analogous elements generally differ either by 7 or by some multiple of 7—the same relationship as the extremities of one or more octaves of music . . . between nitrogen and phosphorus there are 7 elements; between phosphorus and arsenic there are 14; . . .'

Newlands was the first man to give a number to an element, to leave spaces for elements still undiscovered, and even to alter the positions of elements, for example, tellurium and iodine, when the atomic weights placed them in the wrong order.

One of Newlands' arrangements was:

H	Li	Be	B	C	N	O
F	Na	Mg	Al	Si	P	S
Cl	K	Ca	Cr	Ti	Mn	Fe
(*doh*	*ray*	*me*	*fah*	*soh*	*lah*	*te*)

The names below the elements refer to the notes in an octave of music to which Newlands compared the elements that he had arranged in order.

MEYER'S ATOMIC VOLUME CURVE

In 1869 Meyer plotted the atomic volume of each element against its atomic weight. He defined the atomic volume of an element as its atomic weight divided by its density.

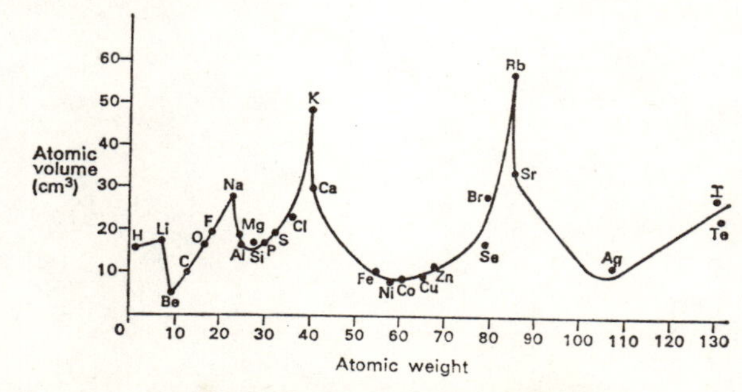

Fig. 10.1 Atomic volume curve

Parts of the graph that he obtained are shown in Fig. 10.1. It illustrates that elements with similar chemical properties appear in similar positions on the waves of the curve. In particular all the alkali metals, lithium, sodium, potassium, and rubidium, are at the tops of the waves; the halogen elements, fluorine, chlorine, bromine, and iodine, are all on ascending parts of the curve. However, in many ways the graph was unsatisfactory because it was not possible to predict how many more elements were yet to be discovered.

MENDELEEV'S PERIODIC TABLE

Mendeleev was a Russian chemist who wrote in 1869: 'The properties of the elements are in periodic dependence upon their atomic weights'. He published a table of the elements and was able to make a number of important deductions from this table. It was not complete because many elements, including the noble gases, had not been discovered. It also contained some errors which now have been corrected, but it did show how the elements were related to each other and how their chemical reactivities changed with increasing relative atomic masses.

The modern version of the Periodic Table is shown overleaf and also at the front of this book.

GROUPS

The modern table is divided into vertical columns called *groups*. There are eight main groups of elements, and the elements in each group all have similar chemical properties as we shall see in later chapters.

We have already seen that the elements in group 1 all form strong alkalis and have a valency of 1. The elements in group 2, the alkaline earth metals, all form bases and have a valency of 2. We also know that the elements in group 7, which are known as the halogens, form acids with hydrogen and salts with metals. The noble gases in group 8 form few compounds.

We can say in general that the elements to the left of the table are metals and that the elements to the right of the table are non-metals because of their chemical properties. The dividing line on the table shows the approximate division of the elements into metals and non-metals.

The most reactive metals are in group 1; the most reactive non-metals are in group 7. It is thus clear that elements at opposite sides of the table react vigorously together. For example, sodium reacts explosively with chlorine; magnesium continues to burn in chlorine gas. Elements in the central groups react less vigorously.

PERIODS

Hydrogen occupies a special place in the table because it is able to react with both metals and non-metals to form hydrides:

	NaH	CaH_2	BH_3	CH_4	NH_3	H_2O	HCl
valency	1	2	3	4	3	2	1

It is therefore placed in the first horizontal row in both group 1 with the alkali metals and in group 7 with the halogens. The first of the noble gases, helium, is also placed in this row.

The horizontal rows are called *periods*. It is found that the most reactive non-metals are in the earlier periods; for example, fluorine is the most reactive of the halogens. On the other hand, the most reactive metals are in the later periods, for potassium is more reactive than sodium.

We can see from the hydrides shown above that the valency of elements in each period varies in a regular manner. The oxides also show this regular variation:

	Na_2O	CaO	Al_2O_3	CO_2	N_2O_5	SO_3	Cl_2O_7
valency of element	1	2	3	4	5	6	7

The noble gases form few compounds.

TRANSITION METALS

In the fourth and succeeding periods, there is an extra group of elements which have very similar properties. They are all metals which form basic oxides, many of which are coloured, and they are called *transition elements* or metals, because their chemical properties are between those of elements in group 2 and those in group 3. Periods 4 and 5 are short periods and contain ten transition elements. Periods 6 and 7 are long periods which contain an extra fifteen elements which have almost identical chemical properties; they are known as the *lanthanides* and *actinides*. The lanthanides were formerly called *rare earth* elements. Only period 6 is shown in the table in this chapter.

We are now able to number the elements in the order in which they fit into the Periodic Table. This number is called the *atomic number* of the element; its significance will be explained in chapter 14.

SUMMARY

Classification of elements to relate the chemical and physical properties of elements to each other and to allow forecast of the existence of other elements.
Lavoisier wrongly divided elements into five different classes.
Dalton prepared the first list of atomic weights.
Dobereiner showed that many elements could be divided into groups of three (triads) by their chemical properties.
Newlands arranged elements in the order of their atomic weights and found that the chemical and physical properties of every eighth element were often similar.
Meyer demonstrated some of Newlands' findings with a graph of atomic volume: atomic weight.
Mendeleev prepared the first periodic table of elements which showed how the properties of elements were repeated.

Periodic Table
Groups of elements: eight vertical columns showing elements whose properties are chemically related.
Periods of elements: seven horizontal rows of elements:

PERIODIC TABLE

	Group 1	Group 2	TRANSITION METALS										Group 3	Group 4	Group 5	Group 6	Group 7	Group 8	
PERIOD 1	1 **H** Hydrogen																1 **H** Hydrogen	2 **He** Helium	PERIOD 1
PERIOD 2	3 **Li** Lithium	4 **Be** Beryllium											5 **B** Boron	6 **C** Carbon	7 **N** Nitrogen	8 **O** Oxygen	9 **F** Fluorine	10 **Ne** Neon	PERIOD 2
PERIOD 3	11 **Na** Sodium	12 **Mg** Magnesium											13 **Al** Aluminium	14 **Si** Silicon	15 **P** Phosphorus	16 **S** Sulphur	17 **Cl** Chlorine	18 **Ar** Argon	PERIOD 3
PERIOD 4	19 **K** Potassium	20 **Ca** Calcium	21 **Sc** Scandium	22 **Ti** Titanium	23 **V** Vanadium	24 **Cr** Chromium	25 **Mn** Manganese	26 **Fe** Iron	27 **Co** Cobalt	28 **Ni** Nickel	29 **Cu** Copper	30 **Zn** Zinc	31 **Ga** Gallium	32 **Ge** Germanium	33 **As** Arsenic	34 **Se** Selenium	35 **Br** Bromine	36 **Kr** Krypton	PERIOD 4
PERIOD 5	37 **Rb** Rubidium	38 **Sr** Strontium	39 **Y** Yttrium	40 **Zr** Zirconium	41 **Nb** Niobium	42 **Mo** Molybdenum	43 **Tc** Technetium	44 **Ru** Ruthenium	45 **Rh** Rhodium	46 **Pd** Palladium	47 **Ag** Silver	48 **Cd** Cadmium	49 **In** Indium	50 **Sn** Tin	51 **Sb** Antimony	52 **Te** Tellurium	53 **I** Iodine	54 **Xe** Xenon	PERIOD 5
PERIOD 6	55 **Cs** Caesium	56 **Ba** Barium	57–71 Lanth-anides	72 **Hf** Hafnium	73 **Ta** Tantalum	74 **W** Tungsten	75 **Re** Rhenium	76 **Os** Osmium	77 **Ir** Iridium	78 **Pt** Platinum	79 **Au** Gold	80 **Hg** Mercury	81 **Tl** Thallium	82 **Pb** Lead	83 **Bi** Bismuth	84 **Po** Polonium	85 **At** Astatine	86 **Rn** Radon	PERIOD 6
PERIOD 7	87 **Fr** Francium	88 **Ra** Radium	89–103 Acti-nides																PERIOD 7

those on the left are metals, those on the right are non-metals. Chemical properties gradually change from very reactive metals to very reactive non-metals.
Transition elements: metals in periods 4, 5, 6, and 7 with similar properties that lie between those of groups 2 and 3.
Reactivity: non-metals with *high* atomic masses are least reactive; metals with *low* atomic masses are least reactive.
Valency of elements within a period changes in a regular manner.

QUESTIONS

1. Give reasons why it is not possible to classify all the known elements into groups of three similar to Dobereiner's triads.
2. Name the eight elements in the second period of the Periodic Table (i.e. between helium and sodium). Name four elements in group 1 and four in group 7.
3. (*a*) Which pair of the following elements are in the same group of the Periodic Table: magnesium, nitrogen, calcium, and sulphur? (*b*) What element in period 3 is likely to react most violently with chlorine? (*c*) Of the eight elements in period 3, which are likely to form a compound of the formula XCl_3 with chlorine? (*d*) Selenium, Se, occurs below sulphur in the Table. What is the probable formula of the hydrogen compound of selenium? (*e*) What is the probable formula of the oxide of silicon (silicon is below carbon in the Table)? (*f*) Are the elements in group 2 and group 6 metals or non-metals?
4. Mendeleev introduced his classification of the elements in 1869. How did his work contribute to the development of chemistry during the next fifty years or so?
5. This question refers to the following Periodic Table of the first fifty-four elements. Some of the elements are shown by letters. (The letters are not the symbols of the elements.)

					1	2				
3 A	4				5	6 E	7	8	9 H	10
11	12				13 D	14	15	16	17	18 I
19	20	21–28	29 C	30	31	32 F	33	34 G	35	36
37 B	38	39–46	47	48	49	50	51	52	53	54

(*a*) Which of the lettered elements is an inert (noble) gas? (*b*) State two of the lettered elements which are in the same group. (*c*) Give the name of each halogen element together with the number of its position in the Table. (*d*) Which one of the lettered elements would you expect to react most violently with chlorine? (*e*) Which one of the lettered elements would you expect to form a compound with the composition 1 mole of the element: 3 moles of chlorine? (*f*) What would you expect the formula of a compound of hydrogen with the element G to be? (Use G as the symbol of the element.) (L.)

6. The second short period is Na Mg Al Si P S Cl Ar. Describe simple experiments you have done with (*a*) the hydroxides and (*b*) the chlorides of sodium, magnesium, and aluminium which illustrate a gradual change in chemical behaviour across a period. Use these trends to predict some of the properties of the chloride and hydroxide of phosphorus. (L.)

11 The Halogens and their Compounds

The halogens are the non-metals in group 7 of the periodic table: fluorine, chlorine, bromine, and iodine.

EXTRACTION AND USES OF COMMON SALT, NaCl

Sea-water contains almost 3 per cent by mass of sodium chloride. The common salt is obtained from sea-water either by evaporation in hot countries, or by letting most of the water separate as ice in cold countries. The remainder of the water is then removed by heating to dryness.

Deposits of rock-salt, a mixture of salt and sand, occur in several regions. The rock-salt has been formed by the evaporation of inland seas; it also contains other halogen salts (halides). It is either mined in the same way as other minerals such as coal and iron ore, or it is dissolved by forcing water through the ground to the rock-salt and then pumping up the salt solution which is called brine. The brine is then evaporated to dryness. Small quantities of common salt are needed by animals. Sodium, chlorine and most of their compounds are made from sodium chloride. Common salt is also used to preserve fish and meat.

To study the properties of sodium and potassium halides

Action of heat Heat sodium chloride crystals in a hard-glass or silica test-tube with a hot flame. Repeat with potassium chloride and potassium iodide.

Action of concentrated sulphuric acid Place a few crystals of sodium chloride, potassium bromide, and potassium iodide in separate test-tubes. Add a few drops of the acid to each. Observe the appearance of the gas formed and of the mixture in the tube. Breathe on the gas and hold a stopper moistened with ammonia solution in the gas.

Action of manganese(IV) oxide and concentrated sulphuric acid Add sodium chloride, potassium bromide, and potassium iodide to separate test-tubes. Add an equal volume of manganese(IV) oxide powder to each and mix well by shaking. Add concentrated sulphuric acid to each tube and warm gently. Test any gas formed with moist litmus paper. Observe the appearance of the gas and of the mixture in the tubes.

Action of silver nitrate solution Make a few cm^3 of a solution in distilled water of each of the halides that were tested above. Add silver nitrate solution to each.

Let the precipitates settle and pour away most of the liquid in each tube. Divide each precipitate into two parts and find if it is soluble in (*a*) dilute nitric acid and (*b*) ammonia solution, by adding these pollutions a little at a time and shaking.

Action of chlorine Dissolve chlorine gas in cold water to make chlorine water. Add the solution to aqueous potassium bromide and then to aqueous potassium iodide. Add tetrachloromethane and shake.

REACTIONS OF CHLORIDES, BROMIDES AND IODIDES

Note that the following symbols are used in a chemical equation to indicate the physical state of a compound:

(g) gas; (l) liquid; (s) solid; (c) crystal; (aq) solution in water.

Heat Sodium chloride crystals break open when heated and make cracking sounds. They melt at 803 °C to form a colourless liquid. The liquid boils at 1430 °C.

Potassium chloride behaves in a similar manner.

Potassium iodide turns brown on heating and also when kept in the light because some iodine is formed.

Melting points are: KCl, 790 °C; KBr, 750 °C; and KI, 705 °C.

Concentrated sulphuric acid Colourless hydrogen chloride, hydrogen bromide and hydrogen iodide are produced; they form misty fumes in damp air.

$NaCl + H_2SO_4 = HCl(g) + NaHSO_4$, sodium hydrogensulphate.

$$Cl^-(aq) + H^+(aq) = HCl(g)$$
$$Br^- + H^+ = HBr(g)$$
$$I^- + H^+ = HI(g)$$

Part of the hydrogen bromide is oxidized by the acid to bromine, which turns the mixture reddish and comes off as a red gas:

$$2HBr + H_2SO_4 = Br_2(g) + SO_2(g) + 2H_2O$$

Most of the hydrogen iodide is also oxidized to iodine which comes off as a violet gas. The mixture in the tube is black.

$$2HI + H_2SO_4 = I_2(g) + SO_2(g) + 2H_2O$$

Manganese(IV) oxide and concentrated sulphuric acid The acid first forms the hydrogen halide (as before) which is then oxidized by the oxide to the halogen. Chlorides form chlorine which bleaches litmus; bromides form bromine as a red gas; iodides form iodine as a violet gas:

$$MnO_2 + 4HCl = Cl_2(g) + MnCl_2 + 2H_2O$$
$$MnO_2 + 4HBr = Br_2(g) + MnBr_2 + 2H_2O$$
$$MnO_2 + 4HI = I_2(g) + MnI_2 + 2H_2O$$

Silver nitrate solution Chlorides form a white precipitate, bromides form a pale yellow precipitate, and iodides form a yellow precipitate:

$$AgNO_3 + KCl = AgCl(s) + KNO_3$$
or $Ag^+(aq) + Cl^-(aq) = AgCl(s)$, silver chloride
$Ag^+(aq) + Br^-(aq) = AgBr(s)$, silver bromide
$Ag^+(aq) + I^-(aq) = AgI(s)$, silver iodide

All of the precipitates are insoluble in dilute nitric acid. Silver chloride is soluble in dilute ammonia solution, silver bromide is slightly soluble, and silver iodide is insoluble.

Chlorine water This forms a red solution with bromides and a brown solution, and possibly black crystals, with iodides. Bromine and iodine are displaced:

$$2KBr + Cl_2 = 2KCl + Br_2(l)$$

or

$$2Br^- + Cl_2 = 2Cl^- + Br_2(l)$$

$$2I^- + Cl_2 = 2Cl^- + I_2(s)$$

Tetrachloromethane (CCl_4) dissolves the bromine to form a reddish solution; it dissolves the iodine to form a violet solution.

To prepare hydrogen chloride

Arrange the apparatus as in Fig. 11.1. Add concentrated sulphuric acid down the thistle funnel to cover the lumps of rock-salt and warm gently. Powdered salt should not be used as it produces excessive froth and some may pass over into the drying bottle. Collect the gas by downward delivery.

$$NaCl + H_2SO_4 \rightleftharpoons HCl(g) + NaHSO_4$$

The equation represents a reversible reaction. In practice, it proceeds to completion on the right because the hydrogen chloride is removed as soon as it is formed and the reverse reaction is then not possible. Sodium sulphate, Na_2SO_4, is not formed in glass apparatus as it is produced only at about 1000°C.

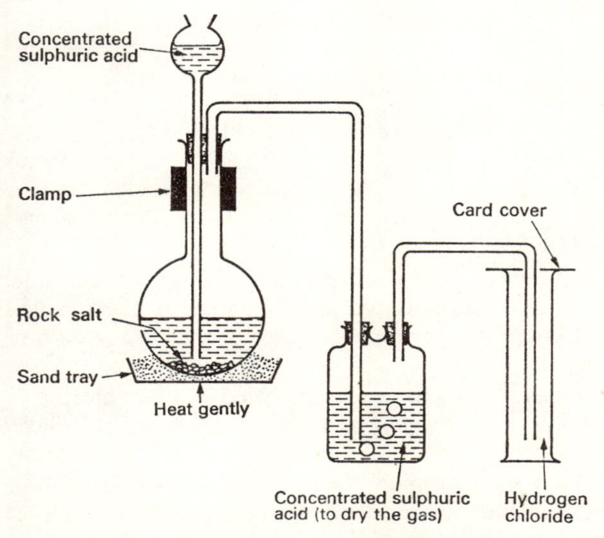

Fig. 11.1 Preparation of dry hydrogen chloride

PROPERTIES OF HYDROGEN CHLORIDE

It is a colourless gas which is denser than air, with a pungent, irritating smell, and a sharp, sour taste. In moist air it forms misty fumes which consist of tiny drops of hydrochloric acid solution. It is very soluble in water and methylbenzene. It turns litmus and Universal Indicator paper red. Liquid hydrogen chloride (b.p. −84°C) has no action on indicators, metals, oxides or carbonates and is therefore not an acid.

Hydrogen chloride neither burns nor supports combustion. Burning sodium and many hot metals decompose the gas and form the chloride and hydrogen:

$$2Na + 2HCl = 2NaCl + H_2$$

$$Fe + 2HCl = FeCl_2 + H_2$$

Fig. 11.2 shows a simple apparatus for the preparation of iron(II) chloride, which forms shiny crystals on the cool parts of the test-tube. This experiment shows that the gas contains hydrogen.

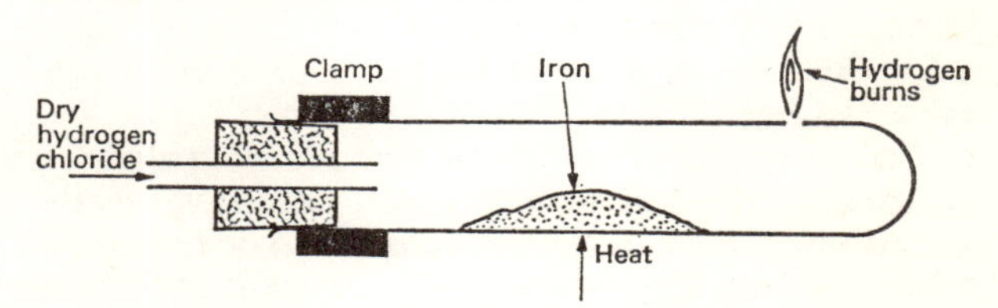

Fig. 11.2 Preparation of iron(II) chloride, $FeCl_2$

Hydrogen chloride gas combines with ammonia to form dense white fumes of ammonium chloride, and it precipitates white silver chloride from silver nitrate solution:

$$NH_3 + HCl = NH_4Cl$$

$$AgNO_3(aq) + HCl(aq) = HNO_3 + AgCl(s)$$

To prepare solutions of hydrogen chloride

Prepare the gas from common salt and sulphuric acid and pass it into water, using an inverted funnel as in Fig. 11.3. In this way the gas dissolves readily because it is in contact with a large area of water. If it dissolves very quickly, water rises up the funnel; thus the water

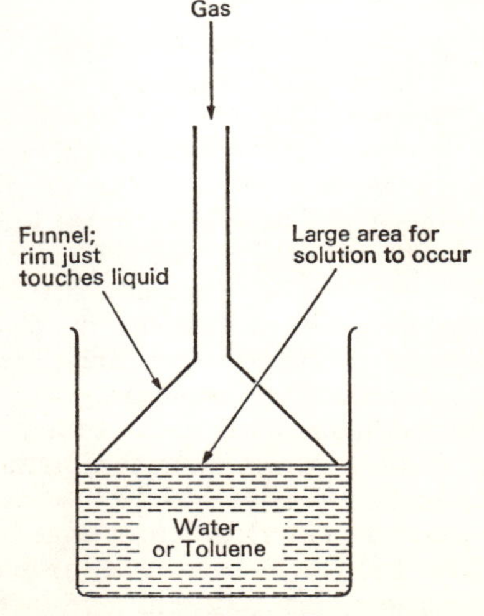

Fig. 11.3 Dissolving hydrogen chloride

level outside drops below the rim and air can enter the funnel. Liquid cannot therefore rise up the funnel and suck back into the acid in the preparation flask. The aqueous solution is hydrochloric acid.

Also prepare a solution in dry methylbenzene (toluene).

1 cm^3 of water dissolves about 500 cm^3 of gas at s.t.p. and much heat is evolved. Aqueous solutions of the other hydrogen halides are called hydrobromic acid, hydriodic acid and hydrofluoric acid.

PROPERTIES OF HYDROCHLORIC ACID

Dilute aqueous hydrochloric acid has a sour taste and it is acidic to indicators. The acid reacts with most metals to form hydrogen, with carbonates and sulphites to form carbon dioxide and sulphur dioxide, and with alkalis and insoluble bases to form salts and water. All of these are typical properties of dilute acids. Refer to p. 33 of chapter 6. The solution contains hydrogen ions.

The concentrated acid is a strong reducing agent. It is oxidized to chlorine by manganese(IV) oxide, potassium manganate(VII), lead(IV) oxide, and red lead or trilead tetraoxide Pb_3O_4. These reactions are used in the preparation of chlorine.

The solution in dry toluene This solution is not acidic to indicators, and does not react with metals, carbonates, sulphites or bases. It has none of the typical properties of a dilute acid because it does not contain hydrogen ions. It forms a white precipitate of ammonium chloride in toluene when ammonia is passed into it.

MANUFACTURE OF HYDROGEN CHLORIDE

It is synthesized by burning hydrogen in chlorine. Some is made by treating common salt with concentrated sulphuric acid. Much hydrogen chloride is a by-product formed when organic compounds are chlorinated, e.g.

$$C_xH_y + Cl_2 = C_xH_{(y-1)}Cl + HCl$$

To determine the formula of hydrogen chloride

1. Place a platinum spiral in one end of a silica combustion tube and pack the rest loosely with glass wool so that gases can pass through easily.
2. Arrange the apparatus as in the diagram. Add 40 cm^3 (excess) of hydrogen to syringe 1 and 20 cm^3 of chlorine to syringe 2.
3. Turn the three three-way taps into position so that the hydrogen can be pushed into the chlorine and form a hydrogen-chlorine mixture. Use a safety screen as this mixture is explosive.
4. Heat the platinum until it is red-hot. Pass the gaseous mixture slowly over the platinum from syringe 2 to syringe 1.
5. Pass the gases from syringe 1 to syringe 2, and then back again to syringe 1, in order to ensure that reaction between the chlorine and hydrogen is complete.
6. Cool the tube with a damp cloth. Measure the volume of the product at room temperature, which contains a mixture of hydrogen chloride and hydrogen.

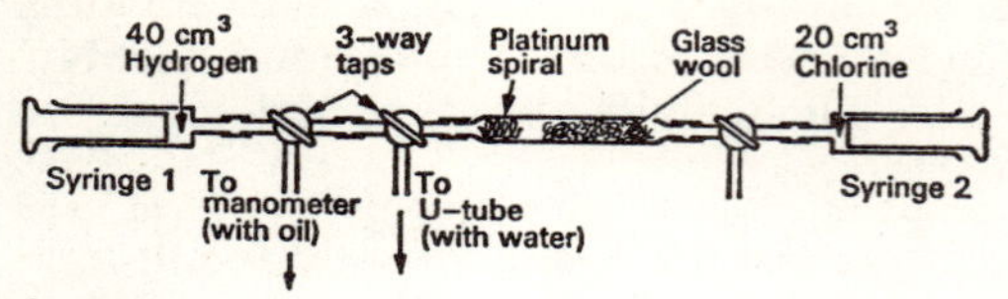

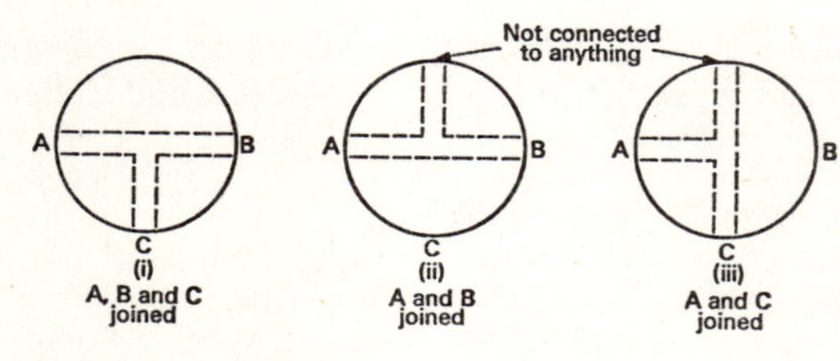

Fig. 11.4 To determine the formula of hydrogen chloride

7. Find the volume of hydrogen chloride by pushing the gases from syringe 1 into the U-tube of water, which dissolves the hydrogen chloride. Use the manometer to ensure that the gas is at atmospheric pressure when the hydrogen chloride is dissolving. Measure the volume of gas left in syringe 1.

Specimen results and calculation

1.	Volume of hydrogen	= 40 cm^3
2.	Volume of chlorine	= 20 cm^3
3.	Volume of gases after reaction with platinum	= 60 cm^3
4.	Volume of gas left after dissolving	= 20 cm^3
3 − 4.	Volume of hydrogen chloride formed	= 40 cm^3
1 − 4.	Volume of hydrogen which reacted	= 20 cm^3

The results show that:

20 cm^3 of hydrogen react with 20 cm^3 of chlorine.

∴ 1000 cm^3 of hydrogen react with 1000 cm^3 of chlorine, and from the densities of the gases we know that:

0·089 g of hydrogen react with 3·17 g of chlorine.

1·008 g of hydrogen react with (1·008 × 3·17)/0·089 = 35·5 g of chlorine,

i.e. 1 mole of hydrogen atoms combines with 1 mole of chlorine atoms, and the empirical formula of hydrogen chloride is HCl.

THE HALOGENS

(F_2, Cl_2, Br_2, and I_2)

Fluorine Potassium hydrogen fluoride, KHF_2, is dissolved in anhydrous hydrogen fluoride and the mixture is electrolysed. Fluorine forms at carbon anodes and hydrogen at steel cathodes. The cell is made of a copper-nickel alloy which is not attacked by the very reactive fluorine.

$$\text{At the anode: } F^- - e = F; \quad 2F = F_2(g)$$

$$\text{At the cathode: } H^+ + e = H; \quad 2H = H_2(g)$$

Hydrogen fluoride itself is a non-electrolyte. Aqueous hydrogen fluoride cannot be used as an electrolyte, because the fluorine formed immediately reacts vigorously with the water. Potassium fluoride is added to liquid hydrogen fluoride to form a conducting solution. The boiling point of the solution is much higher than that of hydrogen fluoride, 19·5 °C. The solution is electrolysed at a temperature of about 80 °C.

Chlorine It is obtained by the electrolysis of concentrated sodium chloride in the manufacture of sodium hydroxide solution and also by the electrolysis of the fused chloride to obtain sodium. Refer to p. 64 of chapter 12.

Chlorine is also obtained by air oxidation of hydrogen chloride, which is a by-product of many chlorinations of organic compounds. The mixture of hydrogen chloride and air is passed over catalysts, e.g. copper salts:

$$4HCl + O_2 = 2Cl_2 + 2H_2O$$

Bromine Sea-water contains about 0·006 per cent by mass of bromine as the bromides of sodium, potassium, and magnesium. The water is acidified with sulphuric acid and the bromine is displaced by chlorine:

$$Cl_2 + 2Br^- = 2Cl^- + Br_2(l)$$

The bromine is then expelled from solution by a stream of air.

Iodine Crude Chile saltpetre is a naturally occurring salt. It consists of sodium nitrate with a little sodium iodate, $NaIO_3$. The iodate remains in solution when the nitrate is crystallized from water. Sodium hydrogen sulphite, $NaHSO_3$, is added to the iodate solution to precipitate the iodine. The equations are too difficult at this stage.

Some seaweeds contain iodides. They are mixed with concentrated sulphuric acid and manganese(IV) oxide and heated:

$$NaI + H_2SO_4 = HI + NaHSO_4$$

$$4HI + MnO_2 = I_2 + MnI_2 + 2H_2O$$

PHYSICAL PROPERTIES OF HALOGENS

	Fluorine	*Chlorine*	*Bromine*	*Iodine*
Appearance	Greenish-yellow gas	Greenish-yellow gas	Dark red liquid	Greyish-black solid
Density of liquid ($g\ cm^{-3}$)	1·11	1·55	3·19	4·9 (solid)
m.p. (°C)	−223	−101	−7	114
b.p. (°C)	−188	−34	59	183

USES OF THE HALOGENS AND THEIR COMPOUNDS

Fluorine Hydrogen fluoride is used to etch glass. The glass is coated with wax and is scratched through the wax with a metal point. The exposed glass is changed by hydrogen fluoride gas. The silica in the glass is attacked:

$$SiO_2 + 4HF = 2H_2O + SiF_4, \text{ silicon(IV) fluoride}$$

Dichlorodifluoromethane, CCl_2F_2, is a volatile unreactive compound used in refrigerators and as a propellant in aerosol sprays. Some other fluorinated and chlorinated hydrocarbons are lubricants. Aluminium fluoride is used in the electrolyte during the extraction of aluminium. The two uranium isotopes, 235 and 238, are separated by gaseous diffusion of their fluorides, UF_6. Tin(II) fluoride, SnF_2, is used in some toothpastes. A little fluorine gas is sometimes dissolved in water supplies in order to reduce teeth decay.

Chlorine is used to bleach paper pulp, cotton, and linen. Water in reservoirs and swimming pools is chlorinated in order to kill germs, such as cholera and typhoid. Most chlorine is used in the manufacture of chlorine-containing carbon compounds. Examples are given below.

Trichloroethene	$CHCl{=}CCl_2$	Industrial solvents and fire extinguishers
Tetrachloromethane	CCl_4	Industrial solvents and fire extinguishers
DDT, dichloro diphenyl trichloroethane	—	Insecticides
'Gammexane', benzene hexachloride	$C_6H_6Cl_6$	Insecticides
TCP, trichlorophenol	—	Antiseptic
Chloroform, trichloromethane	$CHCl_3$	Anaesthetic
Sodium chlorate	$NaClO_3$	Weed killer, also used for removing leaves of cotton plant

Note that chlorine and its compounds are used to 'destroy' dyes, germs, grease, fires, insect pests, and unwanted plants.

Bromine is used mainly to produce dibromoethane, $C_2H_4Br_2$. This is added to petrol in order to remove metallic lead from the engine. The lead is formed from a compound that is added to good petrol in order to avoid 'knocking' or uneven burning of the fuel in the engine. Silver bromide is used on photographic film; some bromides are sedative drugs.

Iodine solution in ethanol is called 'tincture of iodine'; it is used as an antiseptic on wounds. Silver iodide is used on photographic film.

To prepare chlorine

1. FROM HYDROCHLORIC ACID

either Drop the concentrated acid onto solid potassium manganate(VII) using the apparatus in Fig. 11.5. No heat is required.

$$16HCl + 2KMnO_4 = 5Cl_2(g) + 2KCl + 2MnCl_2 + 8H_2O$$

simplified to

$$2HCl + [O] = Cl_2(g) + H_2O$$

or Mix concentrated acid with lumps of manganese (IV) oxide in a flask. Heat the mixture gently in the apparatus shown in Fig. 11.6. Do not use powdered oxide because too much frothing occurs and the reaction is too vigorous.

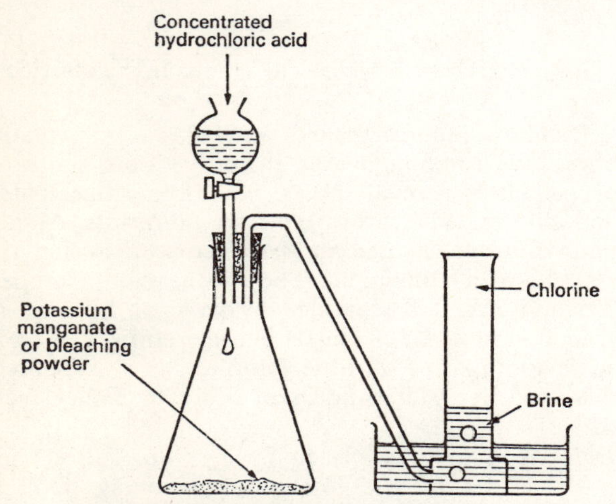

Fig. 11.5 Preparation and collection of chlorine

$$4HCl + MnO_2 = Cl_2(g) + MnCl_2 + 2H_2O$$

2. FROM BLEACHING POWDER

Drop dilute mineral acid onto solid bleaching powder in the apparatus of Fig. 11.5. Do not heat. Any mineral acid may be used.

$$CaOCl_2 + 2HCl = Cl_2(g) + CaCl_2 + H_2O$$

3. FROM COMMON SALT

either Mix common salt with manganese(IV) oxide and add to a flask; see Fig. 11.6. Prepare diluted sulphuric acid by adding the concentrated acid carefully to half its volume of water—*never* add water to acid. Then add the diluted acid to the mixture in the flask using the thistle funnel, and heat gently.

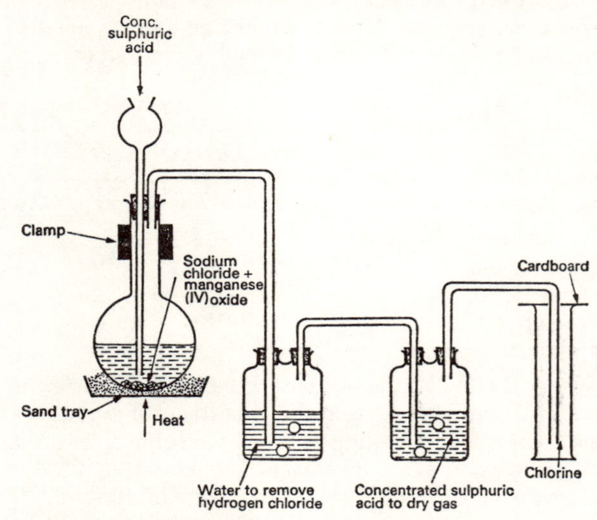

Fig. 11.6 Preparation and collection of dry chlorine

$$H_2SO_4 + NaCl = HCl + NaHSO_4$$
$$4HCl + MnO_2 = Cl_2(g) + MnCl_2 + H_2O$$

or Electrolyse concentrated sodium chloride solution using carbon or platinum electrodes. Chlorine is liberated at the anode and hydrogen at the cathode.

COLLECTION OF CHLORINE

either Collect the dense, greenish-yellow gas in a gas-jar over brine, see Fig. 11.5.

or If the gas is required dry, bubble it through water to remove any acid and then through concentrated sulphuric acid (see Fig. 11.6) or U-tubes of calcium chloride to dry it. Then collect by downward delivery.

Chlorine water Bubble the gas slowly through ice-cold water. The water absorbs about two and a half times its volume of gas and forms a yellow solution.

$$Cl_2 + H_2O \rightleftharpoons HCl + HClO\text{, chloric(I) acid}$$

To study some reactions of chlorine

Hydrogen

Light the gas from a cylinder of hydrogen, and lower the flame into a jar of chlorine.

Hydrocarbons

1. Lower a burning candle into chlorine.
2. Push a burning wax taper into the gas.
3. Hold glass wool in tongs, soak it in petrol, light the petrol, and drop the burning substance into chlorine.

4. Hold glass wool in tongs, soak it in hot turpentine, and drop it into chlorine.

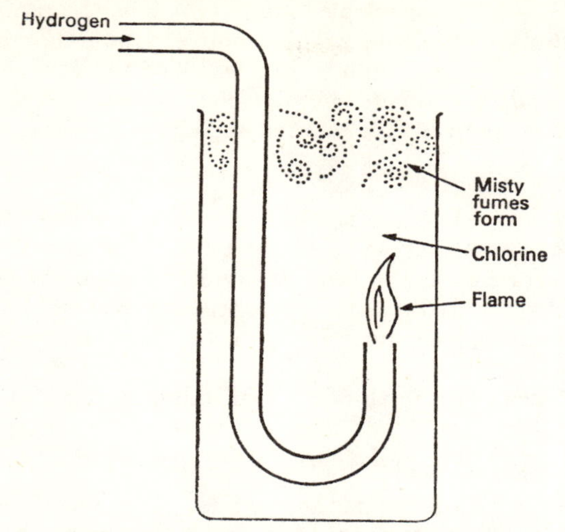

Fig. 11.7 Burning hydrogen in chlorine

Metals

1. Put a piece of asbestos paper on a combustion spoon. Place a small piece of sodium on the paper, heat it directly with a flame until it burns, and then lower it into chlorine.
2. Repeat with burning lithium and burning potassium.
3. Pass dry chlorine over heated iron using an apparatus as in Fig. 11.8.

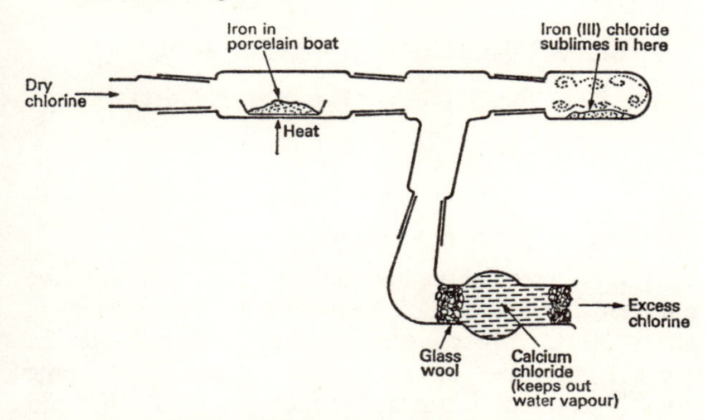

Fig. 11.8 Preparation of iron(III) chloride, $FeCl_3$

4. Repeat with other hot metals, such as magnesium, zinc, and copper.

Non-metals

1. Place dry white phosphorus on a combustion spoon and lower into a jar of chlorine.
2. Repeat with burning red phosphorus.
3. Repeat with burning sulphur and hot carbon.

Reducing agents

Place a gas-jar of chlorine over a gas-jar of ammonia. Remove the covers and let the gases mix. Repeat the test with gas-jars of chlorine and hydrogen sulphide.

Dyes (bleaching)

1. Add moist litmus paper, coloured flowers, and a piece of newspaper with printers' ink and ordinary ink on it into separate jars of moist chlorine.
2. Add dry litmus paper to a jar of chlorine that has been dried by leaving solid calcium chloride in it for at least one hour.

Water

Invert a tall tube of chlorine water in a beaker of water. Leave in sunlight for several hours.

Dilute alkali

Bubble chlorine through dilute sodium hydroxide. Test the product with litmus paper.

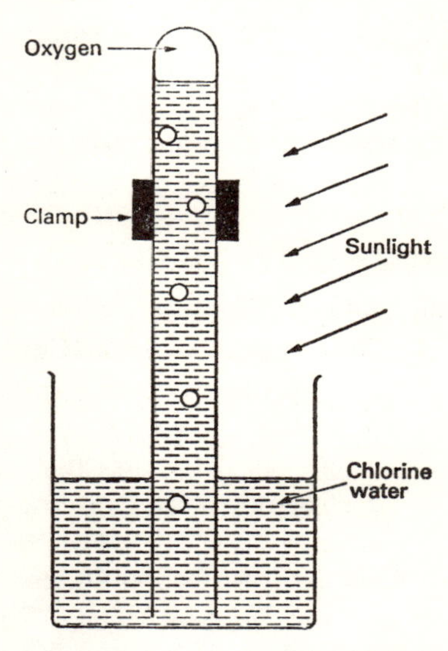

Fig. 11.9 Chlorine water in sunlight

CHEMICAL PROPERTIES OF CHLORINE

Hydrogen A hydrogen-chlorine mixture explodes in direct sunlight or when placed near a flame. The gases react smoothly in diffused light and when passed over heated platinum which catalyses the reaction. Burning hydrogen continues to burn in the gas.

$$H_2 + Cl_2 = 2HCl$$

The violent reaction in sunlight is probably caused by the formation of chlorine atoms when light energy is absorbed:

$$Cl_2 + \text{energy from light} = 2Cl\text{; then}$$

$$H_2 + Cl = HCl + H\text{, and } H + Cl_2 = HCl + Cl\text{, etc.}$$

Hydrocarbons

Wax and petrol continue to burn in the gas. The flame is red. Soot is formed and misty fumes are produced.

$$C_xH_{2y} + yCl_2 = xC \text{ (black)} + 2yHCl \text{ (misty fumes)}$$

Hot turpentine catches fire in chlorine. Substitution products are formed with methane and other alkanes. Alkenes form addition products. Refer to chapter 25 on Organic Chemistry.

Metals Sodium forms white sodium chloride, and iron forms black iron(III) chloride as a sublimate.

$$2Na + Cl_2 = 2NaCl; \quad 2Fe + 3Cl_2 = 2FeCl_3$$

Other metals also form chlorides.

Non-metals White phosphorus bursts spontaneously into flame, and burning red phosphorus continues to burn.

$$2P + 3Cl_2 = 2PCl_3, \text{ phosphorus trichloride}$$
$$2P + 5Cl_2 = 2PCl_5, \text{ phosphorus pentachloride}$$

Sulphur forms a reddish liquid, S_2Cl_2, disulphur dichloride which is pale yellow when pure. Carbon does not react.

Reducing agents Ammonia is oxidized to nitrogen and white fumes of ammonium chloride are also formed.

$$2NH_3 + 3Cl_2 = N_2 + 6HCl; \text{ then}$$
$$NH_3 + HCl = NH_4Cl$$

Hydrogen sulphide is oxidized to sulphur.

$$\underset{}{H_2S} + Cl_2 = \underset{\text{(yellow)}}{S(s)} + \underset{\text{(misty fumes)}}{2HCl(g)}$$

Dyes Dry chlorine does not bleach. The moist gas bleaches many dyes because of the action of chloric(I) acid which oxidizes the dye to a colourless form.

$$\text{Dye} + HClO = HCl + (\text{Dye} + O), \text{ colourless}$$

Printers' ink contains carbon, which cannot be bleached by any substance. Blue litmus is first turned red and then bleached.

Water Sunlight causes the chloric(I) acid present in chlorine water to decompose and form oxygen. The yellowish solution becomes colourless.

$$2H_2O + 2Cl_2 \rightleftharpoons 2HCl + 2HClO$$
$$2HClO = O_2 + 2HCl$$
$$\text{or} \quad 2Cl_2 + 2H_2O = O_2 + 4HCl$$

Dilute alkali Sodium hydroxide absorbs the gas to form a solution containing sodium chloride and sodium chlorate(I), NaClO, which is used as a bleaching agent and as a disinfectant.

$$2OH^- + Cl_2 = Cl^- + ClO^- + H_2O$$

Solid slaked lime also absorbs chlorine and forms bleaching powder.

$$Ca(OH)_2 + Cl_2 = H_2O + CaOCl_2 \text{ (white solid)}$$

PROPERTIES OF BROMINE AND IODINE

Bromine is a dense, dark-red liquid. The liquid corrodes the skin. The vapour is poisonous and attacks the eyes and mucous membranes.

Iodine forms black, lustrous crystals, which are metallic in appearance. It sublimes on heating and forms a violet or purple vapour which is poisonous.

Hydrogen Hydrogen and bromine vapour combine when heated platinum is used as a catalyst. Hydrogen and iodine combine only at high temperatures; the reaction is reversible and incomplete.

$$H_2 + Br_2 = 2HBr \qquad H_2 + I_2 \rightleftharpoons 2HI$$

Hydrocarbons Bromine forms addition products with alkenes; see chapter 25 on Organic Chemistry.

Metals Put about four drops of bromine on glass wool at the bottom of a test-tube. Push iron wool half-way down and heat it, with the tube horizontal. Reaction occurs.

Repeat with three crystals of iodine in place of the bromine.

$$2Fe + 3Br_2 = 2FeBr_3$$

Non-metals Bromine and white phosphorus react explosively. Phosphorus and iodine combine readily.

Water Iodine is only slightly soluble. Bromine forms 'bromine water', which decomposes to form oxygen when exposed to sunlight.

$$Br_2 + H_2O \rightleftharpoons HBr + HBrO, \text{ bromic(I) acid}$$

Bromine water and moist bromine gas are bleaching agents.

Dilute alkali The reactions are similar to that with chlorine, forming sodium bromate(I) and sodium iodate(I).

$$Br_2 + 2NaOH = H_2O + NaBr + NaBrO,$$
$$I_2 + 2NaOH = H_2O + NaI + NaIO,$$
$$\text{i.e.} \quad X_2 + 2OH^- = H_2O + X^- + XO^-$$

PROPERTIES OF FLUORINE

This pale greenish-yellow gas is the most reactive halogen. It combines explosively with hydrogen even at −250 °C. Most heated metals and non-metals, except chlorine, oxygen, and nitrogen, combine directly to form fluorides; for example, CF_4, PF_5, and SF_6. Fluorine displaces the other three halogens from their compounds; it also reacts with water.

$$2Cl^- + F_2 = 2F^- + Cl_2; \; 2F_2 + 2H_2O = 4HF + O_2$$

SUMMARY

Reactions of halides

Conc. sulphuric acid Colourless misty acid fumes in damp air, e.g.

$$NaCl + H_2SO_4 = NaHSO_4 + HCl$$

Bromides give some reddish bromine gas; iodides give violet iodine fumes.

Conc. sulphuric acid with MnO_2 Halogen gas released—bleaches litmus, e.g.

$$MnO_2 + 4HCl = Cl_2(g) + MnCl_2 + 2H_2O$$

Silver nitrate solution Halide precipitated: test solubility in dilute ammonia. AgCl: white, dissolves easily. AgBr: pale-yellow, slightly soluble. AgI: yellow, insoluble.
Chlorine water Displaces iodine and bromine: test solubility in CCl_4. Bromine: red solution, reddish in CCl_4. Iodine: brown solution with a few black crystals, violet in CCl_4.

Hydrogen chloride: colourless gas, denser than air, irritating smell, sharp taste. Misty fumes in damp air. Very soluble in water giving acid solution, and in methylbenzene (neutral solution).
Preparation: from sodium chloride and conc. sulphuric acid.
Properties: does not burn; forms chlorides with heated metals (NaCl, $FeCl_2$); forms NH_4Cl white fumes with ammonia; forms white precipitate with silver nitrate solution.

Hydrochloric acid: dissolve HCl gas through funnel in water.
Properties: acid to indicators; hydrogen released with metals; carbon dioxide with carbonates; sulphur dioxide with sulphites; salts and water with bases; chlorine with oxidizing agents.
Formula: pass chlorine with excess hydrogen over hot platinum.

Manufacture and uses of halogens
Fluorine: electrolyse KHF_2 in anhydrous HF with carbon anode. Its compounds are used to etch glass, and in refrigerators, as lubricants, and in aerosol sprays.
Chlorine: electrolyse conc. sodium chloride or fused sodium chloride, or oxidize HCl gas with air and catalyst. It bleaches materials and sterilizes water; its compounds are solvents, insecticides, antiseptics, anaesthetics, and weed killers.
Bromine: acidify sea-water and pass chlorine. Its compounds are used in photography, as sedatives and to remove lead from petrol engines.
Iodine: crystallize sodium nitrate from Chile saltpetre, then precipitate iodine from remaining solution with sodium hydrogensulphite. Its compounds are used in photography, and iodine dissolved in ethanol is an antiseptic for wounds.

Chlorine: greenish-yellow gas, denser than air, choking smell, poisonous, easily liquefied by pressure.
Preparation: dilute mineral acid on bleaching powder ($CaOCl_2$), or oxidize conc. HCl with $KMnO_4$, MnO_2, or NaCl with H_2SO_4 and MnO_2. Remove HCl fumes with water, dry with conc. H_2SO_4 or $CaCl_2$, and collect over brine or by downward delivery.
Properties
Dissolves in *water* to give chlorine water: HCl and HClO (chloric(I) acid). Mixture is acid but bleaches litmus.
With *hydrogen* forms HCl; reaction mixture explodes in sunlight or near a burning splint.
With *hydrocarbons* (wax, petrol) forms black carbon and HCl, burns with red flame.
With *metals* forms chlorides; e.g. NaCl, $FeCl_3$.
With *non-metals:* P forms PCl_3 and PCl_5; S forms S_2Cl_2; carbon does not react.
With other *reducing agents:* ammonia forms $N_2 + NH_4Cl$ (white fumes); hydrogen sulphide forms yellow sulphur and HCl.
With *alkalis:* NaOH gives NaCl + NaClO, sodium chlorate(I); solid $Ca(OH)_2$ gives bleaching powder, $CaOCl_2$.
With *halide solutions:* bromine and iodine are released from bromides and iodides.
Test for chlorine: moist blue litmus turns red and is then bleached.
Reactivity: fluorine > chlorine > bromine > iodine.

QUESTIONS

1. Sketch the apparatus used to demonstrate the high solubility in water of hydrogen chloride. Describe and explain briefly what happens. What precautions must be adopted when a solution of hydrogen chloride in water is to be prepared in the laboratory?
2. Outline the preparation of chlorine from concentrated hydrochloric acid (*a*) by a chemical reaction, (*b*) by electrolysis. Give the names, formulae, and appearance of the solid products formed when hot iron reacts with (i) hydrogen chloride, (ii) dry chlorine.
3. State three physical properties of chlorine. Describe reactions in which chlorine combines with the hydrogen in other compounds.
4. Outline the preparation of a solution containing sodium chlorate(I). What is the action of chlorine on (*a*) moist slaked lime, (*b*) phosphorus, and (*c*) hydrocarbons?
5. Bromine is a volatile, corrosive liquid which is liable to attack cork and rubber, and which has chemical properties similar to those of chlorine. Describe in detail how you would attempt to obtain a sample of bromine from potassium bromide, giving a careful account of the apparatus you would use. (L.)
6. Explain the following observations: (*a*) dry chlorine has no action on dry litmus paper; (*b*) chlorine turns moist blue litmus paper red and then bleaches it; (*c*) aqueous potassium bromide turns reddish when chlorine is bubbled through it; (*d*) white fumes are formed when chlorine burns in either hydrogen or ammonia; (*e*) a candle burns in chlorine to form white misty fumes and black fumes.
7. Describe the physical and chemical properties of chlorine, bromine, and iodine which indicate that these elements are related.
8. Outline three chemical tests which you would use to distinguish between sodium chloride and sodium iodide. Write equations for the reactions involved.
9. The element fluorine resembles chlorine in some properties, though it is more reactive. (*a*) How would you expect it to react with water? (*b*) Do you expect silver fluoride and calcium fluoride to be soluble in water? (*c*) What difficulties would you anticipate in trying to prepare fluorine from sodium fluoride? (*d*) Write formulae for silver fluoride and copper(II) fluoride, using F as the symbol for fluorine. Explain your answers.

12 The Alkali Metals and their Compounds

The elements in group 1 of the Periodic Table are the alkali metals lithium, sodium, potassium, rubidium, and caesium. They are all good conductors of heat and electricity. They are soft enough to be cut with a knife and their relative densities are low: lithium, sodium, and potassium all float on water. They are very reactive.

OCCURRENCE

Sodium is one of the six commonest metals, but it is never found free. Sea-water contains almost 3 per cent by mass of sodium salts, mainly the chloride, and about 0·1 per cent of potassium salts. Solid sodium chloride occurs in many countries as rock salt, formed by evaporation of seas millions of years ago.

Igneous rocks contain sodium aluminium silicate and potassium aluminium silicate.

The ashes left when plants are burnt contain potassium carbonate, formerly extracted and sold as pot-ash.

Other compounds occurring naturally are:

Chile saltpetre which contains $NaNO_3$ and a little sodium iodate, $NaIO_3$
Sylvine, KCl
Carnallite, $KCl \cdot MgCl_2 \cdot 6H_2O$
Cryolite, Na_3AlF_6, used in the extraction of aluminium.

EXTRACTION OF SODIUM AND POTASSIUM

Davy first obtained each of the metals in about 1807 by electrolysis of the fused hydroxide using platinum electrodes. This process was used for some time to manufacture sodium.

At the cathode: $Na^+ + e \rightarrow Na$
At the anode: $OH^- - e \rightarrow OH$
$4OH \rightarrow O_2 + 2H_2O$

However, the water produced decomposes some of the sodium. Electrolysis of fused sodium chloride is now used because it is cheaper and because no water is produced.

Electrolysis of fused sodium chloride See Fig. 12.1. The cell is a circular steel tank lined with bricks. The anode in the centre of the cell is a huge graphite rod as chlorine does not attack graphite. Around it is a circular steel cathode. The distance between anode and cathode is small in order to reduce the resistance of the cell. A diaphragm of steel gauze is placed between anode and cathode to prevent the liberated sodium and chlorine mixing.

Sodium chloride melts at about 800° C. At this high temperature both chlorine and sodium attack the materials of the cell. The electrolyte is therefore a mixture of sodium chloride and calcium chloride in the ratio 2:3 by mass which melts at a lower temperature of about 600 °C. The electric current maintains the electrolyte at this temperature.

At the cathode:

$$Na^+ + e \rightarrow Na(l); \quad \text{also } Ca^{2+} + 2e \rightarrow Ca$$

At the anode:

$$Cl^- - e \rightarrow Cl; \quad 2Cl \rightarrow Cl_2(g)$$

The molten sodium rises to the top of the electrolyte; it passes into a reservoir. The small amounts of calcium also produced are easily separated as calcium is denser and does not mix with sodium. A cone traps the chlorine.

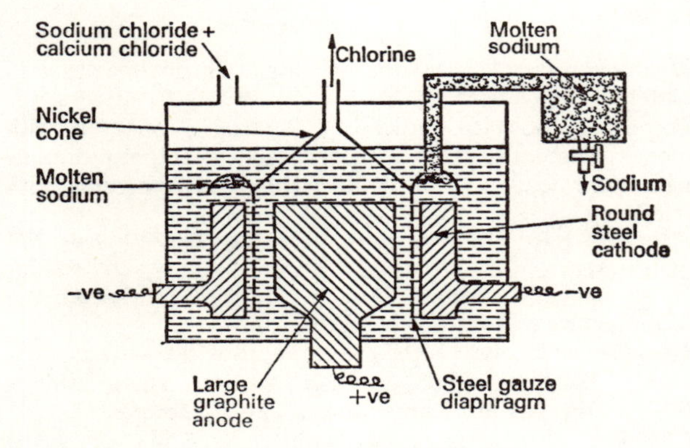

Fig. 12.1 Manufacture of sodium

PROPERTIES OF LITHIUM, SODIUM, AND POTASSIUM

	Lithium	*Sodium*	*Potassium*
Density ($g\ cm^{-3}$)	0·53	0·97	0·86
m.p. (°C)	186	97	62
b.p. (°C)	1336	880	757

Sodium is a soft, silvery white metal. Its melting point, boiling point, relative density, and tensile strength are low compared with most other metals.

Action of air or oxygen Sodium that has been dried with phosphorus(V) oxide for several months does not react with dry oxygen. It is probable that water catalyses this reaction and many similar ones.

Sodium rapidly tarnishes in the moist air of the atmosphere, and a film of oxide forms in seconds. The oxide absorbs water vapour to form the hydroxide, which is slowly converted to carbonate by atmospheric carbon dioxide:

$$4Na + O_2 = 2Na_2O; \; Na_2O + H_2O = 2NaOH(aq);$$
$$2NaOH + CO_2 = H_2O + Na_2CO_3(s)$$

Potassium reacts similarly, but the potassium carbonate is deliquescent and remains as a solution. The metals are stored under kerosine in air-tight containers.

Sodium burns in air or oxygen with a bright yellow flame, forming the oxides Na_2O_2 (sodium peroxide) and Na_2O. Potassium burns with a lilac flame. Lithium burns with a dazzling flame similar to that of magnesium, forming Li_2O.

Action of water Refer to p. 26 in chapter 5. Potassium is most reactive. Lithium decomposes water slowly, like calcium.

Action of ethanol Hydrogen is evolved steadily:

$$2C_2H_5OH + 2Na = H_2 + 2C_2H_5ONa,$$
sodium ethoxide

Action of elements The hot metals combine with chlorine and hydrogen. Lithium also combines with nitrogen.

$$2Na + H_2 = 2NaH, \text{ sodium hydride}$$

$$6Li + N_2 = 2Li_3N, \text{ lithium nitride}$$

Mercury combines with the alkali metals to form liquid amalgams.

HYDROXIDES OF THE ALKALI METALS

Potassium hydroxide and sodium hydroxide are deliquescent and very soluble in water; they are strong alkalis. Lithium hydroxide is only moderately soluble and is only a moderately strong alkali.

Sodium hydroxide is a white solid. Its solution absorbs carbon dioxide from the air and gradually forms solid sodium carbonate-10-water hydrate, $Na_2CO_3 . 10H_2O$. The solution should not be kept in bottles with glass stoppers as the solid carbonate fixes the stoppers tight in the necks of the bottles. Potassium carbonate is deliquescent and does not do this.

The caustic alkalis (KOH and NaOH) precipitate hydroxides from solutions of most metallic salts. The hydroxides of zinc, lead, and aluminium react with excess alkali.

$$Zn(OH)_2 + 2OH^- = Zn(OH)_4^{2-}$$

$$Pb(OH)_2 + 2OH^- = Pb(OH)_4^{2-}$$

$$Al(OH)_3 + OH^- = Al(OH)_4^-$$

Aluminium reacts readily with hot caustic alkali solution, and zinc reacts slowly:

$$2Al + 2NaOH + 6H_2O = 3H_2 + 2NaAl(OH)_4,$$
sodium aluminate

$$Zn + 2NaOH + 2H_2O = H_2 + Na_2Zn(OH)_4,$$
sodium zincate

These formulae are sometimes written $NaAlO_2$ and Na_2ZnO_2 (by removing $2H_2O$ from each formula).

The oxides and hydroxides of lead, zinc, and aluminium are called *amphoteric* compounds because they can act as bases or as acids. They react to form salts both with acids and with alkalis.

Concentrated and fused caustic alkalis slowly react with glass and porcelain, which contain silicon(IV) oxide:

$$2NaOH + SiO_2 = H_2O + Na_2SiO_3, \text{ sodium silicate}$$

Sodium hydroxide is cheaper, and therefore more commonly used than potassium hydroxide. However, potassium hydroxide solution is used to absorb carbon dioxide because (*a*) potassium hydroxide is more soluble in water, and (*b*) potassium carbonate remains in solution in potassium hydroxide whereas sodium carbonate crystallizes from sodium hydroxide solution and hinders further absorption.

USES OF SODIUM HYDROXIDE

It is used to clean machines and metal sheets because it emulsifies oils and grease. It converts fats to soap. It is used in the extraction of aluminium to remove sand from bauxite ore which is impure aluminium oxide. Rayon, drugs, dyes, alkaline bleaching solutions (e.g. NaClO), and many other industrial chemicals are made by its use.

MANUFACTURE OF SODIUM HYDROXIDE

From sodium chloride by electrolysis Saturated brine is electrolysed between graphite anodes and a thin layer of moving mercury which forms the cathode.

At the cathode:

$$Na^+ + e = Na; \quad Na + \text{mercury} = \text{amalgam}$$

At the anode:

$$Cl^- - e = Cl; \quad 2Cl = Cl_2$$

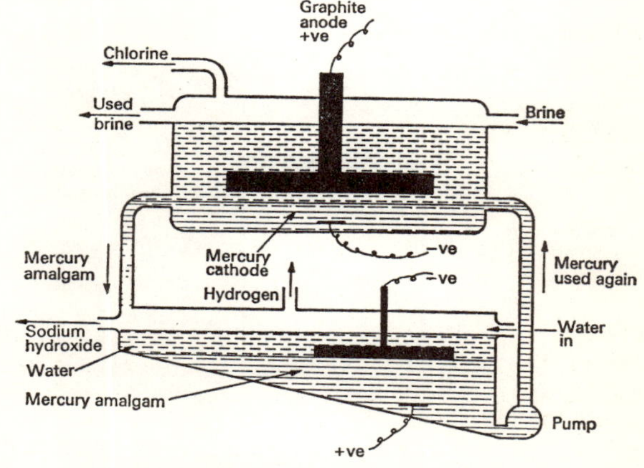

Fig. 12.2 Manufacture of sodium hydroxide

Sodium forms a dilute liquid amalgam with the mercury. The amalgam drops into a second cell containing water in which the amalgam now forms the anode, and carbon or metal forms the cathode. The amalgam, in effect, reacts with the water to form hydrogen, sodium hydroxide solution, and mercury which is used again.

$$2Na\ (amalgam) + 2H_2O = H_2 + 2NaOH + mercury$$

Potassium hydroxide is made in a similar way.

From sodium carbonate Lumps of lime are added to 10 per cent sodium carbonate solution in an iron vessel heated by steam.

$$Na_2CO_3 + Ca(OH)_2 \rightleftharpoons 2NaOH + CaCO_3(s)$$

or $$CO_3^{2-} + Ca^{2+} = CaCO_3(s)$$

Excess lime is used so that the reversible reaction proceeds to the right as much as possible. A sludge of calcium carbonate and excess lime settles. The sodium hydroxide is evaporated to dryness in iron pans, and the molten alkali is poured into iron drums to solidify. It is usually made into sticks or pellets.

SODIUM CARBONATE AND SODIUM HYDROGENCARBONATE

Manufacture from common salt (ammonia-soda or Solvay process) Excess carbon dioxide is passed into a solution of common salt containing ammonia. Ammonium hydrogencarbonate is first formed; this then reacts to form a precipitate of sodium hydrogen carbonate which is only sparingly soluble in brine:

$$NH_3 + H_2O + CO_2 = NH_4HCO_3$$
$$\rightleftharpoons NH_4^+ + HCO_3^-$$
$$Na^+ + HCO_3^- = NaHCO_3(s)$$

(*a*) *Preparation of ammoniacal brine* Ammonia is passed up a tower down which concentrated brine passes slowly.

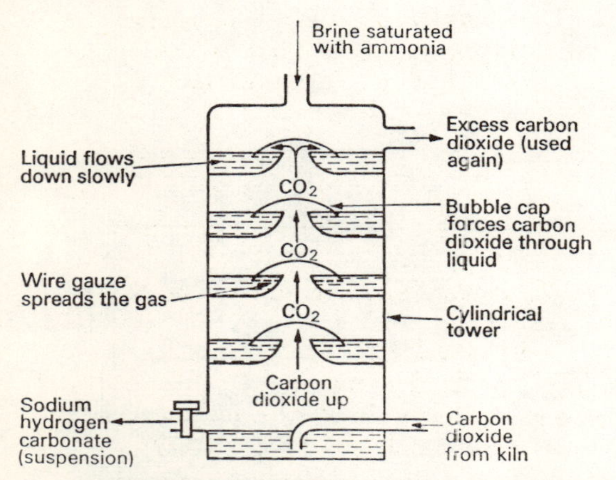

Fig. 12.3 Ammonia-soda or Solvay process for the manufacture of sodium carbonate

(*b*) *Carbonation* The ammoniacal brine is added at the top of a tower up which carbon dioxide passes. There are many compartments in the tower which force the carbon dioxide to bubble through the liquid. Sodium hydrogencarbonate separates as a sludge of fine crystals.

(*c*) *Filtration* The crystals are filtered off, washed to remove ammonium salts and heated to form anhydrous sodium carbonate which is called soda-ash:

$$2NaHCO_3 = Na_2CO_3 + CO_2 + H_2O$$

(*d*) *Ammonia recovery* The liquid from the filters contains ammonium salts. Slaked lime is added and all the ammonia is liberated for use again:

$$2NH_4Cl + Ca(OH)_2 = 2NH_3 + CaCl_2 + 2H_2O$$

The slaked lime is prepared from quicklime (CaO) which is prepared in the lime-kilns which are used to produce the carbon dioxide.

Note that the raw materials are brine and limestone (to produce carbon dioxide) and the complete change may be written:

$$2NaCl + CaCO_3 = Na_2CO_3 + CaCl_2$$

Calcium chloride is the only waste product. The ammonia is used over and over again and only slight losses have to be replaced. The process is continuous.

Potassium carbonate cannot be made by a similar process because the hydrogen carbonate is too soluble. It is obtained by saturating potassium hydroxide with carbon dioxide and then heating the hydrogencarbonate so formed:

$$KOH + CO_2 = KHCO_3$$
$$2KHCO_3 = K_2CO_3 + CO_2 + H_2O$$

PROPERTIES OF THE ALKALI CARBONATES AND HYDROGENCARBONATES

The anhydrous carbonates are white powders. Only Na_2CO_3 forms hydrates: they are the 1-water hydrate, $Na_2CO_3 . H_2O$ and the efflorescent 10-water hydrate $Na_2CO_3 . 10H_2O$. Potassium carbonate is very soluble and deliquesces. Lithium carbonate, like calcium carbonate, is only sparingly soluble.

Action of heat Lithium carbonate decomposes:

$$Li_2CO_3 = CO_2 + Li_2O$$

Potassium and sodium carbonates do not decompose; hydrated sodium carbonate loses its water of crystallization. All the hydrogencarbonates decompose, releasing carbon dioxide and forming the carbonates.

Action of acids Carbon dioxide is evolved. See p. 33 of chapter 6.

Magnesium sulphate solution Add this to solutions of the carbonates and of the hydrogencarbonates. Only the carbonates of the alkali metals form precipitates:

$$Mg^{2+} + CO_3^{2-} = MgCO_3(s), \text{ white}$$

This test distinguishes between the two types of carbonate.

COMPARISON OF THE THREE SODIUM SODAS

	Washing soda $Na_2CO_3.10H_2O$ Colourless crystals	*Baking soda* $NaHCO_3$ White powder	*Caustic soda* $NaOH$ White solid
Air	Effloresces	No change	Deliquesces
Litmus	Moderately alkaline	Weakly alkaline	Strongly alkaline
Acids	Carbon dioxide	Carbon dioxide	No gas evolved
Heat	Water evolved	Carbon dioxide and steam	Melts. No vapour evolved
Magnesium sulphate	White precipitate, $MgCO_3$	No precipitate	White precipitate, $Mg(OH)_2$

USES OF SODIUM CARBONATE AND SODIUM HYDROGENCARBONATE

Sodium hydroxide and glass are manufactured from the carbonate. Sodium carbonate-10-water (washing soda) is used to wash textile fabrics and to make water form a lather easily by removing the carbonates present—see p. 146 in chapter 24.

Sodium hydrogencarbonate is used in baking powder where the heat of baking releases carbon dioxide to make the mixture light, and also in medicines to reduce stomach acidity.

ALKALI METALS COMPARED

Potassium is more reactive than sodium; its salts are more soluble, more likely to be deliquescent, not so hygroscopic and do not form hydrates so readily.

Lithium, the first member of the group, resembles magnesium and calcium in many ways; see chapter 24. The metal is fairly hard and tarnishes only slowly. It decomposes water slowly and burns like magnesium; it also forms a nitride, Li_3N (compare Mg_3N_2). The hydroxide is not caustic as it is only moderately soluble in water. The chloride is hydrated and deliquescent, similar to $CaCl_2$ and $MgCl_2$. The carbonate decomposes on heating, and the hydrogen carbonate, like calcium hydrogencarbonate, exists only in solution.

The properties of the main alkali metals are compared with those of hydrogen and the halogens in the table below.

Flame test

Compounds of the alkali metals can be identified by moistening them with a little concentrated hydrochloric acid, and heating on a nichrome or platinum wire in a Bunsen flame. The flame shows characteristic colours:

Sodium	golden yellow
Potassium	lilac
Lithium	crimson

Halogens	*Hydrogen*	*Alkali metals*
Gas, liquid or solid	Gas	Solids
Vapours, diatomic, X_2	Diatomic, H_2	Vapours monatomic: Na, K, and Li
Non-metals	Non-metal	Metals
Do not burn in oxygen	Burns in oxygen	Burn in oxygen
Form salts with metals	Forms a few hydrides, e.g. NaH, CaH_2	Form alloys with metals and amalgams with mercury

SUMMARY

Alkali metals: lithium, sodium, potassium, rubidium, caesium.

Occurrence: in rock-salt, Chile saltpetre ($NaNO_3$), Sylvine (KCl), Carnallite ($KCl.MgCl_2.6H_2O$) and Cryolite (Na_3AlF_6).

Extraction Sodium: electrolyse fused NaCl mixed with $CaCl_2$ at 600 °C. Potassium: electrolyse fused KOH.

Properties: soft low-density metals with low m.p. and low b.p. Form *oxide* in air which slowly changes to hydroxide and then to carbonate.

Release *hydrogen* from water and ethanol.

Combine with *hydrogen* and *halogens*.

Form *amalgams* with mercury.

Lithium: its chemical properties are more similar to calcium and magnesium than to sodium and potassium.

Reactivity: caesium > rubidium > potassium > sodium > lithium.

Hydroxides

Properties: except LiOH, strong caustic alkalis which are white deliquescent solids. Solutions absorb CO_2; sodium carbonate crystals form, but potassium carbonate remains soluble.

With most *metallic salts*, hydroxide precipitated. Zinc, lead, and aluminium hydroxides redissolve in excess alkali because they are amphoteric.

Hot alkali releases *hydrogen* with zinc and aluminium.

Manufacture of caustic soda: electrolyse brine with mercury cathode and release sodium from amalgam to form alkali with water in another cell, *or* add lime to 10 per cent Na_2CO_3 solution.

Uses: to remove oil and grease from machinery; to make soap; to purify bauxite in manufacture of aluminium; to make drugs, dyes, rayon, and bleaching solutions.

Carbonates and hydrogencarbonates

Manufacture: carbon dioxide in ammoniacal brine—*Solvay process*. Sodium hydrogencarbonate is precipitated; this is heated to give anhydrous carbonate. Recover ammonia with slaked lime, $Ca(OH)_2$. Potassium carbonate is made by saturating KOH with CO_2 and then heating the soluble hydrogencarbonate formed—the Solvay process cannot be used.

PROPERTIES

Heat: lithium carbonate decomposes to CO_2. Sodium carbonate loses water of crystallization. Potassium carbonate is anhydrous and remains unchanged. Hydrogencarbonates release CO_2, giving carbonates.

Acids: CO_2 released.

Test: with aqueous $MgSO_4$, carbonates form precipitate; hydrogencarbonates do not.

USES

Sodium carbonate: manufacture of glass and caustic soda; also as washing soda for textiles and for removing carbonates (softening) in hard water.

Sodium hydrogencarbonate: in baking powder and in medicines for reducing stomach acidity.

Flame test

Compounds of sodium: yellow; potassium: lilac; lithium: crimson.

QUESTIONS

1. Mention similarities between the elements sodium and potassium and between their compounds. In what ways do their nitrates, carbonates, and hydrogencarbonates differ?
2. Outline the ammonia-soda (Solvay) process for the manufacture of sodium carbonate. Explain why potassium carbonate cannot be made in the same way.
3. The element rubidium (Rb), valency 1, resembles sodium and potassium in most of its properties, though it is more electropositive than either. (*a*) How would you expect it to react with water? (*b*) What would be the action of heat on rubidium nitrate and rubidium carbonate? (*c*) Will rubidium carbonate be soluble in water? (*d*) How would you attempt to prepare the metal from rubidium chloride? (L.)
4. Imagine that a new element named trentium has been discovered recently. It is a solid but is easily cut with a knife. When added to water it reacts vigorously and takes fire, forming a caustic, alkaline solution. Answer the following questions concerning Trentium (symbol T). (*a*) How would you store it? (*b*) Is it a metal or a non-metal? Give your reason. (*c*) Where would you place it in the reactivity series? (*d*) What will be its valency? (*e*) Write formulae for its chloride, sulphate and carbonate. (*f*) Write an equation for its reaction with water. (*g*) What do you consider will be the products of decomposition by heat of its nitrate? Give an equation. (*h*) From what compound and by what method do you consider it might have been isolated? (L.)
5. Give one test in each case by which a solution of washing soda could be distinguished from a solution of (*a*) caustic soda, (*b*) sodium hydrogencarbonate. How may washing soda be converted in the laboratory to (*c*) caustic soda solution, (*d*) solid sodium hydrogencarbonate? How and under what conditions does caustic soda react with (*e*) chlorine, (*f*) zinc, (*g*) sulphur dioxide? (L.)
6. Draw and label a diagram of the electrolytic cell used in the manufacture of sodium hydroxide from sodium chloride. How and under what conditions will sodium hydroxide react with (i) carbon dioxide, (ii) sulphur dioxide, (iii) zinc or aluminium?

13 Ions and Electrolysis

UNITS OF ELECTRICITY

One *ampere* is the unit of measurement for the flow of electric current. When one ampere flows for one second a quantity of electricity called one *coulomb* has been used. Silver is deposited on the cathode when an electric current flows through silver nitrate solution between silver electrodes. Accurate experiments show that one coulomb of electricity liberates 0·001 118 g of silver. Therefore one mole of silver atoms (107·87 g) is liberated by 107·87 ÷ 0·001 118, or approximately 96 500 coulombs. The accurate value is 96 487 coulombs. This quantity of electricity is 1 *mole of electrons*, and it is the quantity required to deposit 1 mole of silver atoms. The *Faraday constant* is 96 487 coulombs per mole ($C\ mol^{-1}$).

To find the Faraday constants required to deposit copper

1. Connect the circuit as shown. Switch on the current and adjust the rheostat until a current of 0·5 A or less is flowing. Use a 6-volt d.c. supply.

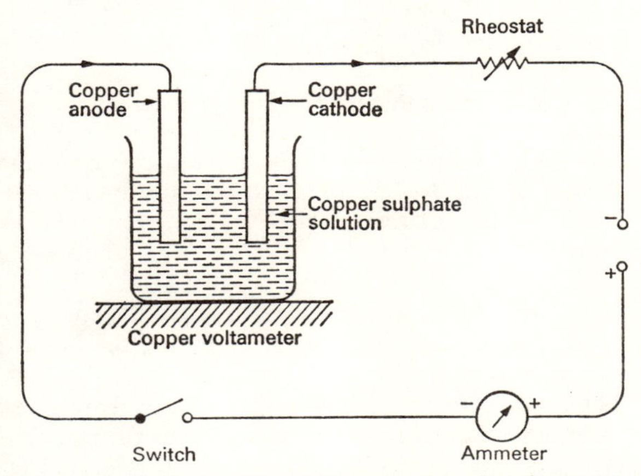

Fig. 13.1 Finding the Faraday constants to deposit copper

2. Switch off. Remove the copper cathode, wash it, and dip in ethanol which evaporates quickly; then leave it to dry.
3. Weigh the clean, dry cathode. Replace it in the voltameter.
4. Switch on the current again. Use the rheostat to ensure that the current is constant during the whole experiment. Pass the current for a known time, e.g. 30 minutes.
5. Switch off. Remove, dry, and weigh the cathode as before.

The increase in its mass is the mass of copper deposited.

1.	Current	$= I$ ampere
2.	Time	$= t$ seconds
3.	Mass of cathode at start	= ?
4.	Mass of cathode at finish	= ?
4 − 3.	Mass of copper deposited	$= w$ grams

w g of copper are deposited by It coulombs

$$\therefore\ 63{\cdot}5 \text{ g of copper are deposited by } \frac{63{\cdot}5}{w} It \text{ coulombs}$$

$$= \frac{63{\cdot}5\, It}{96\,500w}\ \text{C mol}^{-1}$$

To find the Faraday constants required to deposit lead

The experiment with fused lead bromide or iodide described on pp. 17–8 in chapter 3 is repeated and made quantitative.

1. Add lead bromide, to a depth of about 5 mm, to a crystallizing dish and heat until it is just molten. The electrical circuit includes a 12-V d.c. supply, switch, ammeter and rheostat.
2. Place two carbon electrodes in the molten electrolyte.
3. Switch on and adjust the current to a known value, of about 3 A. Pass the current for a known time, about 15 minutes. A bead of lead forms.
4. Switch off, take out the electrodes, and pour the molten electrolyte into a second dish. Leave the lead in the original dish. Allow the lead to cool to room temperature, then take it out and remove any bromide sticking to it.
5. Weigh the lead.

If the current is I ampere, the time t seconds, and the mass of lead w grams, the number of faradays required to deposit 1 mole (207 g) of lead is $(207\ It)/(96\ 500w)$.

The experiments should show that 2 faradays are required to deposit 1 mole of copper or lead. One or more faradays are required to deposit 1 mole of any element. For example, aluminium requires 3 faradays. We can write the ions of these elements as Ag^+, Cu^{2+}, Pb^{2+}, and Al^{3+}.

FARADAY'S LAWS OF ELECTROLYSIS (1830)

Faraday was the first to do quantitative experiments on electrolysis. He summarized his results in two laws.

1. *The mass of a substance liberated during electrolysis is directly proportional to the quantity of electricity passed.*

2. *The quantity of electricity required to deposit 1 mole of any substance is a whole number (1, 2, 3, etc.) of moles of electrons.*

One mole, about 96 500 C, liberates 1 mole of silver atoms, i.e. $6{\cdot}023 + 10^{23}$ atoms.

$$\text{The charge on one silver ion} = \frac{\text{one mole of electrons}}{\text{number of ions}}$$

$$= \frac{96\,500}{6{\cdot}023 \times 10^{23}}\text{C}$$

$$= 1{\cdot}6 \times 10^{-19}\ \text{C}$$

This is the charge of one electron (determined by other methods). Therefore the charge on a silver ion is numerically equal to, but opposite in sign to, that of 1 electron.

An electron is the smallest particle of negative electricity that exists. It is described fully in the next chapter.

IONIC THEORY

This theory was first suggested about a hundred years ago. In its modern form it states that electrolytes contain *ions*. An ion is a positively or negatively charged particle that is formed by loss or gain of one or more electrons by an atom or radical. A negatively charged ion, an anion, moves to the anode and a positively charged ion, a cation, moves to the cathode.

Formula of substance	*Cations*	*Anions*
$AgNO_3$	Ag^+	NO_3^-
$CuSO_4$	Cu^{2+}	SO_4^{2-}
$PbBr_2$	Pb^{2+}	$2Br^-$
H_2SO_4	$2H^+$	SO_4^{2-}

An ion has properties very different from the uncharged particle. For example, sodium and potassium ions do not react with water. Chloride ions and bromide ions are not poisonous like the elements themselves.

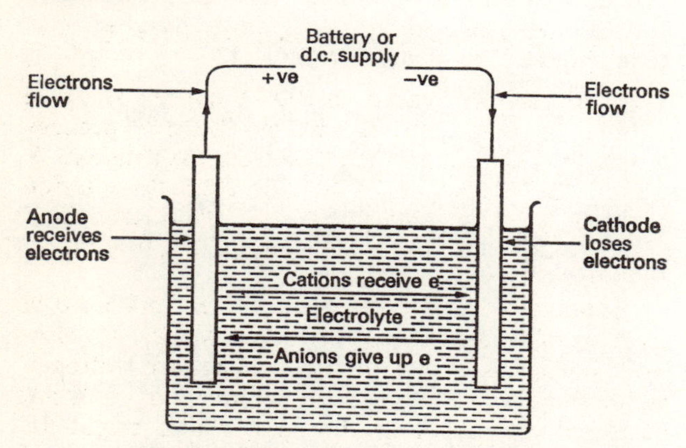

Fig. 13.2 Movements during electrolysis

During electrolysis, electrons flow from the battery to the cathode. The cations move to the cathode and receive electrons, so becoming uncharged particles. The anions take electrons to the anode, and the electrons then flow back to the battery. The current in the wire is a flow of electrons; the current in the electrolyte is a flow of anions and cations.

The movement of ions can be investigated by using an electrolyte of acidified copper chromate, which contains blue hydrated copper ions, $Cu(H_2O)_4^{2+}$, and orange dichromate ions, $Cr_2O_7^{2-}$, produced by the chromate and the hydrochloric acid.

To investigate the migration of coloured ions

1. Prepare copper chromate by mixing equal volumes of M copper sulphate solution and M potassium chromate solution. Filter off and wash the insoluble chromate.
2. Dissolve the copper(II) chromate in the minimum volume of 2M hydrochloric acid. Saturate this solution with urea by stirring for a few minutes; this increases the density of the solution. The solution contains dichromate ions $Cr_2O_7^{2-}$.

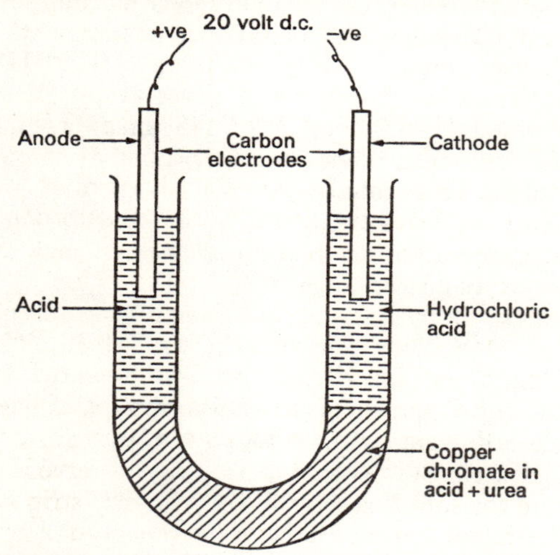

Fig. 13.3 Showing that ions migrate towards electrodes

3. One-third fill a U-tube with 2M hydrochloric acid.
4. Use a pipette to add the urea-acid-chromate mixture to the bottom of the U-tube. The mixture remains at the bottom and there are columns of acid at each side.
5. Dip two carbon electrodes into the acid and connect them to a 20-V d.c. supply. Observe any changes during the next 30 minutes.

The electrolyte around the cathode turns from green to blue, and that around the anode turns orange.

REPLACEMENT OR DISPLACEMENT OF ELEMENTS BY EACH OTHER

We already know from chapter 11 that chlorine displaces bromine and iodine from halide solutions. The essential change is that chlorine, which has a greater affinity for electrons than the other halogens, removes electrons from their ions:

$$Cl_2 + 2Br^- = Br_2 + 2Cl^-; \quad Cl_2 + 2I^- = I_2 + 2Cl^-$$

Fluorine would displace chlorine in the same way. Since fluorine forms negative ions most easily, it is the most *electronegative* of the halogens. Bromine, however, will only displace iodine from iodides and is therefore less reactive and less electronegative than the other two halogens. The series in order of reactivity is:

fluorine chlorine bromine iodine

Dip zinc foil or a piece of iron in copper sulphate solution. Each metal soon becomes reddish-brown owing to a coating of copper. The essential change is that they transfer electrons to copper ions:

$$Zn + Cu^{2+}(aq) = Cu(s) + Zn^{2+}(aq)$$
$$Fe + Cu^{2+}(aq) = Cu(s) + Fe^{2+}(aq)$$

Zinc and iron form positive ions more easily than copper and they are more *electropositive* and more reactive. Displacement of one metal by another can be used to place some metals in the order of their reactivities.

To find, by displacement, the order of reactivities of metals

1. Half fill four test-tubes with about M copper(II) sulphate solution. Do the same with about M lead nitrate or lead ethanoate, iron(II) sulphate, zinc sulphate, and magnesium sulphate, making 20 test-tubes altogether. Place them in racks.
2. Take 4 strips, each about 10 mm long, of copper, lead, iron, zinc, and magnesium. Clean them by rubbing with sandpaper.
3. Add the copper strips to all solutions except copper sulphate, lead strips to all except lead nitrate, etc.
4. Leave the tubes for some time and observe if metal from the solution deposits on the metal strip added. The metal appears as shiny crystals, as a greyish or as a dark, soft deposit.
5. List the metals in a reactivity series.

Aluminium can also be used in this experiment provided the aluminium strip is rubbed with mercury to remove the film of oxide which prevents reaction. Aluminium sulphate and extra test-tubes would also be needed.

ELECTRICAL ENERGY FROM CHEMICAL REACTIONS

A torch cell has a zinc case which contains a carbon rod and white ammonium chloride, together with other chemicals. The zinc and carbon are electrodes and the ammonium chloride is the electrolyte. The cell produces electricity and the zinc is gradually used up. The voltage of the cell is about 1·5 V.

Two different metals dipping in an electrolyte also produce a voltage, or e.m.f. The size of the e.m.f. can be measured and used to place metals in order in a series called the electrochemical or electromotive series (e.m.f. means electromotive force).

To measure the e.m.f. of cells containing two metals

1. Fill a beaker with sodium chloride solution which is the electrolyte.
2. Clean some rods or strips of copper, lead, iron, zinc, and magnesium.
3. Choose the copper rod because copper is the least reactive metal, and place it in the cell. Connect the rod to the positive terminal of a voltmeter.
4. Place the lead rod in the cell and connect it to the negative terminal of the voltmeter. Read the e.m.f. of the cell.
5. Now use each of the other metals in turn in place of the lead.

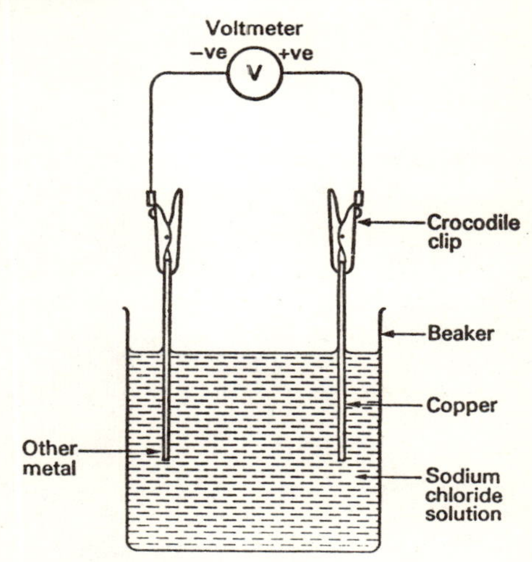

Fig. 13.4 Measuring the e.m.f. of a cell

6. Arrange the metals in an electrochemical series according to the voltage they produce when compared with copper as the standard.

ELECTROCHEMICAL OR ELECTROMOTIVE SERIES

A rod of copper placed in 1·0 M copper sulphate solution becomes positively charged (+0·3 V) because copper ions deposit on the metal from the solution. A rod of zinc placed in 1·0 M zinc sulphate solution becomes negatively charged (—0·8 V) because some of the metal forms positive zinc ions, Zn^{2+}, in solution and leaves electrons on the rod.

The potential difference between a metal and a 1·0 M solution of its ions is its *electrode potential.* All electrode potentials have been measured against hydrogen as the standard. The electrode potentials of gases are measured by bubbling the gas, which might be hydrogen, oxygen, chlorine, and so on, around platinum in a 1·0 M solution of the ions (H^+, OH^-, Cl^-, etc.). Some values of the electrode potentials in volts are given below. The series is known as the electrochemical or electromotive series. Only calcium is in a different place in the reactivity series.

Metals

Lithium	—3·04	Zinc	—0·76
Potassium	—2·92	Iron	—0·44
Calcium	—2·87	Tin	—0·14
Sodium	—2·71	Lead	—0·13
Magnesium	—2·37	Hydrogen	0·00 (standard)
Aluminium	—1·66	Copper	+0·34

The electrochemical series can be continued for non-metals. The list below shows the electrode potentials of elements that are increasingly electronegative.

Non-metals

Oxygen	+0·40	Chlorine	+1·36
Iodine	+0·54	Fluorine	+2·85
Bromine	+1·07		

The results of electrolysis can be explained when the order of preferential discharge of ions in the electrochemical series is known. It is even possible to place negative radicals, such as the sulphate ion (SO_4^{2-}), in this series.

The ions whose potential is closest to hydrogen in the series are the most easily discharged. The lines below show the commonest ions arranged in order; those to the right in each line are the most easily discharged.

K^+	Na^+	H^+	Cu^{2+}	cations
SO_4^{2-}	Cl^-	OH^-		anions

EXPLANATION OF RESULTS OF ELECTROLYSIS

ELECTROLYSIS OF DILUTE SULPHURIC ACID

The following ions are formed:

$$H_2SO_4 \rightarrow 2H^+(aq) + SO_4^{2-}(aq)$$
$$H_2O \rightleftharpoons H^+(aq) + OH^-(aq)$$

The water is only slightly ionized because only about one molecule in several million forms ions. The ionization of water is reversible. If either H^+ or OH^- is removed, more water ionizes.

The reactions that take place are:

$H^+ + e = H; \quad 2H = H_2(g)$ at the cathode

$OH^- - e = OH; \quad 4OH = 2H_2O + O_2(g)$ at the anode

Sulphate ions are not discharged.

Transfer of 4 electrons produces $2H_2$ and O_2. Thus, the volume of hydrogen produced is twice that of oxygen. The acid does not change and the final result is the electrolysis of water:

$$2H_2O + \text{electrical energy} = 2H_2 + O_2$$

ELECTROLYSIS OF COPPER(II) SULPHATE SOLUTION

The following ions are formed:

$$CuSO_4 \rightarrow Cu^{2+}(aq) + SO_4^{2-}(aq)$$
$$H_2O \rightleftharpoons H^+(aq) + OH^-(aq)$$

If both electrodes are copper, the following reactions take place:

$Cu^{2+} + 2e = Cu$ at the cathode

$Cu = Cu^{2+} + 2e$ at the anode

Hydrogen ions are not discharged because they are higher than copper ions in the series.

With copper electrodes, the final result is that copper is transferred from anode to cathode, and the copper sulphate does not change.

A different reaction occurs at the anode if it is made of platinum. Hydroxide ions are discharged and then they react to form oxygen and water:

$$OH^- - e = OH; \quad 4OH = O_2 + H_2O$$

The sulphate ions are not discharged and the platinum anode, unlike copper, does not form ions.

With a platinum anode, copper is deposited on the cathode as before, but oxygen is liberated at the anode. The copper sulphate solution becomes more dilute because of the loss of copper. Finally after prolonged electrolysis, the solution becomes colourless and contains only dilute sulphuric acid which then electrolyses to form hydrogen and oxygen.

The diagram shows what happens during the electrolysis of copper(II) chloride with a platinum anode.

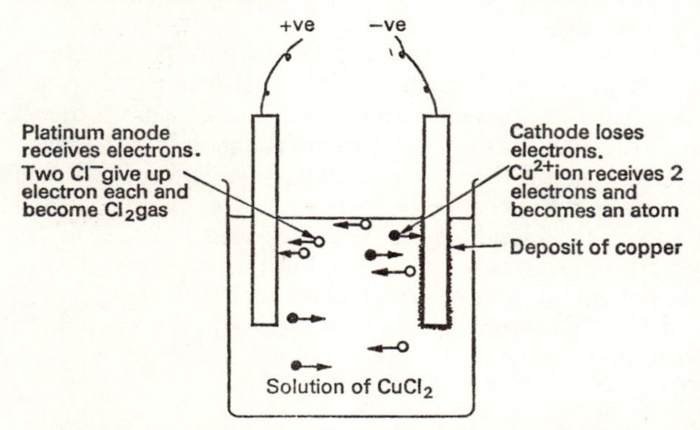

Fig. 13.5 Changes during the electrolysis of copper(II) chloride solution

ELECTROLYSIS OF SOLUTIONS OF SODIUM CHLORIDE

Dilute solution This contains Na^+, H^+, Cl^-, and OH^-. H^+ is discharged at the cathode and OH^- at the anode (as in the electrolysis of dilute sulphuric acid). Hydrogen and oxygen are the products. This is another example of 'the electrolysis of water'.

Concentrated solution The Cl^- ions are now discharged in preference to OH^- because they are present in much greater concentration, or possibly some of both ions are discharged together. Chlorine is formed at the anode (or a mixture of chlorine and oxygen).

Concentrated solution with mercury cathode The Na^+ ions are discharged in preference to H^+. The sodium dissolves in the mercury to form mercury amalgam. This reaction is used in the manufacture of sodium hydroxide. See chapter 12, p. 65.

Note that the concentration of the solution and the materials of the electrodes can affect the course of electrolysis.

Table showing reactivities of metals in relation to their order in the electrochemical series

	Action on metal of				*Action on oxide of*		*Action of heat on*		
	air	*water*	*steam*	*acids*	*hydrogen*	*carbon*	*carbonate*	*nitrate*	
K	Oxides formed even when cold	Violent	Explosive	Explosive	No action	No action	No action	Oxygen and nitrite formed	K
Na									Na
Ca		Moderate	Violent	Violent			Oxides formed plus carbon dioxide	Oxides formed plus oxygen and nitrogen dioxide	Ca
Mg	*most easily* Oxides formed when heated *least easily*	Slow	Rapid	*most easily* Hydrogen evolved *least easily*					Mg
Al		No action	Moderate						Al
Zn			Reversible reaction		Reduced to metals	Reduced to metals			Zn
Fe									Fe
Sn			No action	No action					Sn
Pb									Pb
Cu									Cu
Hg							*most easily*	Metal formed (not oxide)	Hg

SUMMARY

Flow of electricity: the flow of electrons through a conductor or of ions through a solution—demonstrated by electrolysing acidified copper chromate solution between carbon electrodes.
Rate of flow of electricity: measured in amperes.
Quantity of electricity: measured in coulombs.

$$1\ \mathrm{C} = 1\ \mathrm{A}\ \text{for}\ 1\ \mathrm{s}$$

Definition of faraday: the quantity of electricity that liberates 1 mole of silver (107·87 g); equal to 96 500 C approx. *Faraday constant* is 96 487 C mol^{-1}.

Faraday's laws of electrolysis:
1. The mass of a substance liberated during electrolysis is directly proportional to the quantity of electricity passed.
2. The quantity of electricity required to deposit 1 mole of any substance is a whole number of moles of electrons.

Ions are formed by loss or gain of one or more electrons by an atom or by a part (radical) of a molecule.
Cations are positively charged through the loss of one or more electrons; for example, Na^+, Cu^{2+}, Fe^{3+}.
Anions are negatively charged through the gain of one or more electrons: for example, OH^-, Cl^-, CO_3^{2-}, SO_4^{2-}.

Reactivity series: the order in which elements displace the ions of others lower down the series from solution:

magnesium zinc iron lead copper

For example, copper is displaced from solution by all the other elements above and forms a reddish coating. The order of the series is the same as that obtained when we study other chemical reactions.

Electrochemical series
Two different metals, placed in an electrolyte, produce an e.m.f. which can be measured by a voltmeter. Metals can be placed in order according to the e.m.f. they produce compared with copper or hydrogen.
Hydrogen electrode: hydrogen is bubbled round a platinum rod in 1·0 M acid. This electrode is used as the standard in accurate work and its potential is regarded as zero.
Electrode potential of an element is the potential difference between the metal and a 1·0 M solution of its ions—assuming that the hydrogen electrode has a potential of 0·00 V.
Electrochemical (electromotive) series: a list of elements arranged in order of their electrode potentials. The order is the same as that of the reactivity series, except for Ca.

Results of electrolysis
Ions with positive or low negative electrode potentials are the most easily discharged.

Most easily $Cu^{2+} < H^+ < Na^+ < K^+$ Least easily

Most easily $OH^- < Cl^- < SO_4^{2-}$ Least easily

The type of electrode and the concentration of the solution can both affect the results of electrolysis; for example, sodium chloride:
Dilute solution: OH^- discharged rather than Cl^-.
Concentrated solution: Cl^- discharged rather than OH^-.
Mercury cathode: Na^+ discharged rather than H^+.

QUESTIONS

1. Sketch an apparatus suitable for the electrolysis of an aqueous solution of sulphuric acid and measurement of the volumes of gases liberated. Carefully label your drawing and indicate the result observed.

2. Explain how the current passes through dilute sulphuric acid, and account for the products of electrolysis. What is the quantity of electricity in faradays required to liberate 1 mole of each element produced by electrolysis?

3. Explain why an electric current is conducted more easily through a molar solution of hydrochloric acid than through a molar solution of acetic acid.

4. Describe two experiments which give support to the idea of the existence in solutions or liquids of ions.

5. Explain what you understand by the 'electrochemical series of metals'. Arrange the metals zinc, sodium, magnesium, and copper in decreasing order of activity. Justify the order you give by comparing the reactions of these four metals with (*a*) water, (*b*) steam, (*c*) oxygen, and (*d*) dilute hydrochloric acid.

6. State Faraday's laws of electrolysis. The same quantity of electricity was passed through three voltameters, depositing copper in the first, silver in the second, and liberating 200 cm^3 of hydrogen (measured at s.t.p.) in the third. Calculate the weights of copper and silver deposited. (1000 cm^3 of hydrogen at s.t.p. weighs 0·09 g; Cu = 63·6, Ag = 108.) (L.)

7. Arrange the metals *W*, *X*, *Y*, and *Z* in order of their chemical activities by using the following information: (*a*) *X* is displaced from aqueous solutions of its salts by each of the three other metals; (*b*) *Z* displaces metal *W* from an aqueous solution of a salt of *W*, but *Z* has no action on aqueous solutions of salts of *Y*. Explain carefully the reasoning by which you arrive at the order of activities.

8. A current is passed through two voltameters in series, one a copper voltameter with copper electrodes in a solution of copper sulphate, the other with platinum electrodes in a dilute solution of sulphuric acid. After a time, 785 cm^3 of hydrogen, measured dry at 27 °C and 745 mmHg pressure, are collected in one voltameter. What mass of copper will be deposited on the cathode in the other? (Cu = 64, H = 1.) (C.)

9. If one faraday of electricity were passed through dilute sulphuric acid solution, what would be the total volume of gas liberated, measured at s.t.p.? Aluminium reacts slowly with dilute acids and with air, although the metal is high in the activity (electrochemical) series. Suggest a reason for this apparent inactivity. Describe one simple experiment which shows that aluminium is indeed a very reactive metal.

10. Describe the experiment you would do in order to demonstrate that copper ions in aqueous solution each have twice the charge of ions of hydrogen. State clearly what measurements you would take and how you would use them to calculate the result. (Cu = 63·5, H = 1.)

14 Chemical Bonds

PARTICLES SMALLER THAN ATOMS

Between 1896 and 1900 Becquerel and the Curies discovered that atoms of uranium, radium, and certain other elements decompose spontaneously and produce two kinds of particles smaller than the atoms themselves. In 1897 Thompson proved that all atoms contain negatively charged particles called electrons. An atom is electrically neutral. Since it contains electrons with a negative charge, an atom must also contain positively charged particles. In 1920 Rutherford demonstrated clearly the existence of these positive particles, which are called protons. In 1932 Chadwick discovered that atoms also contain particles with no charge. The uncharged particles are called neutrons.

It is now known that all atoms contain electrons and protons and that all atoms except hydrogen atoms also contain neutrons. The charge on an electron is equal and opposite to that on a proton and it is the smallest possible electrical charge that can exist. Details of the three fundamental particles are:

	Relative masses			
	Approx.	*Exact*	*Charge*	*Actual masses*
Electron	1/1840	0·000 549	−1	$9{\cdot}11 \times 10^{-28}$ g
Proton	1	1·007 275	+1	$1{\cdot}673 \times 10^{-24}$ g
Neutron	1	1·008 665	0	$1{\cdot}675 \times 10^{-24}$ g

The mass of a neutron is slightly larger than that of a proton. In some special reactions a neutron splits up into a proton, an electron and energy.

By 1914 Rutherford had proved that an atom consists of a tiny positively charged nucleus surrounded by electrons. Most of the mass of an atom is in its nucleus because the mass of its electrons is negligible.

THE NUCLEI OF ATOMS

Nuclei contain protons and neutrons. These particles are similar except for their electrical properties. They are called *nucleons*. The number of protons in one atom is the *proton number* of the element, and the number of nucleons in one atom is the *nucleon number* of the element. The proton number of an element is now known to be the same as its atomic number in the Periodic Table. The proton numbers of naturally-occurring elements range from one to ninety-two; that of hydrogen is one and that of uranium is ninety-two.

ISOTOPES

A hydrogen atom can have either no neutron or one or two neutrons in its nucleus. Therefore there are three species of hydrogen atoms; they have the same proton number but different nucleon numbers. Since the mass of an atom depends on its nucleon number, the three species of hydrogen atoms have different atomic masses. Chlorine has two species of atoms, and each atom has either eighteen or twenty neutrons in its nucleus. All elements either exist naturally or can be made artificially with different species of atoms. A *nuclide* is a species of atom with the same proton number and nucleon number. Two or more nuclides with the same proton number but different nucleon numbers are called *isotopes* or *isotopic nuclides.*

Hydrogen isotopes

	Protium	*Deuterium*	*Tritium*
Protons	1	1	1
Neutrons	0	1	2
Nucleon number	1	2	3
Relative atomic mass	1·008	2·014	3·016
Symbols	${}^{1}_{1}H$	${}^{2}_{1}D$	${}^{3}_{1}T$

The subscript in a symbol is the proton number or atomic number and the superscript is the nucleon number.

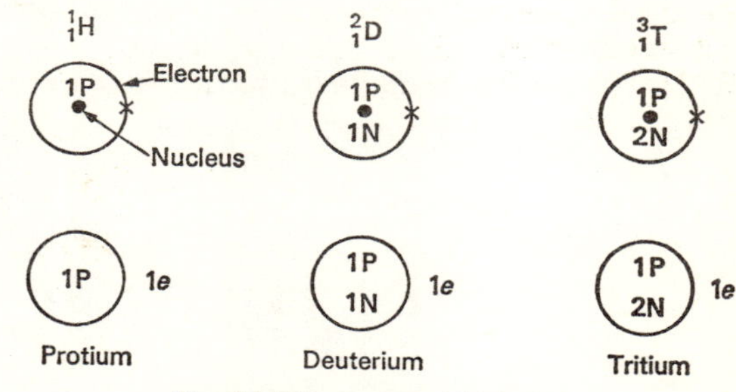

Fig. 14.1 Isotopes of hydrogen

Isotopes of chlorine

Symbols	${}^{35}_{17}Cl$	${}^{37}_{17}Cl$
Protons	17	17
Neutrons	18	20
Nucleon number	35	37
Relative atomic mass	34·981	36·978

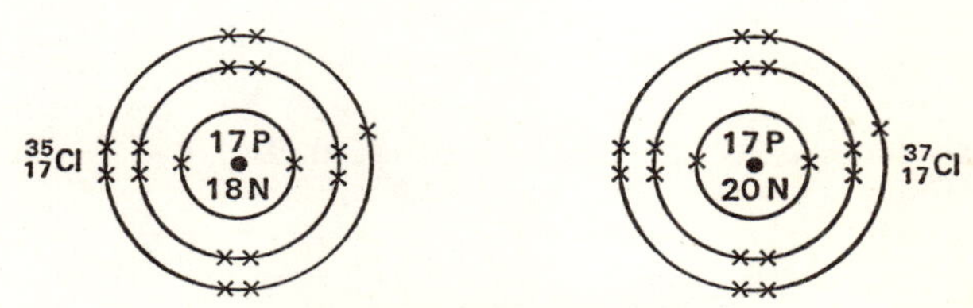

Fig. 14.2 Isotopes of chlorine

Gaseous chlorine contains about three times more chlorine-35 atoms than chlorine-37 atoms. The relative atomic mass of ordinary chlorine is therefore about $[(3 \times 35) + 37]/4 = 35{\cdot}5$ approximately.

ELECTRONS IN ATOMS

The electrons in an atom can move in seven different shells or orbitals, represented by letters *K*, *L*, *M*, *N*, *O*, *P*, and *Q*. An electron in the *K* shell has the least energy and an electron in the *Q* shell has more energy. A *K* electron is nearer to the nucleus than an *L* electron and so on. The arrangements of the electrons in the first twenty elements of the periodic table are as follows.

Period 1 elements A hydrogen atom has one electron in the *K* shell. A helium atom has two electrons in the *K* shell. The *K* shell never has more than two electrons, which are called a *duplet*. The duplet is a very stable arrangement.

Period 2 elements A lithium atom has a duplet in the *K* shell and one electron in the *L* shell, further from its nucleus. Its electron configuration is *K*2, *L*1, which is usually written 2, 1. Atoms of elements after lithium have additional electrons in the *L* shell. A neon atom has eight *L* electrons, which is a stable arrangement called an *octet*.

Period 3 elements The electron configuration of sodium is 2, 8, 1; the atom has a stable duplet, a stable octet, and one electron in the *M* shell. Atoms of elements after sodium have additional electrons in the *M* shell. An argon atom has eight *M* electrons, which form a second stable octet.

The configurations of the first twenty elements are:

Period								
1	H 1							He 2
2	Li 2, 1	Be 2, 2	B 2, 3	C 2, 4	N 2, 5	O 2, 6	F 2, 7	Ne 2, 8
3	Na 2, 8, 1	Mg 2, 8, 2	Al 2, 8, 3	Si 2, 8, 4	P 2, 8, 5	S 2, 8, 6	Cl 2, 8, 7	Ar 2, 8, 8
4	K 2, 8, 8, 1	Ca 2, 8, 8, 2						

FAMILIES OF ELEMENTS

A noble gas has either a stable duplet or a stable octet in its outer shell. The halogens fluorine and chlorine have one electron less than the noble gas which follows them in the periodic classification; an alkali metal such as lithium, sodium or potassium has one electron more than the noble gas it follows. A halogen atom tends to take one more electron so that it has the electron structure of a noble gas. An atom of an alkali metal readily gives up its single electron in order to attain a stable noble gas structure.

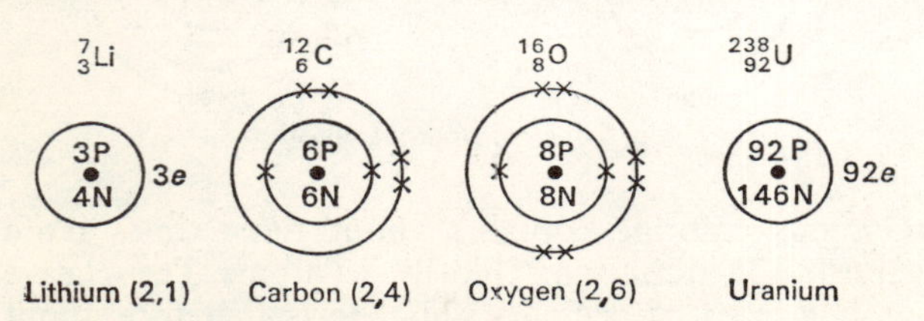

Fig. 14.3 Structures of some atoms

CHEMICAL BONDS

ELECTROVALENT OR IONIC BONDS

A sodium atom tends to lose one electron and a chlorine atom tends to gain one electron. Transfer of one electron from a sodium atom to a chlorine atom enables both to attain stable noble gas structures:

Sodium atom + chlorine atom
2, 8, 1 2, 8, 7
= sodium ion + chloride ion
2, 8 2, 8, 8

The two ions are oppositely charged and therefore attract each other; the bond between them is ionic or electrovalent, and it is caused by the transfer of one electron. Calcium chloride is formed when a calcium atom transfers two electrons, and two chlorine atoms receive one electron each. Calcium oxide is formed when a calcium atom transfers two electrons to an oxygen atom:

Calcium atom + oxygen atom
2, 8, 8, 2 2, 6
= calcium ion + oxide ion
2, 8, 8 2, 8

Simplified electronic formulae do not show the electrons in the complete inner shells; they only show the outer electrons, represented by × or ∘.

Sodium chloride

$Na^{+}Cl^{-} \qquad Na_{\circ} + {}^{\times}\overset{\times\times}{\underset{\times\times}{Cl}}{}^{\times}_{\times} \rightarrow [Na]^{+} \quad \left[{}^{\times}_{\circ}\overset{\times\times}{\underset{\times\times}{Cl}}{}^{\times}_{\times}\right]^{-}$

Calcium chloride

$Ca^{2+}2Cl^{-} \quad Ca^{\circ}_{\circ} + 2{}^{\times}\overset{\times\times}{\underset{\times\times}{Cl}}{}^{\times}_{\times} \rightarrow [Ca]^{2+} \; 2\left[{}^{\times}_{\circ}\overset{\times\times}{\underset{\times\times}{Cl}}{}^{\times}_{\times}\right]^{-}$

Calcium oxide

$Ca^{2+}O^{2-} \qquad Ca^{\circ}_{\circ} + \overset{\times\times}{\underset{\times\times}{O}}{}^{\times}_{\times} \rightarrow [Ca]^{2+} \quad \left[{}^{\circ}_{\circ}\overset{\times\times}{\underset{\times\times}{O}}{}^{\times}_{\times}\right]^{2-}$

Magnesium oxide

$Mg^{2+}O^{2-} \qquad Mg^{\circ}_{\circ} + \overset{\times\times}{\underset{\times\times}{O}}{}^{\times}_{\times} \rightarrow Mg^{2+} \quad \left[{}^{\circ}_{\circ}\overset{\times\times}{\underset{\times\times}{O}}{}^{\times}_{\times}\right]^{2-}$

The sodium chloride crystal Fig. 14.5 shows the arrangement of sodium and chloride ions in a tiny part of a crystal. The part shown in Fig. 14.5a is called a unit cell and it is the smallest part which, by repetition, builds up the entire crystal. The central

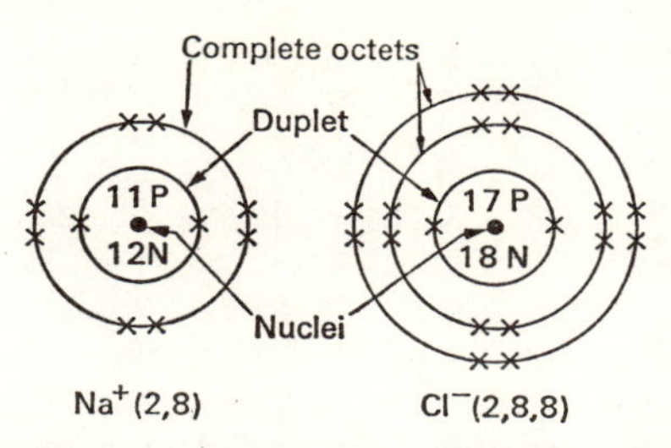

Fig. 14.4 Complete electronic structure of sodium chloride

chloride ion has six sodium ions arranged octahedrally around it, as Fig. 14.5b makes clear (twelve chloride ions are equidistant from it). Similarly, each sodium ion has six chloride ions arranged octahedrally around it and twelve sodium ions are equidistant from it.

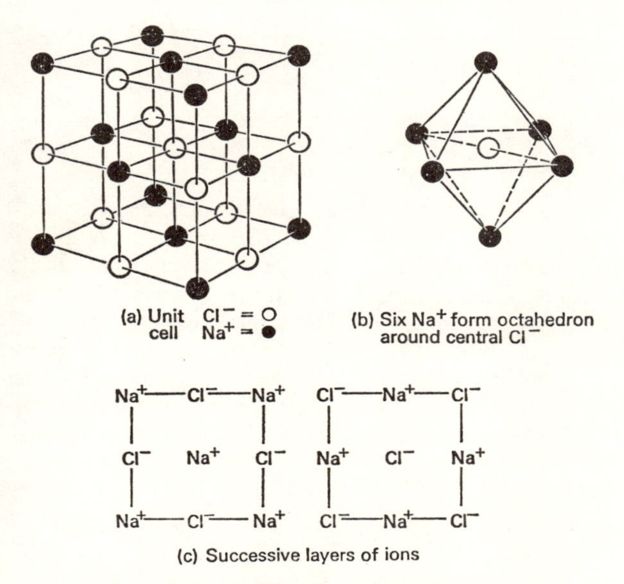

Fig. 14.5 Ions in a sodium chloride crystal

The distance between the centres of neighbouring ions is exactly the same. A sodium chloride molecule does not exist because any ion is joined equally by electrostatic forces to six of the other ions. One crystal is a giant three-dimensional structure held together by attraction between oppositely charged ions.

Figure 14.5 does not show clearly how the ions are packed as it merely represents the centres of the ions but not their sizes. The diameter of a chloride ion is almost twice that of a sodium ion. Fig. 14.6 shows the packing and relative sizes of the ions, which touch each other in a crystal.

The unit cell in Fig. 14.5, consists of eight smaller cubes, each of eight ions with one ion at each corner. The cube is the smallest regular geometrical pattern in the crystal and therefore sodium chloride has a simple cubic structure or lattice.

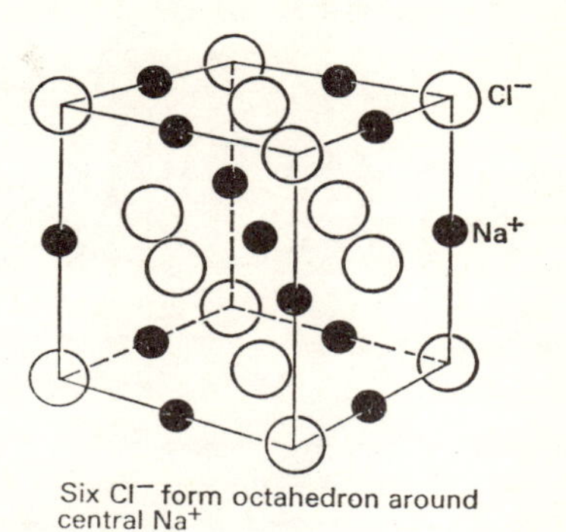

Fig. 14.6 Packing of ions in sodium chloride

Other compounds which consist of simple ions arranged as in sodium chloride are the halides of lithium, sodium, and potassium (e.g. Li^+Br^-, Na^+I^-, K^+Cl^-), the oxides and sulphides of magnesium and calcium (e.g. Mg^{2+} O^{2-}, $Ca^{2+}O^{2-}$), and silver chloride, bromide, and iodide (e.g. Ag^+Cl^-).

COVALENT BONDS

Electrovalent bonds are formed by transfer of electrons. Covalent bonds are formed when two atoms share two electrons.

Hydrogen Two hydrogen atoms combine by each sharing their single electron. The two shared electrons which are called an electron pair or shared-pair, move around both nuclei and hold them together. Since each nucleus has the two electrons, in effect it has the stable helium duplet. The electron pair is a covalent bond and is represented by $H\overset{\times}{\underset{\circ}{}}H$. Both electrons are identical once the bond has been formed. The $\circ$ and $\times$ merely mean that originally they belonged to different atoms. In the simple formula H–H, the line represents two electrons, one from each atom.

Halogen molecules A halogen atom has seven electrons in its outer, incomplete shell. It requires one more to attain the noble gas configuration. Two atoms combine by sharing two electrons, one from each atom, and the electron pair is a covalent bond. Fluorine, chlorine, bromine, and iodine can therefore be represented by X–X which represents:

$$\overset{\times\times}{\underset{\times\times}{{}^{\times}_{\times}X}}\,{}^{\times}_{\circ}\,\overset{\circ\circ}{\underset{\circ\circ}{X^{\circ}_{\circ}}}$$

Figure 14.7 shows the complete structures of the hydrogen molecule and the chlorine molecule.

Double and triple bonds Each oxygen atom in carbon dioxide shares four electrons with the carbon atom, two from each atom. The four electrons are two covalent bonds or a double bond between the atoms. The covalency of oxygen is 2 and the covalency of carbon is 4.

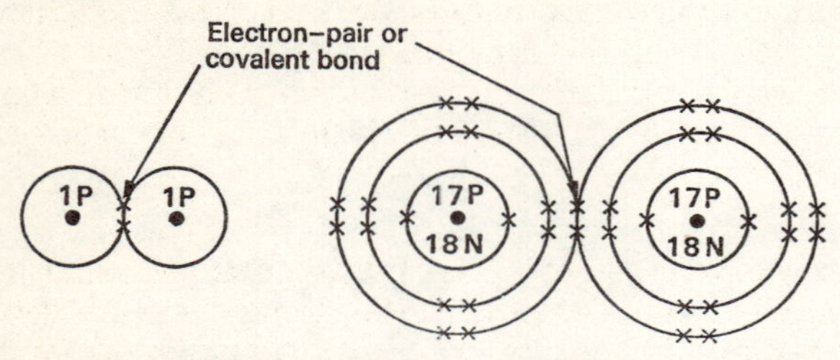

Fig. 14.7 Hydrogen and chlorine molecules with one covalent bond

A nitrogen atom has five electrons in its outer shell. It attains a stable octet by sharing three of these with another nitrogen atom. The three electron pairs of six electrons form a triple bond. Double and triple bonds are common in organic compounds.

Carbon dioxide	O=C=O	××O×o×oC×o×oO××
Nitrogen	N≡N	×N×o×o×oNo
Ethene (ethylene)	H H / C=C / H H	H H / C×o×oC / H H
Ethyne (acetylene)	H—C≡C—H	H×oC×o×o×oC×oH

Hydrogen chloride is a covalent compound when it is a gas or liquid. It is also covalent in methylbenzene but in aqueous solution it dissociates into ions:

H×oCl×× $[H]^+$ $[\times_o Cl \times\times]^-$

Hydrogen chloride HCl   Hydrochloric acid $H^+ . Cl^-$

The other halogen halides behave similarly.

SHAPES OF SOME SIMPLE MOLECULES

Unlike ionic bonds, covalent bonds are directional in character. The electron-pairs tend to be as far as possible from one another. The four bonds of carbon are arranged tetrahedrally around the carbon atom, and this arrangement is explained in chapter 25. The angle between two adjacent bonds is about 109½°.

The ammonia molecule has three covalent bonds, each of two shared electrons. The other two unshared electrons around the nitrogen atom are called a *lone pair*. The lone pair repels the shared-pairs and the angle between two adjacent bonds is about 108°, which is slightly less than the angle between tetrahedral bonds. The bond angle in water is only 104½° because the two lone pairs in a water molecule repel the shared-pairs between the oxygen atom and the hydrogen atoms.

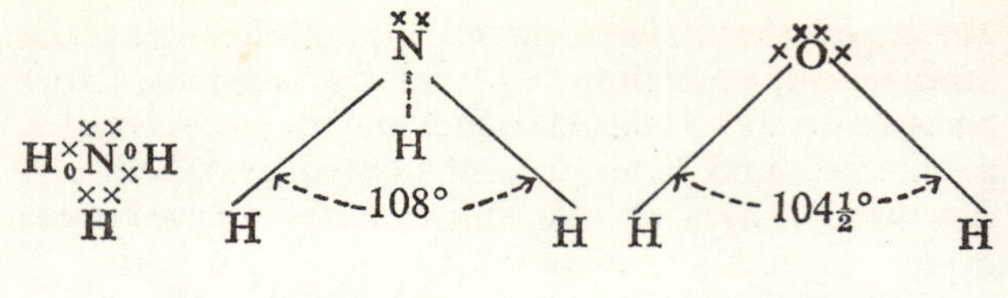

Three covalent bonds
One lone pair

Two covalent bonds
Two lone pairs

SHAPES OF GRAPHITE AND DIAMOND

Graphite crystals consist of layers of carbon atoms, and each layer is a giant structure of atoms. In effect a layer is a giant molecule in two dimensions only. Each carbon atom is at the corner of a regular hexagon and is the same distance from three similar atoms. The four atoms are linked by covalent bonds. The distance between layers is twice the distance between atoms in a layer. The layers are not joined by covalent or ionic bonds, and therefore the force of attraction between them is small. As a result, graphite is a soft substance and the layers easily slide over each other.

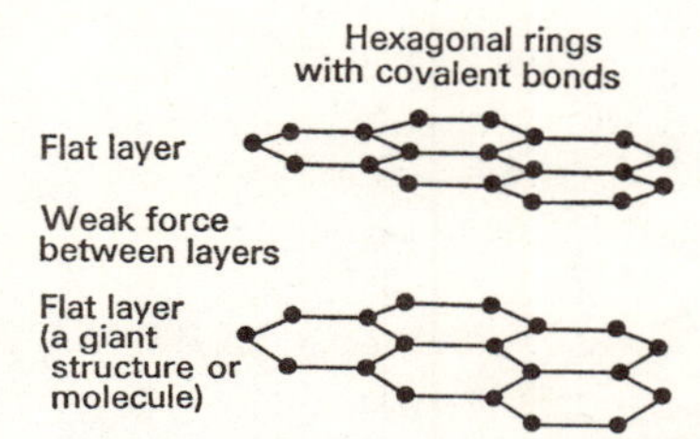

Fig. 14.8 Arrangement of atoms in graphite

Diamond consists of carbon atoms in which each atom is joined to four similar atoms by covalent bonds arranged tetrahedrally; the angle between adjacent bonds is about 109½°. Diamond is therefore a giant structure of atoms, and a crystal of diamond may be regarded as a giant molecule in three dimensions. Four atoms are arranged tetrahedrally around a central carbon atom.

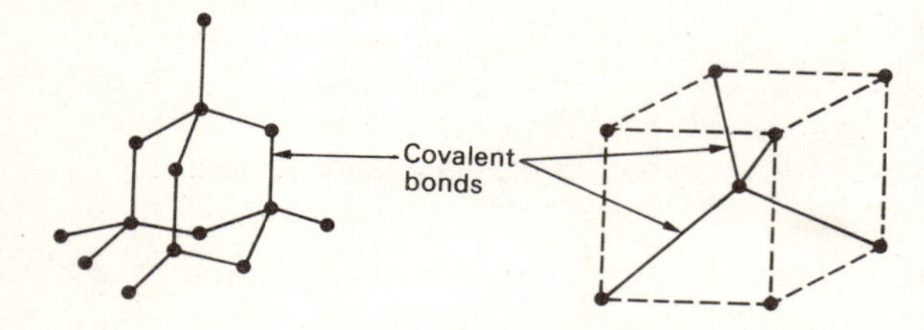

Fig. 14.9 Arrangement of atoms in diamond

IONIC COMPOUNDS AND SIMPLE COVALENT COMPOUNDS COMPARED

Ionic compounds, e.g. Na^+Cl^- and $Ca^{2+}O^{2-}$, are usually hard solids because the electrostatic forces between the ions are strong and non-directional. They contain no molecules. Molecules of covalent compounds do exist and have definite shapes because a covalent bond is directional. Most covalent compounds are gases, liquids or soft solids at ordinary temperatures; graphite and diamond are two exceptions.

Ionic compounds are electrolytes because their ions can move when the compounds are melted or dissolved in suitable solvents. Covalent compounds are non-electrolytes unless, like hydrogen chloride and ammonia, they react with water. Ionic compounds are usually, but not always, soluble in water and insoluble in organic solvents such as petrol, benzene, etc., whereas covalent compounds are usually insoluble in water and soluble in organic compounds.

The energy required to break the bonds in ionic crystals is high and therefore the melting points are high. The melting points and boiling points of most covalent compounds are low. The molar latent heat of fusion of a solid is the energy required to change 1 mole of the solid into liquid at a constant temperature. The molar latent heat of vaporization of a liquid is the energy required to change 1 mole of the liquid into gas at a constant temperature. The molar latent heats of ionic compounds are high and those of most covalent compounds are low.

EXAMPLES OF DIFFERENT TYPES OF BOND

MOLECULAR CRYSTALS

A sodium chloride crystal consists of ions in a giant structure. A graphite crystal consists of atoms in a giant structure of layers of atoms, and a diamond crystal consists of atoms in a giant structure in three dimensions. Since the forces between the ions or atoms are great, the substances are solids with high melting points and boiling points. Much energy is needed to separate the ions in sodium chloride or the atoms in graphite and diamond so that they can move freely in the liquid or gaseous states.

When iodine crystals were heated (chapter 2) the solid easily changed to a gas, and sulphur crystals also melted readily. The iodine molecule consists of two atoms, joined by a covalent bond. The forces holding the molecules together in the crystal are small and they are easily broken when heat energy is supplied.

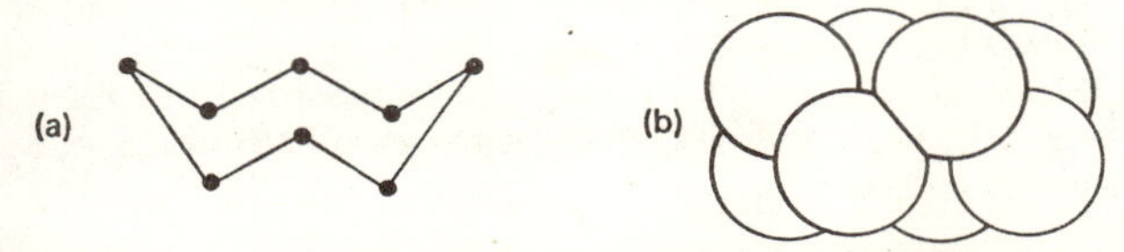

Fig. 14.10 Arrangement of atoms in a sulphur molecule, S_8

The sulphur molecule consists of eight atoms arranged in a ring. The atoms are firmly held together by covalent bonds, but the molecules in a crystal are not firmly held. When the sulphur melts to an amber liquid, the rings of eight atoms are still present. At higher temperatures (about 160 °C) they break up and form long chains of atoms.

The crystals of iodine and sulphur are examples of molecular crystals. Ice, solid carbon dioxide, sugar, and many other compounds also form molecular crystals. Their melting points are relatively low. Since molecular crystals and the liquids they form when melted consist of molecules they do not conduct electricity.

METALLIC CRYSTALS

An atom of magnesium or calcium has two electrons in its outer incomplete shell. It must transfer these electrons to another atom in order to attain the stable noble gas structure. Clearly one magnesium atom cannot transfer two electrons to another magnesium atom. Metals usually have high melting points and high boiling points and these facts indicate that they have a giant structure. The solids are good conductors and therefore contain mobile electrons. Magnesium is a giant structure of magnesium ions, held together by electrostatic forces, and the outer electrons of the atoms are free to move more or less anywhere. Usually the metallic ions are packed together very closely and this fact accounts for the high densities of metals. Sodium and potassium are examples of metals with exceptional properties. They are light elements with densities less than that of water and their melting points are below 100 °C. Mercury is the only metal which is liquid at ordinary temperatures.

CRYSTALS OF COVALENT COMPOUNDS

There are three types of crystals in which covalent bonds are present. They are:

1. *Molecular crystals* such as ice, solid carbon dioxide, sugar, and triiodomethane, CHI_3. The forces between the molecules are weak and they are similar to the weak forces between the molecules of gases. Molecular crystals are easily melted and their molar latent heats of fusion are low.
2. *Giant layer structures* such as graphite. Each layer is a two-dimensional large molecule (macromolecule). The melting points and molar latent heats of fusion of substances with these structures are very high because the covalent bonds which hold the atoms together are powerful.
3. *Giant three-dimensional structures* such as diamond. Strong covalent bonds hold all the atoms together, and each crystal is one giant structure. The melting points and molar latent heats of fusion are very high because every atom is joined to one or more others by covalent bonds.

THE STRUCTURE OF MATTER

The noble gases and mercury vapour consist of single atoms which are stable. All other substances we study consist of either atoms or ions. The molecules of sulphur S_8, phosphorus P_4, water, carbon dioxide and so on contain several atoms. Giant covalent structures consist of atoms but giant ionic structures consist of ions. Free ions exist when ionic compounds are in the liquid state or in solution.

NUCLEAR CHANGES LIBERATE GREAT ENERGY

Tremendous energy is liberated when nuclei break up in radioactive elements and nuclear fission and when nuclei combine in nuclear fusion. Matter is destroyed in these processes, and it is now known that the destruction of one kilogram of matter yields 9×10^{16} joule or 2×10^{13} kilocalorie of energy.

Bombardment of uranium-235 with neutrons causes nuclear fission; one nucleus splits up into two and releases two to three free neutrons which in turn may split up more uranium nuclei:

Uranium-235 + neutron = two nuclei + 2 or 3 neutrons. About 0·1 per cent of mass is destroyed in this fission.

In nuclear fusion, two or more atoms or nuclei combine to form one particle. The sun and stars produce their energy by fusion reactions; for example,

four hydrogen atoms = one helium atom

Almost 1 per cent of the mass is destroyed in this fusion and it is transformed into energy. The energy of hydrogen bombs depends on fusion reactions.

MODERN FORM OF DALTON'S ATOMIC THEORY

Every statement made by Dalton has been shown to be inexact. The statements, however, apply to ordinary chemical reactions because the nuclei of the atoms do not change and the energy changes are relatively small; the changes in mass corresponding to these energy changes are so small that they cannot be detected by ordinary chemical balances.

A modern statement of the theory is:

1. Matter consists of atoms; atoms consist of electrons, protons, and neutrons.
2. Atoms can be created and destroyed in radioactive changes and in nuclear reactions, but not in chemical reactions.
3. All atoms of an element contain the same number of protons. They may differ in mass because they may contain different numbers of neutrons which do not affect their chemical properties.
4. Chemical combination usually occurs between small, whole numbers of atoms; it can occur between very large numbers of atoms, especially with carbon compounds called organic compounds; see chapters 25 and 26.

SUMMARY

Fundamental particles of matter. Electron (negative charge), proton (positive charge), neutron (no charge). Protons and neutrons are called nucleons. An atom consists of a nucleus surrounded by one or more electrons.

Proton number: the number of protons in one atom of an element. Its value is the same as the atomic number.
Nucleon number: the number of protons and neutrons in one atom of an element.

Isotopes or isotopic nuclides

Two or more species of atoms with the same proton number but different nucleon numbers. Examples: protium, deuterium, and tritium; chlorine isotopes $^{35}_{17}Cl$ and $^{37}_{17}Cl$.
Nuclide: species of atoms with the same proton number and the same nucleon number.

Electron configurations

Naturally-occurring atoms have between one and ninety-two electrons around the nucleus. The electrons are in seven shells *K* to *Q*.
Duplet of electrons: two electrons in the *K* shell, as in helium and all other atoms except hydrogen.
Octet of electrons: eight electrons in the *L* shell or in the *M* shell. Examples: neon 2,8; argon, 2,8,8; sodium 2,8,1.

Chemical bonds

Electrovalent or ionic bond: formed by the transfer of one or more electrons from one atom or radical to another, usually from metallic to non-metallic elements. Examples: Na^+Cl^-, $Mg^{2+}O^{2-}$.
Sodium chloride crystal: a giant structure which consists of the ions Na^+ and Cl^-; there are no molecules.
Covalent bond: formed by the sharing of two electrons, one from each atom. Two covalent bonds, of four shared electrons, is a double bond, and three covalent bonds is a triple bond.
Shapes of molecules: the four covalent bonds in methane are arranged tetrahedrally, at an angle of $109\frac{1}{2}°$. The angles between bonds in ammonia and water are less owing to repulsion of the bonds by the lone pairs of electrons. Graphite has a giant layer structure, and diamond has a giant three-dimensional structure; both structures consist of atoms.
Ionic compounds: usually they are hard solids, with high melting and boiling points and high molar latent heats.
Simple covalent compounds: usually are gases or liquids, with low melting and boiling points and low molar latent heats. Examples: iodine, sulphur, sugar, water.
Metallic crystals: giant structures of ions, and some electrons are free to move. Metals are good conductors and have high melting and boiling points and high molar latent heats. Sodium, potassium and mercury are exceptions.

Nuclear reactions

Fission: the decomposition of nuclei into two or more parts. Great energy is released by fission of U-235.
Fusion: the combination of two or more nuclei to form one nucleus. Example: four hydrogen nuclei form one helium nucleus.

QUESTIONS

1. Mention briefly the evidence which indicated the existence of particles smaller than atoms.
2. Explain briefly the arrangements of electrons in orbitals for elements of atomic number between 1 and 20 (i.e. up to calcium).

3. Chlorine (atomic number 17) has two naturally-occurring isotopes, of mass numbers 35 and 37. If the relative atomic mass of chlorine is 35·5, what is the ratio by atoms of the isotopes in chlorine gas? Explain your answer.
4. By reference to the electronic structures of sodium and calcium, show what the formation of ions by metals involves. Mention two reactions of each of these metals which support your statements, and write suitable equations for these reactions.
5. A molecule of sodium chloride does not exist. Explain this statement by describing the structure of a sodium chloride crystal.
6. Describe briefly the crystal structures of graphite and diamond. How do they account for a difference in properties of the two substances?
7. How do the differences in the electronic structures affect the physical properties of the following pairs of substances: (i) calcium chloride and carbon tetrachloride; (ii) copper and ice? (L.)
8. 'The position of an element in a Group in the Periodic Table and its chemical properties are related to the number of outer electrons which the atom of the element contains.' Discuss, with illustrative examples, this statement with respect to the following elements: hydrogen, oxygen, sodium, neon, chlorine. (Atomic numbers: H = 1, O = 8, Na = 11, Ne = 10, Cl = 17.) (L.)
9. The table below shows some properties of some elements and compounds.

   From this list select: (*a*) a substance that could be a salt; (*b*) a substance that has a 'giant molecular' structure; (*c*) a substance which is liquid at room temperature; (*d*) a substance which has an ionic lattice structure.

   What precautions would you take in storing the following? Give your reasons. (*e*) Substance *U*. (*f*) Substance *S*.

   (*g*) The products of burning two of the substances in air would combine to form a salt. Which two? (L.)

Substance	*Melting point* °C	*Boiling point* °C	*Electrical conductivity*		*Effect of air*
			Solid	*Liquid*	
P	−112	−107	Poor	Poor	No reaction
Q	801	1413	Poor	Good	No reaction
R	97·5	880	Good	Good	Burns when heated
S	44	280	Poor	Poor	Ignites, even at room temperature
T	Not known	4200	Poor	—	Burns when strongly heated
U	−110	46·3	Poor	Poor	Burns readily on slight warming

15 Kinetic Theory. The Gaseous State

KINETIC THEORY OF GASES

The experiments on the diffusion of solutes, liquids, and gases and on the Brownian movement described in chapter 8 show that the particles of these substances are in constant motion. The kinetic theory of gases, developed in the mid-nineteenth century, states that gases consist of molecules which are moving rapidly in every direction. The molecules change direction when they collide with each other and, since they are perfectly elastic, no energy is lost on collision. The molecules also collide with the walls of any containing vessel and this bombardment is the cause of the pressure of the gas. However, the molecules are sufficiently small and far apart not to interfere with each other's movements.

The speed and therefore the energy of the molecules increases as the temperature rises. Hydrogen molecules at 0 °C move at about 1838 metres per second. Measurements taken on oxygen gas give the following values at 0 °C and atmospheric pressure:

Diameter of a molecule of oxygen
$= 3 \times 10^{-10}$ m

Average distance between centres of molecules
$= 3 \times 10^{-9}$ m

Average speed $= 460$ m/s

Distance travelled between collisions
$= 10^{-7}$ m

Number of collisions in one second
$= 4 \times 10^{9}$

Liquids When a gas is cooled sufficiently, and sometimes when great pressure is exerted on a gas, it changes to a liquid. The molecules still move at random but they are much closer together. Most molecules remain within the liquid, but some at the surface can escape. Molecules that escape from the liquid form a vapour, but molecules also pass back from the vapour into the liquid.

In a closed vessel, the numbers of molecules moving out of and into the liquid finally become equal and a dynamic or kinetic equilibrium is established. The vapour exerts a certain definite pressure, known as the *saturated vapour pressure.* If the temperature is raised, more molecules of liquid have sufficient energy to escape and the saturated vapour pressure therefore rises. When it equals the external pressure, which is about 760 mmHg if it is atmospheric pressure, the liquid boils. A large number of the particles in ionic liquids are ions and not molecules.

Solids When a liquid is cooled sufficiently, it changes to a solid. The speed of its molecules or ions become so slow that the forces of cohesion can hold the particles in fixed positions, although they may continue to vibrate, oscillate, and rotate.

Particles escape from the surfaces of solids; those escaping from solids such as iodine, naphthalene, ice, and solid carbon dioxide are so numerous that they exert an appreciable vapour pressure. In fact, the vapour pressure of solid iodine reaches 760 mmHg before the solid melts, and therefore iodine sublimes on heating.

THE GAS LAWS

The volume of a gas depends upon its pressure and its temperature. In 1662, Boyle stated that the volume V of a given mass of gas at constant temperature is inversely proportional to the pressure p. Expressed mathematically:

$$V \propto 1/p \qquad \text{—equation 1}$$

and $pV =$ a constant at constant temperature

The volume of a gas also depends on its temperature. The absolute unit of temperature, the *kelvin,* is the fraction 1/273 of the temperature of the melting point of pure ice. This statement is not exact. More accurately, the kelvin is the fraction 1/273·16 of the temperature of the *triple point* of water. The triple point of a substance defines the temperature and pressure at which solid, liquid, and vapour can all exist together. The temperature of the triple point of water is 0·01 °C.

The melting point of ice is 0 °C or 273·15 K, and the boiling point of water at atmospheric pressure is 100 °C or 373·15 K. Ignoring the numbers after the decimal point:

Temperature in kelvin = 273 + temperature in °C
∴ 10 °C = 283 K; −183 °C = 90 K; 298 K = 25 °C

In 1787 Charles stated that the volume of a given mass of gas at constant pressure is directly proportional to its absolute temperature T.

$$V \propto T \qquad \text{—equation 2}$$

and $V/T =$ a constant at constant pressure

Combining equations 1 and 2 gives the *gas equation*:

$$pV/T = \text{a constant}$$

For a given mass of gas, if a volume V_1 at pressure p_1 and absolute temperature T_1 changes to a volume V_2 at pressure p_2 and absolute temperature T_2, then $(p_1V_1)/T_1 = (p_2V_2)/T_2$. If the temperatures are t_1 °C and t_2 °C, the gas equation is

$$(p_1V_1)/(273 + t_1) = (p_2V_2)/(273 + t_2).$$

The volume and density of a given mass of gas can have a wide range of values, depending on its temperature and pressure. Therefore, they are usually stated at *standard temperature and pressure* (s.t.p.) which is a temperature of 0 °C (273 K) and a pressure of 760 mmHg.

MOLAR VOLUME OF A GAS AT S.T.P.

By experiment it is known that

0·089 g of hydrogen at s.t.p. occupy 1000 cm^3

∴ 2·016 g of hydrogen at s.t.p. occupy

$$\frac{2{\cdot}016}{0{\cdot}089} \times 1000\ cm^3 = 22\,400\ cm^3 \text{ (approx.)}$$

Thus, 1 mole of hydrogen at s.t.p. occupies 22 400 cm^3.

This is the molar volume of hydrogen at s.t.p. The molar volumes of other gases at s.t.p. can be calculated similarly; it is found that the molar volume of all gases is approximately 22 400 $cm^3\ mol^{-1}$. The exact value of the molar volume of a gas at s.t.p. is 22 414 $cm^3\ mol^{-1}$ or 0·022 414 $m^3\ mol^{-1}$.

AVOGADRO'S PRINCIPLE

In 1811 Avogadro suggested that there are equal numbers of molecules in the same volumes of different gases at the same temperature and pressure. Modern reasoning supports this theory.

1 mole of a gas at s.t.p. occupies 22 400 cm^3

∴ 6×10^{23} molecules of gas at s.t.p. occupy 22 400 cm^3.

Changes of temperature and pressure from s.t.p. change the volume of all gases equally; i.e. 22 400 cm^3 of any gas becomes some other constant value if the gas laws are obeyed. Therefore, one mole of any gas occupies the same volume at the same temperature and pressure; this is an alternative way of stating the Avogadro principle. The importance of the principle is that we can write 'molecules' instead of 'volumes' in reactions involving gases; for example,

1 volume of hydrogen + 1 volume of chlorine = 2 volumes of hydrogen chloride

∴ 1 molecule of hydrogen + 1 molecule of chlorine = 2 molecules of hydrogen chloride

GAY-LUSSAC'S LAW

In the early nineteenth century Gay-Lussac observed that gases react in volumes which are in simple ratio to each other. His law states that the volumes of gases which react are in simple whole-number ratio to each other and to the volumes of any gaseous products, provided the volumes are measured at constant temperature and pressure. Some examples of his law are:

2 volumes of hydrogen + 1 volume of oxygen = 2 volumes of steam

1 volume of nitrogen + 3 volumes of hydrogen = 2 volumes of ammonia

Sulphur (solid) + 1 volume of oxygen = 1 volume of sulphur dioxide

1 volume of methane + 2 volumes of oxygen = 1 volume of carbon dioxide + 2 volumes of steam

By applying the Avogadro principle, the word molecule can be substituted for volume in each of the above statements.

ATOMICITY OF THE HYDROGEN MOLECULE

We have seen from the experiment on p. 58 in chapter 11 that

1 volume of hydrogen + 1 volume of chlorine = 2 volumes of hydrogen chloride

∴ 1 molecule of hydrogen + 1 molecule of chlorine = 2 molecules of hydrogen chloride

by applying the Avogadro principle.

One molecule of hydrogen chloride contains half a molecule of hydrogen, but it must contain a whole number of hydrogen atoms.

½ molecule of hydrogen = 1, 2, or 3, etc. atoms of hydrogen.

∴ the formula for hydrogen is H_2 or H_4 or H_6, etc.

The formula for hydrogen chloride is HCl or H_2Cl or H_3Cl, etc. assuming that the chlorine formula is Cl_2.

Since hydrogen chloride forms only one sodium salt, its formula is probably HCl because H_2Cl should form two salts, NaHCl and Na_2Cl, and H_3Cl should form three salts. This means that the formula for hydrogen is H_2 and the hydrogen molecule is diatomic.

$$\underset{1\text{ volume}}{H{-}H} + \underset{1\text{ volume}}{Cl{-}Cl} = \underset{2\text{ volumes}}{H{-}Cl + H{-}Cl}$$

Note In chapter 11 the formula of hydrogen chloride was deduced by using the densities and atomic masses of hydrogen and chlorine instead of using the Avogadro principle.

To find the formula of water (steam)

This was found on p. 48 in chapter 9 by passing hydrogen over copper(II) oxide and weighing the water formed. The atomic masses of hydrogen and oxygen were needed. The following method uses the volume composition of steam and the Avogadro principle.

The explosion tube is kept at a constant temperature of 138 °C by vapour from boiling pentanol. Add 20 cm³ of hydrogen and 10 cm³ of oxygen to the explosion tube, measuring each gas at atmospheric pressure. The mercury levels in the explosion tube and in the levelling tube should be equal. Reduce the pressure by lowering the levelling tube in order to reduce the violence of the explosion. Pass a spark through the gases. Allow the temperature to become steady and observe the volume of gas left. Now allow to cool to room temperature and note the volume of any gas left.

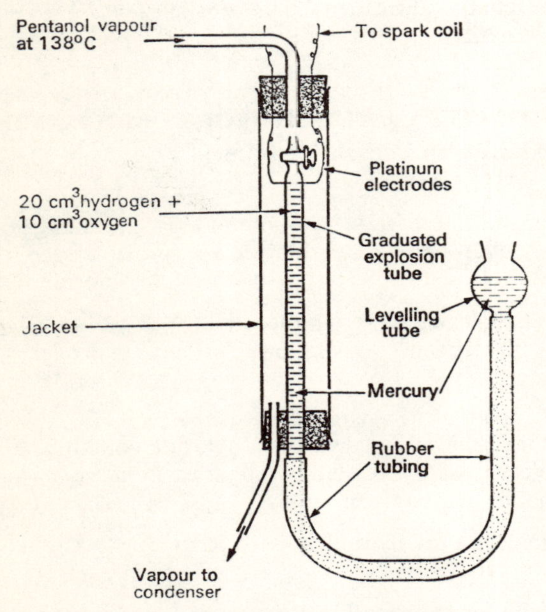

Fig. 15.1 To determine the formula of steam

No gas is left at the end, which shows that all the hydrogen and oxygen reacted. The volume of the steam is 20 cm³.

	20 cm³ of hydrogen	+ 10 cm³ of oxygen	= 20 cm³ of steam
∴	2 volumes of hydrogen	+ 1 volume of oxygen	= 2 volumes of steam
∴	2 molecules of hydrogen	+ 1 molecule of oxygen	= 2 molecules of steam
i.e.	4 atoms of hydrogen	+ 2 atoms of oxygen	= 2 molecules of steam

assuming that hydrogen and oxygen are both diatomic. The formula of steam is therefore H_2O.

ATOMICITY OF OXYGEN

One molecule of steam contains half a molecule of oxygen, but it must contain a whole number of oxygen atoms.

∴ $\frac{1}{2}$ molecule of oxygen contains 1, 2, or 3, etc. atoms of oxygen, and the formula for oxygen is O_2 or O_4 or O_6, etc.

If the formula is O_4 or O_6, one volume of oxygen should sometimes form four or six volumes of gaseous product in a reaction; for example,

$$O_4 + X_4 \text{ (or } 2X_2 \text{ or } 4X) = 4XO \text{ (4 volumes)}$$

$$O_6 + 2X_3 \text{ (or } 3X_2 \text{ or } 6X) = 6XO \text{ (6 volumes)}$$

In practice one volume of oxygen never forms more than two volumes of gaseous product, and the formula is therefore probably O_2.

$$O_2 + X_2 \text{ (or } 2X) = 2XO \text{ (2 volumes)}$$

or

$$O_2 + Y_4 \text{ (or } 2Y_2) = 2Y_2O \text{ (2 volumes)}$$

A similar argument also applies to the atomicity of hydrogen. One volume of hydrogen never forms more than two volumes of gaseous product, and therefore H_2 is a better formula than H_4 or H_6, etc.

SUMMARY

Kinetic theory: gases consist of molecules moving rapidly in all directions. The pressure of a gas in a vessel is a measure of the number and magnitude of molecular collisions with the walls of the vessel. The temperature is a measure of the speed of the molecules.
Liquids: the gas molecules are slowed down sufficiently to be close enough to influence each other's movement. Some molecules escape from the surface of the liquid—evaporation. If a sufficient number of molecules escape to give a vapour pressure equal to that of atmosphere pressure, the liquid boils.
Solids: the molecules can only vibrate, oscillate or rotate.

Gas laws
Boyle's law: at constant temperature $p \propto 1/V$.
Charles' law: at constant pressure, $V \propto T$, the absolute temperature.
Triple point: the temperature and pressure at which the solid, liquid, and gas of a substance can all exist together.
Standard temperature and pressure: s.t.p. 0 °C (273 K) and 760 mmHg—a standard condition for measuring the volume of a gas.

Avogadro's principle: equal volumes of all gases at the same temperature and pressure contain the same number of molecules. Therefore substitute moles for volumes in gaseous reactions, provided the volumes are at the same temperature and pressure.
Molar volume: 22 400 cm³ mol⁻¹ at s.t.p.—the volume at s.t.p. occupied by 1 mole of any gas.
Gay-Lussac's law: gases react in volumes which are in simple ratio to each other.
Atomicity: the number of atoms in one molecule of a gas. Find by experiment and Avogadro's principle; for example, H_2, O_2, Cl_2.

QUESTIONS

1. If 50 cm^3 of oxygen contain x molecules, what number of molecules is contained in (*a*) 100 cm^3 of hydrogen, (*b*) 25 cm^3 of air, if the temperature and pressure of the three gases are the same? What assumptions did you make in order to give your answers?
2. By what factor must you multiply the volume of 1mole of oxygen at 20 °C in order to determine what its volume would be at 200 °C, if the pressure remains the same? If 1 mole of gaseous nitrogen dioxide, N_2O_4, undergoes the same temperature change, the observed expansion is much greater. Account for this fact.
3. The densities of hydrogen and of oxygen at s.t.p. are 0·09 and 1·43 g per 1000 cm^3. Calculate from these figures their relative molecular masses, and explain your reasoning.
4. By how many grams is 5600 cm^3 of sulphur dioxide heavier than the same volume of carbon dioxide (both gases at s.t.p.)? 10 000 cm^3 of an oxide of nitrogen, N_2O, are passed over heated copper and the gas formed is collected. If all volumes are measured at s.t.p., what is the volume of the gas collected and the mass of copper oxide formed? ($S = 32$, $C = 12$, $O = 16$, $N = 14$, $Cu = 64$.) (C.)
5. Describe an experiment to show the volume composition of either hydrogen chloride or steam. Show what deductions may be made from the experiment you have chosen concerning the atomicity of either hydrogen or oxygen. State the law on which you base your deduction. (C.)
6. When phosphine gas is decomposed at a high temperature, four volumes of phosphine form one volume of phosphorus vapour, P_4, and six volumes of hydrogen. Deduce the formula of phosphine and mention what assumptions you make.
7. In 1807 Dalton wrote: 'Are there the same number of particles of any gas in a given volume and under a given pressure? No: nitrogen and oxygen mixed in equal volumes give half the number of particles of nitrogen monoxide in the same volume.' (Modern names are given in the quotation.) Explain the error in Dalton's reasoning and how Avogadro accounted for the experimental fact.
8. Explain how the following observations are accounted for by the kinetic theory: (*a*) gases are readily compressed; (*b*) gases are perfectly miscible with one another; (*c*) gases exert pressure; (*d*) smoke particles exhibit Brownian motion.
9. The mass of 1000 cm^3 at s.t.p. of oxygen, ammonia, and carbon dioxide are respectively 1·43, 0·765, and 1·97 g. Show that there is a simple relationship between these values and the relative molecular masses of the gases. What is the probable molecular mass of a gas whose density at s.t.p. is 0·718 g dm^{-3}?

16 Chemical Equations

Aqueous lead nitrate and aqueous potassium iodide react together to form a yellow precipitate. If we assume the formulae of the two compounds, the left-hand side of the chemical equation for the reaction can be written in several possible ways:

$$Pb(NO_3)_2 + KI =$$
$$Pb(NO_3)_2 + 2KI =$$
$$2Pb(NO_3)_2 + KI = \quad \text{and so on.}$$

Most nitrates are soluble and therefore the precipitate cannot be potassium nitrate or a similar compound. It is likely to be lead iodide or a similar compound. If we assume the charges on the ions of lead and iodine, the left-hand side of the equation can be written ionically as follows:

$$Pb^{2+}(aq) + 2I^-(aq) =$$
$$Pb^{2+}(aq) + I^-(aq) + OH^-(aq) =$$

The first of these two ionic equations assumes that the precipitate is lead iodide, PbI_2, and the second assumes that it is a basic compound such as Pb(OH)I. These examples make clear that we must know the number of molecules or ions which react with each other before we can write a balanced equation.

The following experiments illustrate ways of finding this information.

To determine the equations for reactions in solution

LEAD NITRATE AND POTASSIUM IODIDE

Principle Known volumes of 1·00 M solutions of the reactants are mixed and the heights of the precipitates are observed. M lead nitrate solution contains 331 g of

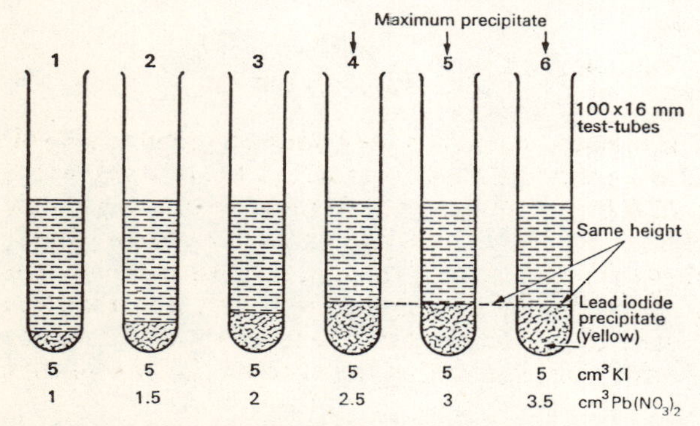

Fig. 16.1 Finding the lead nitrate—potassium iodide equation

nitrate per dm^3 and M potassium iodide solution contains 166 g of iodide per dm^3. Equal volumes of the solutions contain equal numbers of solute molecules. (Actually the solutes are ionized but we can ignore that fact for the purpose of this experiment.)

1. Add 5 cm^3 of M potassium iodide solution to each of either six or nine test-tubes. The 100 × 16 mm size of tube is suitable.
2. Use a 5 cm^3 graduated pipette to add different volumes of M lead nitrate to the tubes: 1 cm^3 to tube 1, 1·5 cm^3 to tube 2, and so on up to either 3·5 cm^3 to tube 6 or 5 cm^3 to tube 9.
3. Stir each mixture with a glass rod for ten seconds. Add a few drops of ethanol which makes the precipitate settle more readily, and then centrifuge the tubes.
4. Place the tubes in a rack and measure the heights of the six or nine precipitates. The height is a maximum in several tubes. Note the minimum volume of lead nitrate solution which produced this maximum height of precipitate. Let this minimum volume be x cm^3.

$$\frac{\text{Volume of potassium iodide solution}}{\text{Volume of lead nitrate solution}} = \frac{5}{x}$$

$$\therefore \quad \frac{\text{Molecules of potassium iodide}}{\text{Molecules of lead nitrate}} = \frac{5}{x}$$

The left-hand side of the equation is:

$5KI + Pb(NO_3)_2 =$ if $x = 1\ cm^3$,

$2KI + Pb(NO_3)_2 =$ if $x = 2{\cdot}5\ cm^3$,

$KI + Pb(NO_3)_2 =$ if $x = 5\ cm^3$, and so on.

Any two solutions that react to form a dense precipitate can be used in the above experiment. For example, barium chloride or nitrate and a soluble sulphate, which form white barium sulphate.

COPPER(II) SULPHATE AND IRON FILINGS

1. Weigh a test-tube.
2. Add about 0·5 g of non-greasy, dry iron filings. Weigh again.
3. Dissolve about 3 g of copper sulphate crystals by warming them in a test-tube one-third full of distilled water. This amount of copper sulphate is much more than is needed to react with the iron.
4. Add the hot copper sulphate solution to the test-tube containing the iron. Stir gently with a glass rod. Copper is precipitated and heat is evolved.
5. Centrifuge the mixture or let the copper powder settle. Remove the solution of excess copper sulphate and iron(II) sulphate with a teat pipette.
6. Wash the copper with distilled water and then with ethanol or propanone, removing each liquid with a teat pipette. Warm the test-tube and copper in a beaker of boiling water to remove the last traces of liquid and dry the copper.
7. Dry the test-tube, let it cool, and weigh.

1.	Mass of test tube	= ?
2.	Mass of tube + iron filings	= ?
3.	Mass of tube + copper	= ?
2 − 1.	Mass of iron	= x gram
3 − 1.	Mass of copper displaced	= y gram

x g of iron displace y g of copper,

$$\therefore \text{ 56 g of iron displace } \frac{56y}{x} \text{ g of copper,}$$

$$\text{i.e. 1 mole of iron displaces } \frac{56y}{63{\cdot}5x} \text{ moles of copper.}$$

If this number is 1, the equation is

$$Fe(s) + CuSO_4(aq) = Cu(s) + FeSO_4(aq)$$

or, ionically,

$$Fe(s) + Cu^{2+}(aq) = Cu(s) + Fe^{2+}(aq)$$

In the above experiment, zinc can be used instead of iron and lead nitrate solution can be used instead of copper sulphate.

SODIUM CARBONATE AND HYDROCHLORIC ACID

M hydrochloric acid, which contains 36·5 g of acid per dm³, and M sodium carbonate, which contains 106 g of the anhydrous carbonate per dm³, are required. Equal volumes of the two solutions contain equal numbers of molecules. The acid is added to 5 cm³ of the carbonate until no more effervescence occurs.

1. Add 5 cm³ of M sodium carbonate to a small flask.
2. By using a graduated pipette or a burette, add the acid 1 cm³ at a time. Shake well after each addition and note if bubbles of carbon dioxide are formed.
3. Continue to add the acid until it causes no further effervescence. Record the maximum volume of acid which causes effervescence. Let this volume be x cm³.

$$\frac{\text{Volume of M carbonate}}{\text{Volume of M acid}} = \frac{5}{x}$$

$$\therefore \quad \frac{\text{Molecules of carbonate}}{\text{Molecules of acid}} = \frac{5}{x}$$

The left-hand side of the equation is,

$HCl + Na_2CO_3 =$ if $x = 5$,

$2HCl + Na_2CO_3 =$ if $x = 10$, and so on.

To measure the volume of gas evolved when calcium carbonate and hydrochloric acid react

1. One mole of calcium carbonate is 40 + 12 + 48 = 100 g. Weigh 0·20 g of the carbonate (0·002 mole) and add it to a small test-tube which can drop easily into a larger one.
2. Add hydrochloric acid to the larger tube. Saturate the acid solution with carbon dioxide by adding to it a little calcium carbonate, but not the carbonate already weighed. Arrange the apparatus as shown in Fig. 16.2. The weighed carbonate and acid must not mix yet. Note the reading of the gas syringe.
3. Turn the large test-tube until the carbonate and acid react completely. Twist the piston gently to ensure that it is not sticking and therefore the carbon dioxide is at atmospheric pressure. Note the reading of the syringe again; the difference between the reading is the volume, v cm³, of carbon dioxide evolved. Therefore:

$$\text{0·2 g of carbonate form } v \text{ cm}^3 \text{ of carbon dioxide,}$$

$$\therefore \text{ 100 g of carbonate form } 500v \text{ cm}^3 \text{ of carbon dioxide,}$$

i.e. 1 mole of carbonate forms $500v$ cm³ of carbon dioxide.

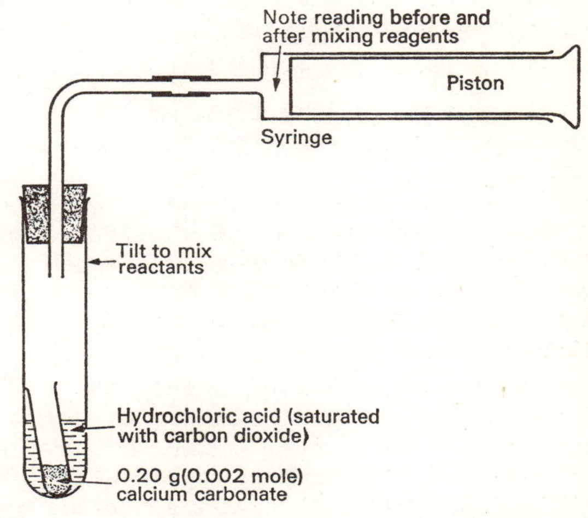

Fig. 16.2 Measuring the volume of gas evolved from a carbonate-acid mixture

If room temperature, t °C, and atmospheric pressure, p mmHg, are measured, this volume of $500v$ cm³ can be converted to s.t.p. (as described on p. 82). Substituting in the formula:

$$\frac{p_1V_1}{T_1} = \frac{p_2V_2}{T_2}$$

$$\text{we have } \frac{p \times 500v}{(273 + t)} = \frac{760 \times \text{volume at s.t.p.}}{273}$$

$$\text{Volume at s.t.p.} = \frac{p \times 500v \times 273}{(273 + t) \times 760}$$

If the experiment has been done accurately, this volume is either the molar volume at s.t.p. of a gas, which is 22 400 cm³, or some simple multiple or sub-multiple of it such as 2 × 22 400 cm³ or ½ × 22 400 cm³. In the above experiment with calcium carbonate, the volume at s.t.p. should be about 22 400 cm³, and therefore 1 mole of carbonate reacts to form 1 mole of carbon dioxide. This means that the equation for the reaction is:

$$CaCO_3 + \text{acid} = CO_2 + \text{other products}$$

Other experiments must be done before the complete equation can be written.

The above experiment can be done with a metal instead of a carbonate. Hydrogen is evolved. It is found that 1 mole of zinc, magnesium or iron forms hydrogen whose volume, calculated at s.t.p., is 22 400 cm^3 approximately. The equation for the reaction is therefore:

$$M + acid = H_2 + other\ product$$

(M = Zn or Mg or Fe)

BALANCING EQUATIONS

The following equations are *not* correct:

$$H_2 + O_2 = H_2O; \quad Cu + H_2SO_4 = CuSO_4 + H_2$$

The first equation is incorrect because one atom of oxygen appears to have been destroyed. The second equation is incorrect because copper does not react with sulphuric acid to form hydrogen.

Balancing an equation means making the number of each kind of atom or ion on the left side equal to the number on the right side.

Write balanced equations by:

(*a*) writing correct formulae for the reactants on the left and for the products on the right, and

(*b*) using whole numbers in front of the formulae where necessary. Sometimes it is easier to balance by using fractions and then multiplying the whole equation to remove the fractions. Two examples are given below. In each case, (*b*) is the balanced equation.

(*a*) $H_2O_2 = O_2 + H_2O$ (*a*) $P + O_2 = P_4O_{10}$
(*b*) $H_2O_2 = \frac{1}{2}O_2 + H_2O$ (*b*) $4P + 5O_2 = P_4O_{10}$
i.e. $2H_2O_2 = O_2 + 2H_2O$

MEANING OF EQUATIONS

$$Mg(s) + H_2SO_4(aq) = H_2(g) + MgSO_4(aq)$$

Mg(s)	H_2SO_4(aq)	H_2(g)	$MgSO_4$(aq)
24 g	98 g or 1000 cm^3 M	2 g or 22 400 cm^3 at s.t.p.	120 g

This equation means that under suitable conditions, solid magnesium reacts with aqueous sulphuric acid to form hydrogen gas and aqueous magnesium sulphate. Several facts about each substance and about the whole reaction are known from the equation.

1. Magnesium is a solid, and is an element.
2. One molecule of sulphuric acid contains 2 atoms of hydrogen, 1 atom of sulphur and 4 atoms of oxygen.
3. One mole (98 g) of acid contain 2 g of hydrogen, 32 g of sulphur and 64 g of oxygen.
4. Similarly the composition of a molecule of magnesium sulphate is known and its composition by mass is known.
5. Hydrogen is diatomic and is a gas.
6. One atom of magnesium reacts with 1 molecule of acid to form 1 molecule of hydrogen and 1 molecule of magnesium sulphate.
7. The masses of reactants and products, including the volumes of gases at s.t.p. and the volumes of M aqueous solutions, can be deduced. They are given under the above equation.
8. The heat change can be given as explained in chapter 17.

The equation does not give the rate of reaction. It does not state if heat or a catalyst is required but sometimes these two facts are written over the = sign or the ⇌ sign in a reversible reaction.

SUMMARY

Chemical equations show the formulae of reactants and of products, and the number of particles (molecules, atoms, or ions) that react and that are formed.

Relative numbers of particles involved in reactions are found by:

1. Measuring the heights of precipitates formed when various volumes of M solutions react. Examples: $Pb(NO_3)_2$ and KI, $BaCl_2$ or $Ba(NO_3)_2$ and a soluble sulphate.
2. Measuring the mass of product displaced by a known mass of reactant. Examples: Fe or Zn and solutions of $CuSO_4$ or $Pb(NO_3)_2$.
3. Measuring the volumes of solutions of known molarity that react. Example: Na_2CO_3 and HCl.
4. Measuring the volume, converted to s.t.p., of gas evolved from a known mass of reactants. Examples: CO_2 from Na_2CO_3 or $CaCO_3$ and HCl, H_2 from metal and acid.

Balancing chemical equations: the same number of every kind of atom or ion must appear on each side of a chemical equation.

Information from a complete chemical equation: it gives the state of the reactants and products, atomicities of elements, formulae of compounds, relative numbers and masses of reacting particles, volumes at s.t.p. of any gases, volumes of M solutions that react, and the heat change. It may state if heat or a catalyst is necessary. It states nothing about the rate of reaction.

QUESTIONS

1. Describe how you would use molar barium chloride and molar sodium carbonate to find the relative number of reacting ions in the reaction between the two substances.
2. What is the volume at s.t.p. of gas liberated by the action of heat on 5·05 g of potassium nitrate? The equation is: $2KNO_3 = 2KNO_2 + O_2$. (K = 39, N = 14, O = 16.)
3. What mass of calcium carbonate is required to produce 336 cm^3 of carbon dioxide measured at 819 K and a pressure of 1520 mmHg? (Ca = 40, O = 16, C = 12.)
4. $3Fe + 4H_2O = Fe_3O_4 + 4H_2$. From this equation calculate (*a*) the maximum volume of hydrogen, measured at s.t.p., that could be obtained by using 56 g of iron, (*b*) the mass of iron oxide that would be formed at the same time. (Fe = 56, H = 1, O = 16.) (L.)

5. The following mixtures of gases are exploded and allowed to cool. What will be the volume and composition of the resulting gas in each case, all volumes being measured at laboratory temperature and pressure? (*a*) 10 cm^3 of carbon monoxide and 10 cm^3 of oxygen; (*b*) 15 cm^3 of hydrogen and 10 cm^3 of oxygen; (*c*) 4 cm^3 of methane, CH_4, and 20 cm^3 of oxygen. (C.)

6. Describe, with the help of a diagram, how you would determine the volume of oxygen produced at atmospheric pressure when some potassium chlorate is heated to constant weight. In such an experiment, the chlorate lost 0·370 g and 278 cm^3 of oxygen (measured at 760 mmHg pressure and 20 °C) were evolved. Calculate from these results the molecular mass of oxygen. (C.)

7. Excess sodium hydroxide solution was added to a solution of silver nitrate and the precipitate was filtered off and dried. 0·232 g of this substance yielded on heating 0·216 g of silver and 0·016 g of oxygen. What is the formula of the precipitate? Write an equation for the reaction between the two solutions. (Ag = 108, O = 16·0.) (L.)

8. The mineral azurite is a compound of copper carbonate, $CuCO_3$, and copper hydroxide, $Cu(OH)_2$. When 1·72 g of azurite are heated, 224 cm^3 of carbon dioxide at s.t.p. and 0·090 g of water are evolved, and a residue of copper oxide is left. Find the formula of the azurite. Sketch an apparatus you would use to determine the mass of water formed in this experiment and explain how you would use it. (Cu = 63·5, C = 12, O = 16, H = 1.) (L.)

9. Nitrogen combines with lithium, Li, and the nitride reacts with water giving off ammonia and leaving a metallic hydroxide. When 1·75 g of lithium nitride is treated with water, 1180 cm^3 of ammonia at 17 °C and 765 mmHg pressure is evolved. (*a*) Calculate the empirical formula of lithium nitride. (*b*) What is the valency of lithium in this compound? (*c*) Write the equation for the action of water on lithium nitride. (Li = 7·00, N = 14·0.) (L.)

17 Heats of Reactions

ENERGY

The energy of a substance is its capacity for doing work. Gunpowder, atom bombs, petrol, oil, and coal contain much chemical energy, which can be used to do work. In addition to chemical energy there is heat energy, electrical energy, light energy, sound energy and mechanical energy.

The various types of energy are either energy of motion, called *kinetic energy*, or energy of position or condition, called *potential energy*. Kinetic energy includes the energy of molecules which are in motion and the energy of any moving object, such as moving water, moving air, cars and rockets in motion, and so on. Potential energy includes the chemical energy in fuels and other substances, which is energy of condition, and the energy of substances raised above the earth's surface, such as the water at the top of a waterfall, which is energy of position.

Each kind of energy can be changed into any other kind. For example, the kinetic energy of moving water can be used to turn a dynamo and produce electrical energy; the chemical energy of oil or gas can be used to produce heat energy.

UNITS OF ENERGY

The unit of energy or work is the *joule*, J. The joule is the energy used when a force of one newton (1 N) moves a distance of one metre. A joule is sometimes called a metre-newton. The force on a 100-g mass is about one newton; if you lift a 100-g mass through a height of one metre you do about one joule of work.

The kilocalorie, kcal, is a unit of heat energy, and it is equal to 1000 calories. The calorie and kilocalorie were widely used at one time but their use is gradually being abandoned. A *kilocalorie* is the quantity of heat required to raise the temperature of one kilogram of water through 1 °C. A *calorie* is the quantity of heat required to raise the temperature of one gram of water through 1 °C. The calorie and kilocalorie are directly related to the joule and kilojoule:

1 calorie (1 cal)

= 4·185 5 joules (4·185 5 J)

1 kilocalorie (1 kcal)

= 4·185 5 kilojoules (4·185 5 kJ)

Simpler figures are used for ordinary work:

1 calorie

= 4·19 or 4·2 joules (1 cal = 4·19 or 4·2 J)

1 kilocalorie

= 4·19 or 4·2 kilojoules (1 kcal = 4·19 or 4·2 kJ)

ENERGY REQUIRED TO HEAT, MELT, AND BOIL SUBSTANCES

Heat energy is required to raise the temperature of substances, to melt solids and to boil liquids. Heat also melts solids and boils liquids without causing a rise in temperature; this heat is called *latent heat* which means 'hidden' heat. For example, heat melts ice at 273 K and heat boils water at 373 K if the pressure is atmospheric.

Various definitions and values are given below. Pupils studying physics will be familiar with them. Chemistry students should understand them but do not have to remember them; they are here for reference only. Remember that a rise of temperature of 1 °C equals a rise of one kelvin, 1 K.

Specific heat capacity of a substance is the energy required to raise the temperature of 1 kg of the substance through 1 °C or 1 K. The unit of specific heat capacity can be expressed in different ways, as follows:

joule per kilogram kelvin (J/kg K or $J\ kg^{-1}\ K^{-1}$)

joule per kilogram degree C (J/kg °C)

The specific heat capacity for water is:

$4{\cdot}185 \times 10^3\ J\ kg^{-1}\ K^{-1}$ or $4{\cdot}185\ kJ\ kg^{-1}\ K^{-1}$

and simple values are

$4{\cdot}2 \times 10^3\ J\ kg^{-1}\ K^{-1}$ or $4{\cdot}2\ kJ\ kg^{-1}\ K^{-1}$

Specific latent heat of fusion of a solid is the energy required to change 1 kg of the solid to liquid at the same temperature. The unit of specific latent heat is:

joule per kilogram or J/kg or $J\ kg^{-1}$

The specific latent heat of ice is $3{\cdot}36 \times 10^5\ J\ kg^{-1}$ (and formerly its value was given as 80 calories per gram or 80 kilocalories per kilogram).

Specific latent heat of vaporization of a liquid is the energy required to change 1 kg of the liquid to gas at the same temperature. The unit of specific latent heat of vaporization is the same as for the latent heat of fusion. The specific latent heat of vaporization of water is $2{\cdot}25 \times 10^6\ J\ kg^{-1}$ (and formerly its value was 537 calories per gram or 537 kilocalories per kilogram).

Molar latent heats, mentioned on p. 79, are more useful in chemistry than specific latent heats. A molar latent heat refers to 1 mole of substance whereas specific latent heat refers to 1 kg. In other words, molar latent heat is the heat required to melt or boil $6{\cdot}023 \times 10^{23}$ atoms or molecules of the substance, which is 1 mole.

The molar latent heats of ice and water refer to 18 grams of each substance because 1 mole of each substance is 18 g. For example, the specific heat of ice is $3{\cdot}36 \times 10^5\ J\ kg^{-1}$, which equals $3{\cdot}36 \times 10^2\ J\ g^{-1}$. The molar latent heat of ice is therefore $18 \times 3{\cdot}36 \times 10^2\ J\ mol^{-1}$.

To measure molar latent heats of ice and water

1. Light a medium Bunsen flame, about 5 cm high, and do not alter it during the experiment. It forms heat at a steady rate.
2. Take about 25 g of ice at its melting point, dry it with filter papers or cloth, and put it in a thin metal can.
3. Place the can and ice on a tripod without a gauze, put the Bunsen under it, and start a stopclock.
4. Observe the time taken until all the ice has melted. Stir the ice well all the time.

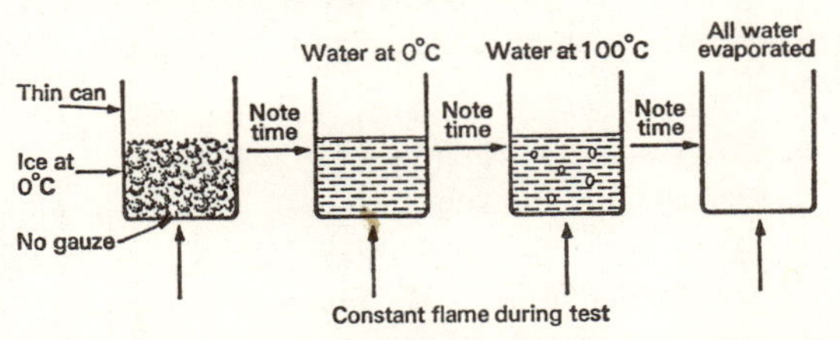

Fig. 17.1 Measuring the molar latent heats of ice and water

5. Now observe the time taken for the water at 0 °C to start to boil.
6. Observe the time taken for the boiling water to evaporate completely.

SPECIMEN RESULTS AND CALCULATION

1. Time to melt ice = 4 min
2. Time to heat the water from 0 to 100 °C = 5 min
3. Time to change the boiling water to steam = 28 min

$$\frac{\text{Heat gained by ice}}{\text{Heat gained by water}} = \frac{4}{5};$$

$$\frac{\text{Heat gained by steam}}{\text{Heat gained by water}} = \frac{28}{5}$$

Since the masses of ice, water, and steam are the same,

$$\frac{\text{Heat gained by 1 g of ice}}{\text{Heat gained by 1 g of water}} = \frac{4}{5}$$

1 g of water heated through 100 °C gains

$$4{\cdot}2 \times 100 = 420 \text{ joules}$$

$\therefore$ Heat gained by 1 g of ice $= 4 \times 420/5 = 336$ joules

The heat gained by 1 g of water when it changes to steam is calculated in the same way. It is $420 \times 28/5$ joules.

The heat required to melt one mole of ice

$$= 18 \times 336 \text{ joules} = 6048 \text{ joules}$$

This is the molar latent heat of ice. The molar latent heat of vaporization of water is calculated similarly.

MOLAR LATENT HEATS AND STRUCTURE OF MATTER

The particles of elements and compounds with giant structures, such as graphite, diamond, ionic compounds, and metals, are held together by strong forces. Great energy is needed to overcome these forces and permit the particles to move freely as in a liquid. Therefore their molar latent heats are high. For the same reason, the melting points of these substances are high, as the particles must have great kinetic energy before they can overcome the cohesive forces holding them in fixed positions. The molar latent heat of ice is also high, and this fact indicates that there are special forces (which do not concern us at this stage) between ice particles.

The particles of liquids which contain ions, such as fused ionic compounds and fused metals, are also held by strong forces. Therefore their molar heats of vaporization and their boiling points are high.

The particles in molecular crystals, such as iodine and sulphur, are held together by weak forces. Their molar heats of fusion and vaporization and their melting- and boiling-points are therefore low.

ENERGY CHANGES IN CHEMICAL REACTIONS

Energy as heat, light, and sound is either evolved or absorbed in most chemical changes. Heat is evolved in the reactions between water and anhydrous copper sulphate, oxygen and many elements, chlorine and many elements, sulphur and many metals, and so on. Life depends on the energy liberated during respiration, when oxygen reacts with glucose. Plants need the energy of sunlight before photosynthesis can occur and produce starch. Modern civilized life depends on the energy liberated when coal, petrol, and oil burn.

The *principle of conservation of energy* states that energy is neither created nor destroyed in chemical reactions. It is merely transformed, e.g. chemical energy is changed to heat energy or vice-versa. Electrical energy is changed to chemical energy during electrolysis, and chemical energy is changed to electrical energy in various cells and accumulators.

HEAT OF REACTION

This is the heat change when the number of moles of the reactants indicated by the chemical equation have reacted completely. Its symbol is ΔH. H is the symbol for heat content of a substance.

$\Delta H = H_2$ (of products) $- H_1$ (of reactants). If heat is evolved in a reaction, the products have less heat content than the reactants; H_2 is less than H_1 and ΔH is negative.

$$Cu(s) + S(\text{rhombic}) = CuS(s)$$
$$\Delta H = -48{\cdot}5 \text{ kJ g-equation}^{-1}$$

This equation means that 48·5 kilojoules are evolved when 63·5 g of copper combine with 32 g of rhombic sulphur to form 95·5 g of copper(II) sulphide and the sulphide has cooled to room temperature. Reactions in which heat is evolved are called *exothermic reactions*. When a reaction is exothermic its ΔH value is negative.

$$H_2(g) + I_2(s) = 2HI(g)$$
$$\Delta H = +52{\cdot}0 \text{ kJ g-equation}^{-1}$$

This equation means that 52 kJ are absorbed when 2 g of hydrogen gas combine with 254 g of iodine to form 256 g of hydrogen iodide. Reactions in which heat is absorbed are called *endothermic reactions*. When a reaction is endothermic its ΔH value is positive.

Energy level diagrams can be used to show energy changes in a simple manner. The diagrams for the two reactions given above are as follows:

$Cu(s) + S(s)$

Exothermic $\Delta H = -48{\cdot}5$ kJ g-eq^{-1}

$CuS(s)$

$2HI(g)$

Endothermic $\Delta H = +52{\cdot}0$ kJ g-eq^{-1}

$H_2(g) + I_2(s)$

The diagrams show that the heat content of copper sulphide is less than the heat content of the elements from which it is made, but the heat content of hydrogen iodide is greater than the heat content of its elements.

Energy diagrams should be drawn to a suitable scale. If 1 mm represents 1 kJ, the arrows in the above diagrams would be 48·5 mm and 52 mm long respectively.

HEAT OF COMBUSTION

This is the heat change when 1 mole of the substance is burnt completely in oxygen. The word 'completely' is necessary to exclude partial combustions such as carbon to carbon monoxide or methane to carbon and carbon monoxide. Carbon dioxide is the product of complete combustion in the first reaction below.

$$C(s) + O_2(g) = CO_2(g);$$
$$\Delta H = -390 \text{ kJ g-equation}^{-1}$$
$$\text{or } -390 \text{ kJ/g-equation}$$

$$H_2(g) + \tfrac{1}{2}O_2(g) = H_2O(l);$$
$$\Delta H = -285 \text{ kJ g-equation}^{-1}$$
$$\text{or } -285 \text{ kJ/g-equation}$$

The hydrogen equation is written with 1 mole of hydrogen and therefore only $\frac{1}{2}O_2$ is required to balance it. These two equations mean that the heat of combustion of carbon is -390 kJ per mole (kJ mol^{-1}) and of hydrogen is -285 kJ mol^{-1}. The minus sign indicates that heat is evolved and the reactions are exothermic.

The combustion of fuels provides energy for homes and factories. The 'slow combustion' of digested food inside the body provides us with heat and other energy. Experiments have shown that fats, sugars, starches, and other foods produce the same energy when they oxidize slowly in the body during respiration as they do when they burn rapidly in oxygen or air. The heat energy of a fuel or of a food is called its *calorific value*. This is defined as the heat evolved when 1 g or 1 kg of the solid or liquid fuel or food is completely burnt in oxygen. The calorific values of three types of food are approximately:

Carbohydrate 16 kJ per g;
Fat 38 kJ per g;
Protein 16 kJ per g.

The heat of combustion of an alcohol or petrol can be determined with moderate accuracy by the following method.

To measure the heat of combustion of ethanol

1. Use ethanol as the fuel in a simple lamp as shown in Fig. 17.2. Weigh the lamp, complete with stopper, wick, and ethanol.

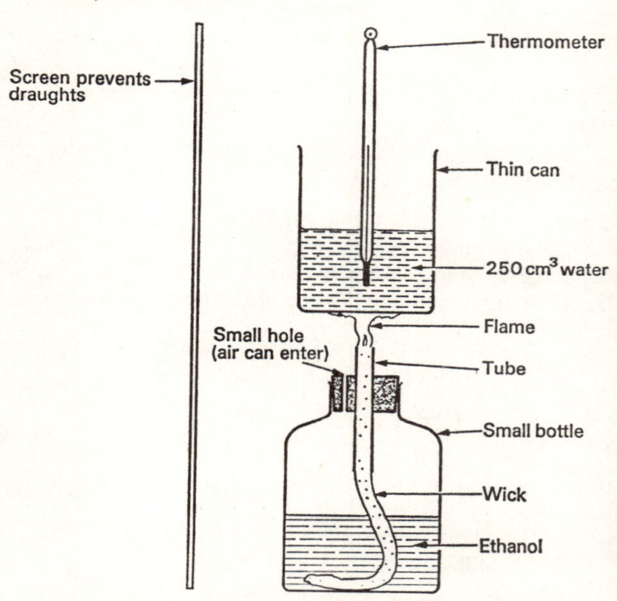

Fig. 17.2 Measuring the heat of combustion of ethanol

2. Add 250 cm^3 (about 250 g) of water to a thin can. Observe the temperature of the water.
3. Light the lamp and place it directly under the can so that most of the flame touches the metal and most of the heat produced passes through the metal into the water.
4. When the temperature rise is about 30 °C, put out the flame, stir the water well, and note the highest temperature reached.
5. Weigh the lamp.

1.	Mass of lamp at start	=
2.	Temperature of water at start	=
3.	Temperature of water at finish	=
4.	Mass of lamp at finish	=
1 − 4.	Mass of ethanol burned	= w g
3 − 2.	Rise in temperature of water	= t °C

1 mole of ethanol, C_2H_5OH,

$$= 24 + 5 + 16 + 1 = 46 \text{ g}$$

$\therefore$ $(w)/46$ mole of ethanol were burned.

Heat gained by the water

$$= \text{mass} \times \text{rise in temperature} \times 4{\cdot}2 \text{ joules}$$

$$= 250 \times t \times 4{\cdot}2 = 1050t \text{ joules}$$

$(w)/46$ mole of ethanol produces $1050t$ joules

$\therefore$ 1 mole of ethanol produces $(1050t \times 46)/w$ joules

This is the heat of combustion of ethanol. Since part of the heat produced by the ethanol flame warms the can and the air, this value is usually less than the correct one.

HEAT OF SOLUTION

This is the heat change when one mole of solute dissolves in a large volume of solvent (usually water). The words 'a large volume of' are essential because the heat change usually varies with the quantity of solvent used, and there is a heat change (called the heat of dilution) when a concentrated solution is diluted. The heat of solution applies when addition of more solvent causes no further heat change; the solution is then said to be 'at infinite dilution' and the symbol (aq) is used to denote it.

$$HCl(g) + aq = HCl(aq) \qquad \Delta H = -71 \text{ kJ mol}^{-1}$$

$$NH_3(g) + aq = NH_3(aq) \qquad \Delta H = -34 \text{ kJ mol}^{-1}$$

$$NH_4Cl(s) + aq = NH_4Cl(aq) \qquad \Delta H = +16{\cdot}7 \text{ kJ mol}^{-1}$$

Two energy changes take place when an ionic compound A^+B^- dissolves in water. Energy is absorbed in order to separate the ions from the crystal structure of the solid. This energy is usually between 500 and 1000 kilojoules per mole. Energy is evolved when the ions become hydrated to form $A(H_2O)_x^+$ and $B(H_2O)_y^-$. The difference between the two energy changes is the heat of solution in water.

To measure some heats of solution

In these simple experiments, 1 mole of solvent is added to water to produce 1 kg of solution. The heat required to raise the temperature of the final solution through 1 °C is 1 kcal or 4·2 kJ and therefore the calculation is very easy.

Sulphuric acid Add 53·5 cm^3 of concentrated sulphuric acid, which is 98 g or 1 mole of acid, to 900 cm^3 of water, at a known temperature, in a large container of thin glass or plastic. Stir well and note the temperature change, t °C.

The heat of solution is therefore about $4{\cdot}2t$ kJ mol^{-1} or t kcal mol^{-1}. The final solution is approximately 1·0 M because it contains 1 mole of acid in about 1 litre of water.

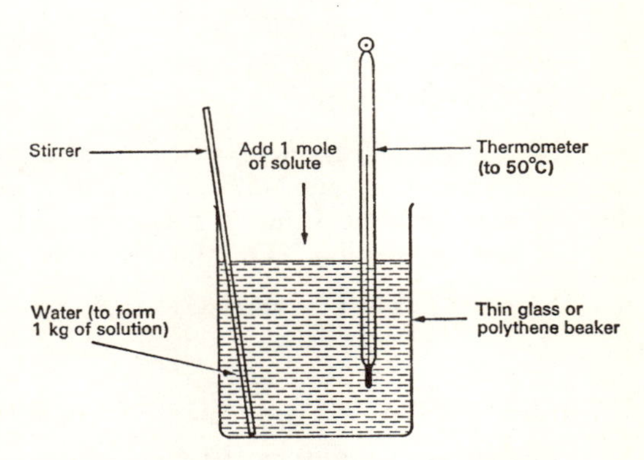

Fig. 17.3 Measuring heat of solution

The experiment can be done with one-tenth of the above quantities, using a small polythene beaker of negligible heat capacity. The total energy change is, of course, only one-tenth of that using the larger quantities, but the mass of solution is only one-tenth also. The temperature change is therefore the same. The experiment with larger quantities may be more accurate as heat losses to the air, thermometer, container, etc. are a smaller fraction of the total heat change.

Ammonium nitrate Make 80 g (1 mole) of the nitrate into a fine powder. Place 920 cm^3 of water in a thin vessel, read its temperature, add the nitrate, stir well, and observe the final temperature. The final solution is about 1·0 M concentration.

Sodium thiosulphate Add 248 g of the crystals (1 mole of $Na_2S_2O_3 . 5H_2O$) to 750 g of water, stir well, and observe the temperature change. The final solution is about 1·0 M concentration.

Energy level diagrams are:

$H_2SO_4(l) + aq$

$\Delta H = -71$ kJ g-equation^{-1}

Exothermic ↓ M H_2SO_4

M NH_4NO_3

Endothermic ↑

$\Delta H = +25$ kJ g-equation^{-1}

$NH_4NO_3(s) + aq$

HEAT OF NEUTRALIZATION

This is the heat change when acid which produces one mole of hydrogen ions is neutralized by alkali, and both the acid and alkali are in dilute solution. Some values are:

Acid	*Alkali*	*kilojoules*
HCl(aq)	NaOH(aq)	−57
HNO_3(aq)	KOH(aq)	−57
$\frac{1}{2}H_2SO_4$(aq)	NaOH(aq)	−61
HF(aq)	NaOH(aq)	−69
HCN(aq)	KOH(aq)	−12
HNO_3(aq)	$\frac{1}{2}Ca(OH)_2$	−58

The heat of neutralization of a strong acid by a strong alkali is usually about −57 kJ. The ionic theory readily explains why the value is a constant. If the acid is HX and the alkali is YOH the change is

$$HX + YOH = H_2O + YX,$$

or ionically,

$$H^+ + X^- + Y^+ + OH^- = H_2O + Y^+ + X^-,$$

i.e. $H^+(aq) + OH^-(aq) = H_2O(l)$

The reaction that occurs is the formation of 18 g of water from 1 g of hydrogen ions and 17 g of hydroxide ions.

Acids such as hydrocyanic or ethanoic (acetic) acid are weak acids and most of their molecules have to form ions before the neutralization can occur. The heat of ionization changes the heat evolved when they are neutralized.

To measure heats of neutralization

Add 500 cm³ of 2 M hydrochloric acid or 2 M nitric acid or M sulphuric acid (all of which contain 1 mole of hydrogen ions) to a thin glass or polythene vessel. Read its temperature. Add 500 cm³ of 2 M sodium hydroxide or potassium hydroxide (which will just neutralize the acid) at the same temperature. Stir well and observe the maximum temperature change, t °C.

Since the product is fairly dilute, its heat capacity can be regarded as that of water. Its mass is about 1 kg and therefore a rise of 1 °C means that 4·2 kJ of heat have been evolved. The heat of neutralization is therefore $-4{\cdot}2t$ kJ mol⁻¹.

To measure the heat of precipitation of silver chloride

Let 25 cm³ of 0·5 M silver nitrate and 25 cm³ of 0·5 M sodium chloride stand until they are both at room temperature. Read the temperature. Mix the solutions in a small, thin polythene bottle and observe the maximum temperature change, t °C. This change would be the same if 500 cm³ of each solution had been used.

$$Ag^+(aq) + Cl^-(aq) = AgCl(s)$$

25 cm³ M/2 + 25 cm³ M/2 = 1/80 mole
Change = t °C

∴ 500 cm³ M/2 + 500 cm³ M/2 = 0·25 mole
Change = t °C

∴ 500 cm³ 2 M + 500 cm³ 2 M = 1 mole
Change = $4t$ °C

ΔH for the reaction is $-4t \times 4{\cdot}2$ kJ mol⁻¹.

To measure the heat of displacement of copper

Repeat the previous experiment by adding 50 cm³ of M/5 copper(II) sulphate solution to 1 g (i.e. excess) of either zinc filings or non-greasy iron filings in a small bottle. Let the temperature change be t °C.

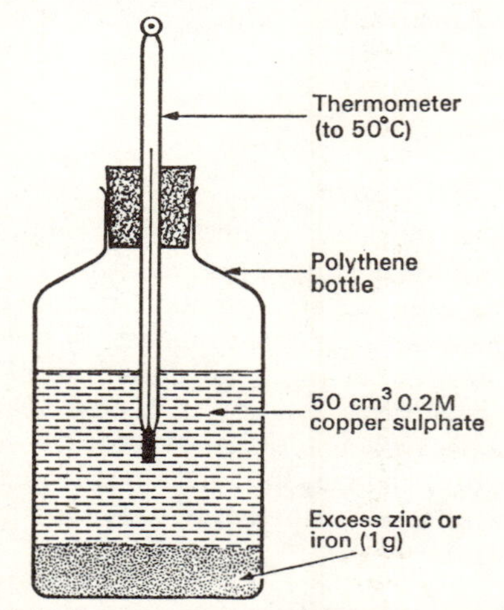

Fig. 17.4 Measuring heat of displacement of copper

$$Cu^{2+}(aq) + Zn(s) = Cu(s) + Zn^{2+}(aq)$$

50 cm³ M/5	0·01 mole	Change = t °C
∴ 1000 cm³ M/5	0·20 mole	Change = t °C
∴ 1000 cm³ M	1·00 mole	Change = $5t$ °C

ΔH for the reaction is $-5t \times 4{\cdot}2$ kJ mol⁻¹.

In the last two experiments, observe that increasing the volume of the same solution would not alter the temperature change. The energy change would increase, but so would the mass of solution to absorb this energy, and the temperature rise would be constant. Heat losses to the air, thermometer, etc. are proportionately greater when small volumes are used, and therefore larger volumes should give more accurate results.

BOND ENERGIES AND HEATS OF REACTION

Any chemical change involves the rearrangement of atoms. Chemical bonds between some atoms are broken and new bonds between these or other atoms are made. Energy is absorbed when chemical bonds are broken and evolved when they are made. The energy of a bond is the energy change in the reaction:

$$X\ (atom) + Y\ (atom) = X\text{—}Y; \quad \Delta H = -x \text{ kJ mol}^{-1}$$

For example:

$$H + H = H\text{—}H; \quad \Delta H = -433 \text{ kJ mol}^{-1}$$

$$H_2 = H + H; \quad \Delta H = +433 \text{ kJ mol}^{-1}$$

The energy of the H—H bond equals the energy absorbed when one mole of hydrogen molecules is split into free atoms, and this energy can be measured by spectroscopic means which do not concern us. Other bond energies are given below in kJ per mole.

C—C	C—O	C—H	O—O
344	344	414	495

H—O	Cl—Cl	H—Cl
460	242	430

These are the energies required to break the bonds and therefore they are a measure of the relative stabilities of the bonds. The energy of a compound is largely the energy of its bonds. Its kinetic energy, resulting from movements of its molecules, atoms or ions, is usually small compared with the bond energy.

Bond energies in ethanol A molecule of ethanol (CH_3CH_2OH) contains five C—H, one C—C, one C—O and one O—H bonds. Its total bond energy is therefore

Five C—H = 5 × 414 = 2070 kJ
One C—C = 344 kJ
One C—O = 344 kJ
One O—H = 460 kJ

making a total of 3218 kJ mol^{-1}. The heat required to form one mole of ethanol from its atoms is 3200 kJ mol^{-1} which is close to the total bond energy.

Heat of the hydrogen–chlorine reaction The experimental value is given by the equation:

$$H_2(g) + Cl_2(g) \rightarrow 2HCl(g); \quad \Delta H = -175 \text{ kJ g-equation}^{-1}$$

The energy change, in terms of bond energies, is

One H—H bond broken = 433 kJ absorbed

One Cl—Cl bond broken = 242 kJ absorbed

Two H—Cl bonds made = 2 × 430 = 860 kJ evolved

The total heat evolved is 175 kJ g-equation^{-1}, which agrees with the experimental figure in the equation above. This close agreement does not apply to every reaction.

ELECTRICAL ENERGY FROM CHEMICAL REACTIONS

A zinc rod placed in zinc sulphate solution becomes negatively charged because zinc ions form and electrons remain on the rod. A copper rod placed in copper sulphate solution becomes positively charged because copper ions change to atoms by acquiring electrons from the rod.

$$Zn(s) = Zn^{2+}(aq) + 2e; \quad Cu^{2+}(aq) + 2e = Cu(s)$$

These two reactions are used in a Daniell cell, and electrons flow from the zinc to the copper through the wire connecting them. The total change is merely a transfer of electrons from copper ions to zinc atoms:

$$Zn(s) + Cu^{2+}(aq) = Zn^{2+}(aq) + Cu(s) + \text{energy}$$

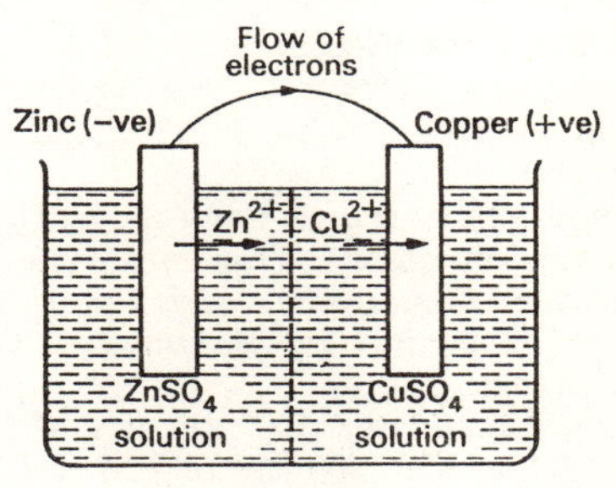

Fig. 17.5 Daniell cell

A voltmeter joined to the two metal rods reads about 1·1 V which is the e.m.f. of the cell. All of this voltage is not available to produce useful electrical energy. Part of it, called *lost volts*, is used up and produces heat in moving the anions and cations inside the cell.

$$\text{Total energy of reaction} = \text{Useful electrical energy} + \text{Heat energy (lost volts)}$$

i.e. $\Delta H = \Delta G + \text{wasted energy}$

SUMMARY

Energy is the capacity for doing work. It may be chemical, heat, electrical, and mechanical energy, and so on.
Units of energy: a *joule* is a metre-newton—the energy used when a force of 1 N moves a distance of 1 m.
Kilocalorie: the energy required to heat 1 kg of water through 1 °C. 1 kcal = 1000 cal = about 4·2 kJ = about 4200 J.

Latent heat is the heat required to melt solids or boil liquids at a constant temperature. *Specific latent heats* refer to 1 kg of the substance; example: ice 3·36 × 10^5 J kg^{-1}.
Molar latent heat refers to 1 mole of substance; it is heat required to melt or boil the same number of particles (6 × 10^{23}) of the substances. Molar latent heats of ionic compounds and substances with giant structures are high because of strong forces between the particles; molar latent heats of most covalent compounds are relatively low because of the weak forces between the molecules.

Energy of reactions (ΔH)

Heat of reaction: the heat change when the number of moles of reactants indicated by the chemical equation have reacted completely; measured in kJ/g-equation (sometimes written kJ g-equation^{-1}).

Exothermic reaction: heat is evolved and ΔH is negative.

Endothermic reaction: heat is absorbed and ΔH is positive.

Heat of combustion: the heat change when 1 mole of substance is burnt completely in oxygen.

Calorific value of fuel or food: the heat evolved when 1 g (sometimes 1 kg) of fuel or food is completely burnt in oxygen.

Heat of solution: the heat change when 1 mole of substance dissolves in a large volume of solvent.

Heat of neutralization: the heat change when acid which produces 1 mole of H^+ is neutralized by alkali (acid and alkali must be in dilute solution). For a strong acid and strong alkali it is −57 kJ.

Measuring heats of reactions

Heat of combustion of ethanol: heat water in a can by burning ethanol in a lamp. Calculate the heat produced by 1 mole of ethanol.

Heat of solution. Sulphuric acid: add 1 mole of acid to about 900 cm³ of water and observe temperature rise. *Ammonium nitrate or sodium thiosulphate:* add 1 mole of solid to enough water to form 1 kg of solution. Observe the temperature change.

Heat of neutralization: mix 500 cm³ of 2 M HCl or 2 M HNO_3 or M H_2SO_4 with 500 cm³ of 2 M NaOH or KOH. Observe the temperature rise.

Heat of precipitation: mix equal volumes of 0·5 M $AgNO_3$ and 0·5 M NaCl and observe temperature change.

Heat of displacement: add excess Zn or Fe to 50 cm³ of 0·2 M $CuSO_4$. Copper is displaced.

Bond energies and heats of reaction

Bond energy: the energy absorbed when 1 mole of X—Y is split into free atoms X and Y. The energy of a compound is mainly in its bonds. The energy change in a reaction is produced by the breaking and making of bonds.

QUESTIONS

1. What is meant by the heat of combustion of a substance? Outline one method by which it can be measured.
2. When 1 mole of ammonium nitrate is dissolved in water to form 1000 cm³ of solution, the fall in temperature is 6 °C. What is the heat of solution in kcal mol^{-1} and kJ mol^{-1}? What would be the fall in temperature when 0·1 mole of the nitrate is dissolved in water to form 100 cm³ of solution?
3. 50 cm³ of 0·5 M silver nitrate and 50 cm³ of 0·5 M sodium chloride are mixed. If the heat of precipitation of silver chloride is −58 kJ mol^{-1}, show clearly that the temperature change should be about 3·5 °C.
4. Explain why the heat of neutralization of a strong acid by a strong alkali is reasonably constant. What is the meaning of the word 'strong' in this statement?
5. Outline how you would measure the approximate heat of neutralization of a weak organic acid by sodium hydroxide solution.
6. The average bond energies associated with the C—H bond, the C—C bond, the C—O bond and the O—H bond are respectively 414, 344, 344, and 460 kJ mol^{-1}. Use these values to calculate the theoretical heat of formation of ethanol, (CH_3CH_2OH), and methanol (CH_3OH).
7. Use the following values of heats of reactions to calculate the average bond energy of the O—H bond.

$$2H_2(g) + O_2(g) = 2H_2O(g) \qquad \Delta H = -487 \text{ kJ g-eq}^{-1}$$
$$4H = 2H_2(g) \qquad \Delta H = -870 \text{ kJ g-eq}^{-1}$$
$$2O = O_2(g) \qquad \Delta H = -495 \text{ kJ g-eq}^{-1}$$

   What is the meaning of the negative sign associated with each value of ΔH?
8. The heat evolved when certain volumes of acid were added to 100 cm³ of 2 M sodium hydroxide were measured and the results were:

Volume of acid used (cm³)	60	80	100	120	140
Heat evolved (kJ)	5·6	8·0	10·4	10·4	10·4

   (*a*) Estimate the volume of acid solution which neutralizes 100 cm³ of the sodium hydroxide. (*b*) How many kilojoules are evolved when 90 cm³ of acid are added to the hydroxide? (*c*) What is the mass of sodium hydroxide in 1000 cm³ of the 2 M solution? (*d*) Estimate the heat evolved when 1 mole of calcium hydroxide in solution is neutralized by the same acid. (*e*) When 50 cm³ of 4 M sodium hydroxide was used instead of the 100 cm³ of 2 M solution, the heat evolved was slightly more than the values given above. How do you explain this fact?

18 Rates of Reactions and Dynamic Equilibria

Precipitation reactions, which are the result of combination between ions, appear to take place instantaneously. Reactions between mineral acids and metals, bases, carbonates or sulphites are completed in a few minutes. The corrosion of iron and other metals takes place over a period of months or years. Mixtures of hydrogen and oxygen or hydrogen and chlorine react explosively when a flame is put to them, but under ordinary conditions they react at a rate so slow that it cannot be measured. It is clear that rates of reactions can vary enormously. We can now study the effects of various factors on the rates.

To study the effect of concentration on rates of reactions

Sodium thiosulphate(VI), $Na_2S_2O_3$, and acid react to form a precipitate of sulphur. Equal volumes of the same acid are added to different concentrations of thiosulphate solutions. The times taken for the same amount of sulphur to form are observed.

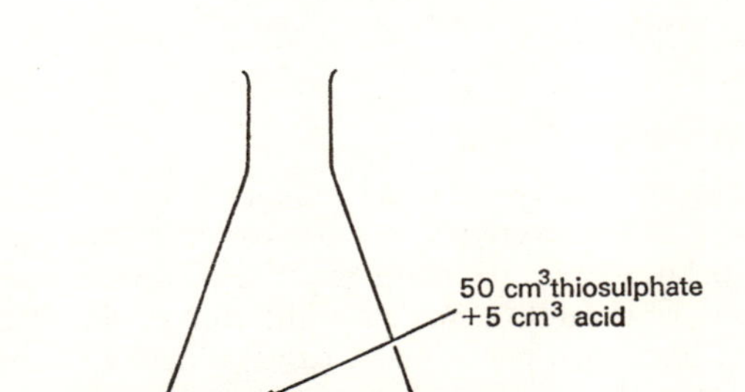

Fig. 18.1 To measure how the rate of a reaction depends on concentration of a reactant

1. Prepare sodium thiosulphate solution of concentration about 40 g per 1000 cm³ (40 g dm⁻³).
2. Take five conical flasks and add 50, 40, 30, 20, and 10 cm³ of the solution to them. Add water to make the total volume 50 cm³ in each flask. The ratio of the concentrations of the thiosulphate is 5:4:3:2:1.
3. Start a stopclock just as 5 cm³ of 2 M hydrochloric acid are added to one flask. Shake the flask quickly and then place it on white paper with a black mark on it.
4. Look through the solution from above. The sulphur precipitate gradually obscures the black mark. Stop the clock when the mark first becomes invisible.
5. Repeat with the other four flasks.

If the reaction is fast, the time for the mark to become invisible is short, i.e. the rate of reaction is inversely proportional to the time, or rate of reaction is proportional to 1/time ($time^{-1}$). Plot a graph of concentration of the thiosulphate solution against $time^{-1}$ and draw any conclusion you can from it.

To study the effect of particle size on rates of reactions

The same volume of the same acid reacts with calcium carbonate chips of different sizes. The rate at which carbon dioxide is evolved is determined by measuring the loss in mass of the reacting vessel and contents.

1. Add 40 cm³ of dilute hydrochloric acid to a small conical flask. Add 20 g of marble chips of medium size and place a loose plug of glass wool in the mouth of the flask to hold back acid spray.
2. Weigh the whole apparatus on a direct-reading balance. Start a stopclock at the same time.
3. Observe the time taken to lose successive amounts of 0·10 g, and stop after ten to fifteen minutes.
4. Repeat the whole test with smaller marble chips and then with tiny chips.

The rate of reaction is inversely proportional to the time taken to lose a definite mass, that is, $time^{-1}$ or 1/time. The rate is slower with the larger particles.

Effect of concentration of acid Modify the experiment by using the same marble chips, but using 40 cm³ of 2 M, M and 0·5 M acid in separate tests. Deduce from the results how the rate of reaction varies with the concentration of the acid. The rate is slower with the more dilute acid.

To study the effect of temperature on rates of reactions

We shall use the reaction where sulphur is precipitated from sodium thiosulphate and hydrochloric acid. The times taken to produce the same quantity of sulphur precipitate at different temperatures are observed.

1. Add 10 cm³ of thiosulphate solution and 40 cm³ of water to a small conical flask. Warm to 20 °C if necessary.
2. Add 5 cm³ of dilute hydrochloric acid and start a stopclock. Shake well and place the flask as before, over a black mark on white paper.
3. Observe the time taken before the mark becomes invisible when viewed from above through the solution.
4. Repeat the whole test at 30 °C, 40 °C, etc. Warm the thiosulphate solution slightly above the desired temperature because the acid cools it when added. Then measure the exact temperature after adding the acid.

Plot a graph of temperature of the reactants against 1/time ($time^{-1}$), which is proportional to the rate of reaction.

KINETIC THEORY AND RATES OF REACTIONS

Collisions between molecules, atoms or ions occur before most chemical reactions can occur. Particles A and B must collide before C and D can be formed in the reaction

$$A + B = C + D$$

Every collision does not produce a reaction, but the rate of reaction between A and B is proportional to the number of collisions in one second. The diagram shows that the number of collisions is proportional to the number of molecules of A and B present; that is, the number of collisions is proportional to the concentrations.

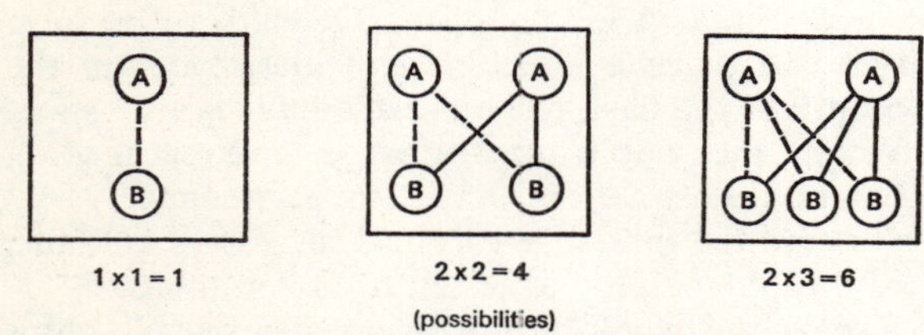

Fig. 18.2 Possibilities of collisions between reacting particles

A rise in temperature increases the speeds of the particles. Therefore there will be more collisions per second. Rate of reaction should therefore increase with temperature. Not only are there more collisions at higher temperatures but more energy is involved at each collision, and therefore the proportion of effective collisions (those that produce reaction) also increases rapidly.

The effective concentration of a solid reactant depends on its surface area. The greater the area, the more chance there is of a liquid or gaseous reactant colliding with the solid reactant. Rate of reaction should therefore be proportional to surface area; that is, the rate of reaction is inversely proportional to the size of the solid particles. (Remember that a 2 mm cube can form eight 1 mm cubes, which have double the surface area of the original 2 mm cube—8×6 mm^2 as compared with 4×6 mm^2.)

To study the effect of a catalyst on rates of reactions

Manganese(IV) oxide catalyses the decomposition of hydrogen peroxide into oxygen and water. The rate of reaction is measured by the rate of formation of oxygen.

1. Add 2 cm^3 of 20-volume hydrogen peroxide solution to a flask and then dilute with 48 cm^3 of water. Connect the flask to a gas syringe. Observe that practically no oxygen is formed.
2. Add a little manganese(IV) oxide to the peroxide and insert the stopper of the flask at once. Observe the volume of oxygen every minute or half-minute.

Effect of concentration Repeat the experiment with 1 cm^3, 4 cm^3, etc. of the 20-volume peroxide, each being diluted so that the volume is 50 cm^3.

Plot a graph of volume of peroxide added (which is proportional to its concentration) against volume of oxygen formed per minute (which is proportional to the rate of reaction).

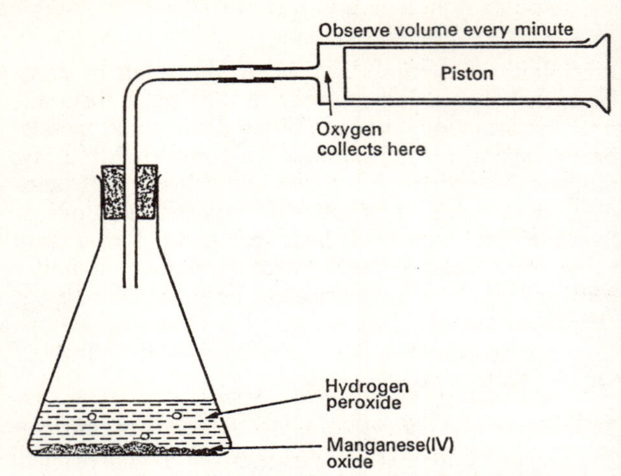

Fig. 18.3 To find how the rate of reaction depends on concentration, temperature, and presence of a catalyst

Effect of temperature Use 2 cm^3 of peroxide solution, diluted to 50 cm^3 with water, at temperatures of 20 °C, 30 °C, etc. In each test add about the same mass of manganese(IV) oxide. Find how the volume of oxygen formed per minute varies with the temperature of the hydrogen peroxide.

CATALYSIS

A catalyst usually increases the rate of a chemical reaction. A few catalysts decrease the rate; for example, food preservatives decrease the rate of decomposition of food, and dilute acids slow the rate of decomposition of hydrogen peroxide. A catalyst does not change in mass and does not change chemically although it may change physically. For example, lumps of manganese(IV) oxide change to powder when added to hydrogen peroxide. Small amounts of certain impurities may stop catalytic action and their effect is called 'poisoning of the catalyst'. Compounds of arsenic 'poison' platinum used in the reaction between sulphur dioxide and oxygen to form sulphur(VI) oxide.

Many reactions occur only when the particles of reactants have sufficient energy. A catalyst speeds a reaction by reducing the amount of energy required. We need not discuss how it does this.

DYNAMIC EQUILIBRIUM

The decomposition of calcium carbonate is a reversible reaction because the products of reaction can react and reform the carbonate:

$$CaCO_3(s) \rightleftharpoons CaO(s) + CO_2(g)$$

When it is heated strongly in the open air, all the carbonate decomposes because the carbon dioxide moves away and cannot react with the calcium oxide. When it is heated in a closed vessel, the carbonate decomposes until a certain pressure of carbon dioxide is reached. After that, no more carbonate seems to decompose and the reaction is at equilibrium. However, reaction does not cease. The carbonate still decomposes, and the calcium oxide and carbon dioxide still combine. The two rates of reaction are equal. The equilibrium is a dynamic one. The final mixture is called an *equilibrium mixture*.

Most reversible reactions proceed to completion in one direction if a reactant is removed. When steam is passed over heated iron, all the iron changes to iron oxide:

$$3Fe(s) + 4H_2O(g) \rightarrow Fe_3O_4(s) + 4H_2(g)$$

The hydrogen is swept away as soon as it is formed and therefore the reverse reaction has no chance to occur. Similarly, all the oxide changes to iron when it is heated in a stream of hydrogen, which sweeps away the steam and prevents the opposing reaction. If the reactants are heated in a closed vessel, both forward and reverse reactions occur and finally an equilibrium is reached when the rates of forward and reverse reactions are equal. The number of atoms of iron formed in one second equals the number reacting in one second. The equilibrium is dynamic.

Bromine and water react reversibly:

$$\underset{\text{(reddish)}}{Br_2(l) + H_2O(l)} \rightleftharpoons \underset{\text{(products colourless)}}{H^+(aq) + Br^-(aq) + HBrO(aq)}$$

This reaction is used in the following experiment.

To study the effect of concentration on a reversible reaction

1. Add about 1 cm^3 of liquid bromine (it is a dangerous, corrosive liquid) to about 100 cm^3 of distilled water in a beaker. Shake or stir well. Bromine water forms.
2. Place 10 cm^3 of the bromine water in a beaker on white paper so that its colour is easily observed.
3. Add bench sodium hydroxide solution, a few drops at a time, and note any colour change. The alkali removes $H^+(aq)$ from the reaction mixture.
4. Now add bench hydrochloric acid or sulphuric acid, a few drops at a time, and note any colour change. The acid increases the concentration of $H^+(aq)$ in the reaction mixture.
5. Repeat steps 3 and 4 several times.

The alkali makes the mixture become colourless and the acid causes the colour to return. A little bromine is still present even when the mixture is colourless.

A RADIOACTIVE TRACER SHOWING DYNAMIC EQUILIBRIUM

Lead iodide can be obtained in a radioactive form. It contains radioactive iodine as part of the iodine present. The radioactive molecules are easy to trace; in effect, they are lead iodide molecules with a label on them.

Radioactive lead iodide crystals were added to a saturated solution of ordinary lead iodide. The concentration of the iodide did not increase, but the solution became radioactive. These facts proved that lead iodide molecules passed from the solid iodide into solution, and therefore iodide molecules must have passed from solution to the solid in order to maintain a constant concentration in the solution. There is a dynamic equilibrium between the undissolved and the dissolved lead iodide:

$$PbI_2(s) \rightleftharpoons PbI_2(aq)$$

SUMMARY

Rate of a reaction depends on concentration of reactants, temperature, presence of a catalyst, and the surface area of solid reactants.
Kinetic or collision theory of reactions: reaction can only occur when particles of reactants collide with sufficient energy.
Increased concentration increases the number of collisions in unit time.
Increased temperature increases the energy of reactant particles.
Catalysts can increase or decrease the rate of reaction by decreasing or increasing the energy required for reaction to occur.
Increased surface area of solid reactants increases the chances of collision of reactant particles.

Dynamic equilibrium occurs when the rate of the forward reaction of a reversible reaction equals the rate of the reverse reaction. The forward reaction is complete if one or more products are removed and the reverse reaction is assisted by removing one or more reactants.
Reversible reactions are indicated by the sign $\rightleftharpoons$.

Effect of conditions on rate of reaction
Concentration: add a given volume of HCl to aqueous sodium thiosulphate of various concentrations. Observe when precipitated sulphur hides a black mark on paper.

Add a given volume of HCl of various concentrations to equal masses of the same marble chips. Observe the times taken for a certain loss in mass to occur.

Add a given volume of H_2O_2 of various concentrations to equal masses of manganese(IV) oxide. Observe the rates of formation of oxygen.
Concentration on reversible reaction: add, alternately, acid and alkali to bromine water; observe the colour changes.
Temperature: use HCl and thiosulphate, of constant concentrations, at different temperatures.

Add equal volumes of the same H_2O_2 at different temperatures to equal masses of manganese(IV) oxide. Observe the rates of formation of oxygen.
Particle size (surface area): add equal volumes of the same HCl to marble chips of varying sizes.

QUESTIONS

1. The decomposition of hydrogen peroxide solution is catalysed by manganese(IV) oxide and by copper(II) oxide. Outline a method by which you would determine which substance is the better catalyst at room temperature.
2. You are required to determine the mass of 1 cm^3 of oxygen at s.t.p. Draw a diagram of the apparatus you would use to collect and measure the oxygen evolved when approximately 0·1 g of potassium chlorate is decomposed by heat. (The volume is less than 100 cm^3.) (C.)
3. Explain carefully how you would carry out the above experiment, obtain the necessary data, and calculate the result. (C.)
4. How do changes in temperature affect the rate of a reaction? In your answer refer to the kinetic theory and to the breaking and making of bonds between atoms.
5. A given volume of standard sodium thiosulphate solution was mixed with the same volume of dilute hydrochloric acid at different temperatures. The times for the appearance of a sulphur precipitate were noted. The results were:

Temperature °C	25	35	44	61	68
Time (s)	60	25	16	6	4
1/time	0·017	0·040	0·062	0·17	0·25

Plot a graph of the temperature of the reactants against 1/time. State and explain what you can deduce from the graph.

6. 0·20 g of zinc powder was added to 50 cm^3 of standard sulphuric acid in a flask. The time taken for the zinc to disappear was noted. The experiment was then repeated with the same mass of zinc, the same volume of acid and at the same temperature, but with acids of different concentrations. Readings were:

Molarity of acid	0·5	0·8	1·0	1·3	1·5
Time (s)	250	80	50	20	15

Draw a graph showing molarity of the acid on the horizontal axis and 1/time (i.e. $time^{-1}$) on the vertical axis.
(*a*) What is the effect of concentration on the rate of reaction between zinc and the acid?
(*b*) Estimate the time it would take for 0·20 g of zinc to react completely with 0·25 molar acid.
(*c*) The temperature rise in the experiments is approximately the same. Account for this fact.
(*d*) What criticisms do you have of this method of investigating how the rate of this reaction depends on the molarity of the acid?
(*e*) The experiment was repeated with 0·20 g of zinc foil instead of zinc powder. How do you expect the readings of time to change, and why?
(*f*) The experiment was repeated at a temperature 10 °C higher than the original temperature. How do you expect the readings of time to change, and why?

7. The reaction between hydrogen peroxide and iodide ions, in acid solution, can be represented by the equation:

$$H_2O_2(aq) + 2I^-(aq) + 2H^+(aq) = 2H_2O(l) + I_2(aq)$$

The reaction, which is rather slow, can be followed by the darkening of the solution due to the formation of iodine, which remains in solution in the presence of iodide ions. Copper(II) ions, $Cu^{2+}(aq)$, are said to act catalytically on the reaction. Describe in detail how you would test the truth of this statement experimentally. (L.)

8. You wish to investigate the effect of varying the proportion of carbon dioxide in air on the rate at which iron rusts. Draw and label a diagram of the apparatus you would use. Indicate what readings and measurements you would take but do not otherwise describe the method.

19 Acids, Bases and Salts

IONIC THEORY OF ACIDS AND BASES

The ionic theory states that acidity is due to hydrogen ions H^+, and alkalinity is due to hydroxide ions OH^-.

An acid is a compound which, when dissolved in water, forms hydrogen ions as the only positively charged ions (cations).

An alkali is a compound, which, when dissolved in water, forms hydroxide ions as the only anions.

A base is a compound which contains either hydroxide ions or oxide ions, OH^- or O^{2-}; it reacts with hydrogen ions to form water and a salt is also formed.

Acid $\quad HCl \rightarrow H^+ + Cl^-$

$$H_2SO_4 \rightarrow H^+ + HSO_4^-$$
$$\rightarrow 2H^+ + SO_4^{2-}$$

Alkali $\quad NaOH \rightarrow Na^+ + OH^-$

Base $\quad 2H^+ + CuO \rightarrow H_2O + Cu^{2+}$

$$2H^+ + Cu(OH)_2 \rightarrow 2H_2O + Cu^{2+}$$

The typical properties of dilute acids are really those of hydrogen ions, which are illustrated by the following equations:

Metals $\quad Mg + 2H^+ = Mg^{2+} + H_2(g)$

Carbonates $\quad CO_3^{2-} + 2H^+ = CO_2(g) + H_2O$

$$HCO_3^- + H^+ = CO_2(g) + H_2O$$

Sulphites $\quad SO_3^{2-} + 2H^+ = SO_2(g) + H_2O$

$$HSO_3^- + H^+ = SO_2(g) + H_2O$$

Bases $\quad OH^- + H^+ = H_2O(l)$

$$O^{2-} + 2H^+ = H_2O(l)$$

In the last two reactions, hydrogen ions of an acid combine with hydroxide or oxide ions of a base to form water, and the reactions are neutralizations.

Neutral, acidic, and alkaline solutions Pure water and other neutral solutions contain equal numbers of hydrogen and hydroxide ions. 1000 cm^3 of a neutral solution contains 10^{-7} mole of hydrogen ions and 10^{-7} mole of hydroxide ions; the pH of the solution is said to be 7. An acidic solution contains a greater concentration of hydrogen ions, e.g. 10^{-6} or 10^{-5} mole in 1000 cm^3, and its pH is less than 7. An alkaline solution contains less hydrogen ions than water and its pH is greater than 7. If the concentration of hydrogen ions is 10^{-9} mol dm^{-3}, the pH of the alkaline solution is 9. Refer to p. 32.

STRONG AND WEAK ACIDS AND ALKALIS

Take four test-tubes and half-fill them with molar solutions of sulphuric acid, ethanoic acid, sodium hydroxide solution and ammonia solution. Now add two or three drops of Universal Indicator to each solution and compare the colours formed.

Sulphuric acid is strongly acidic and ethanoic acid is weakly acidic. Sodium hydroxide is strongly alkaline and ammonia solution is weakly alkaline.

According to the ionic theory, *strong acids and alkalis are completely or almost completely ionized in solution*; they exist only or mainly as ions and not as molecules. *Weak acids and alkalis are only slightly ionized*; they exist mainly as molecules and not as ions. In dilute ethanoic acid only one out of about every three hundred molecules is ionized.

BASICITY OF ACIDS

One molecule of an acid can form one or more hydrogen ions; the number of hydrogen ions it forms is called the *basicity* of the acid.

Hydrochloric acid $\quad HCl = H^+ + Cl^-$ Monobasic

Sulphuric acid $\quad H_2SO_4 = 2H^+ + SO_4^{2-}$ Dibasic

Phosphoric acid $\quad H_3PO_4 = 3H^+ + PO_4^{3-}$ Tribasic

Ethanoic acid $\quad CH_3COOH = H^+ + CH_3COO^-$ Monobasic

Ethanoic acid contains four hydrogen atoms in its molecule, but since only one can form a hydrogen ion, the acid is monobasic. The basicity of an acid indicates the number of sodium salts it can form. For example, hydrochloric acid forms one, NaCl; sulphuric acid forms two, $NaHSO_4$ and Na_2SO_4; and ethanoic acid forms only one, CH_3COONa.

THE HYDROGEN ION

In the above discussion the hydrogen ion, H^+, is regarded as a single proton. It is now known that this does not exist in aqueous solutions which contain the hydrated forms H_3O^+, $H_5O_2^+$, $H_7O_3^+$, and even $H_9O_4^+$. Therefore the ion is often written as $H^+(aq)$ and called the hydrated hydrogen ion, or as H_3O^+, the *oxonium* ion. Whenever you read H^+ in this book assume that it means either $H^+(aq)$ or H_3O^+.

The four chemical equations in the section above can therefore be written:

$$HCl + H_2O = H_3O^+ + Cl^-$$

Monobasic

$$H_2SO_4 + 2H_2O = 2H_3O^+ + SO_4^{2-}$$

Dibasic

$$H_3PO_4 + 3H_2O = 3H_3O^+ + PO_4^{3-}$$
Tribasic

$$CH_3COOH + H_2O = H_3O^+ + CH_3COO^-$$
Monobasic

BRÖNSTED–LOWRY THEORY OF ACIDS

The simple ionic theory given above is restricted to aqueous solutions. The modern theory applies to acids and alkalis in water and other solvents. It states that an acid is a *proton donor* and a base is a *proton acceptor*. An acid or a base can be a molecule, a radical, or an ion.

When an acid loses a proton it becomes a base, and when a base gains a proton it becomes an acid. Thus, in the examples below, acid 1 becomes base 1, and base 2 becomes acid 2. A strong acid forms a weak base, and a weak acid forms a strong base.

$$\text{Acid 1} + \text{Base 2} \rightleftharpoons \text{Acid 2} + \text{Base 1}$$

$$HCl + H_2O \rightleftharpoons H_3O^+ + Cl^-$$

$$H_2SO_4 + 2H_2O \rightleftharpoons 2H_3O^+ + SO_4^{2-}$$

$$NH_4^+ + H_2O \rightleftharpoons H_3O^+ + NH_3$$

$$HCl + CH_3COOH \rightleftharpoons CH_3COOH_2^+ + Cl^-$$

$$H_2O + H_2O \rightleftharpoons H_3O^+ + OH^-$$

The relationship between an acid and its base is:

$$\underset{\text{(proton-donor)}}{\text{Acid}} \rightleftharpoons \underset{\text{(proton)}}{H^+} + \underset{\text{(proton-acceptor)}}{\text{Base}}$$

In water: $\text{Acid} + H_2O \rightleftharpoons H_3O^+ + \text{Base}$

The compounds are called a conjugate acid and a conjugate base; the two compounds are called a *conjugate pair*. The word conjugate means that the acid and base have one common constituent and that they exist together in equilibrium.

The Brönsted–Lowry theory shows that water can act as an acid or as a base, that the ammonium ion is an acid, and that ethanoic acid is a base when hydrogen chloride is dissolved in it. Methylbenzene and benzene do not accept protons at all and therefore hydrogen chloride is not acidic when dissolved in these solvents:

$$HCl + \text{methylbenzene or benzene} = \text{no reaction}$$

SALTS

Salts are formed when either all or part of the ionizable hydrogen of an acid is replaced by another cation. Examples are NaCl from HCl; $NaHSO_4$ and Na_2SO_4 from H_2SO_4; $Ca(HCO_3)_2$ and $CaCO_3$ from H_2CO_3. Salts which contain no ionizable hydrogen are called *normal salts* and salts which contain ionizable hydrogen are called *acid salts*.

Examples of acid salts are:

$$NaHSO_4 \rightarrow Na^+ + H^+ + SO_4^{2-}$$
$$KHCO_3 \rightarrow K^+ + H^+ + CO_3^{2-}$$

These acid salts are not acids because they form other cations in addition to hydrogen ions. Acids form hydrogen ions and no other cations.

SOLUBILITIES OF SALTS

Soluble salts are first prepared in aqueous solution and crystals are obtained by crystallization. Insoluble salts are precipitated. Therefore it is useful to know which salts are soluble in water and which are insoluble. Of the salts we study:

All sodium, potassium, and ammonium salts are soluble.

All nitrates are soluble.

All chlorides are soluble except lead chloride (which is soluble in hot water) and silver chloride.

All sulphates are soluble except those of lead, barium, and calcium. Calcium sulphate is slightly soluble.

All carbonates are insoluble except those of sodium, potassium, and ammonium.

GENERAL METHODS OF PREPARING SALTS

There are six common methods of preparing salts.

1. The action of an acid on a metal. For example, sulphuric acid on zinc, iron or magnesium.
2. The action of an acid on an insoluble oxide or hydroxide. For example, nitric acid on lead oxide or hydroxide.
3. The action of an acid on an insoluble carbonate. For example, hydrochloric acid, sulphuric, or nitric acid on calcium or copper carbonates.
4. The action of an acid on a soluble base (alkali) or on a soluble carbonate. For example, the mineral acids on the caustic alkalis or ammonia solution or sodium carbonate.
5. The precipitation of an insoluble salt, such as lead chloride, silver chloride or barium sulphate.
6. The synthesis of a salt, such as iron(II) sulphide or iron(III) chloride from its elements. See p. 13 in chapter 2 and p. 61 in chapter 11.

To prepare iron(II) sulphate from iron

1. Add about 30 cm^3 of cold 4 M sulphuric acid (which is about twice as concentrated as the dilute acid) to a large beaker whose capacity is about 250 cm^3. Frothing occurs during the preparation and the reaction mixture overflows readily if a small beaker is used. The ordinary dilute acid reacts too slowly with iron.
2. Add iron filings to the acid. Bubbles of hydrogen are evolved. If the reaction is too slow, warm the mixture gently but do not boil. If all the iron reacts, add more. All the acid must be used up and excess metal must remain.

$$Fe + H_2SO_4 = H_2(g) + FeSO_4(aq), \text{ pale green}$$

3. Filter off the metal and collect the pale green filtrate in an evaporating basin or beaker.
4. *Crystallization* Evaporate the filtrate until it is concentrated enough to crystallize on cooling. Test from time to time by removing a few drops in a tube and noting if crystals form as they cool. Prolonged evaporation turns the liquid brown owing to atmospheric oxidation.

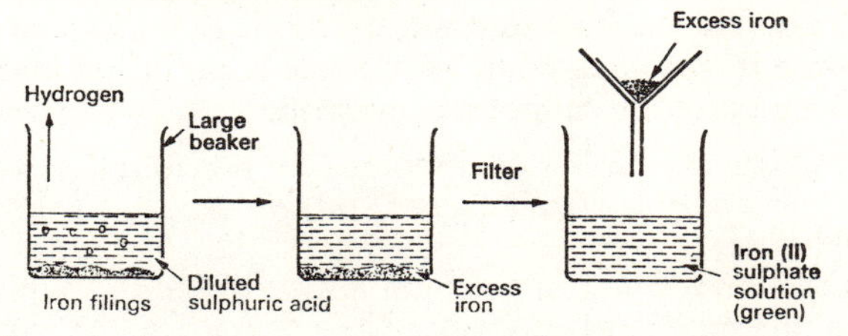

Fig. 19.1 Preparation of iron(II) sulphate, $FeSO_4 . 7H_2O$

5. Cover the vessel containing the concentrated solution and leave to stand at least twenty-four hours. Crystals form.
6. Filter off the green crystals ($FeSO_4 . 7H_2O$, green vitriol) and dry them with filter paper.

Zinc sulphate crystals (white vitriol, $ZnSO_4 . 7H_2O$) are prepared in a similar way. A few drops of copper sulphate solution make the reaction proceed quickly.

To prepare lead(II) nitrate from lead(II) oxide or hydroxide

1. Add about 50 cm^3 of dilute nitric acid to a beaker and warm it gently but do not boil.
2. Add lead oxide or lead hydroxide a little at a time from the end of a spatula. At first the base reacts and forms a clear solution. Add more base until it no longer reacts and all the acid has been used up.
3. Filter off the excess base. Obtain white crystals of anhydrous lead nitrate as described in paras. 4 to 6 above.

$$PbO + 2HNO_3 = H_2O + Pb(NO_3)_2$$

Copper sulphate crystals (blue vitriol, $CuSO_4 . 5H_2O$) are prepared in a similar manner from dilute sulphuric acid and copper oxide or hydroxide. They can also be prepared from sulphuric acid and copper carbonate. Fig. 19.2 shows the stages of its preparation.

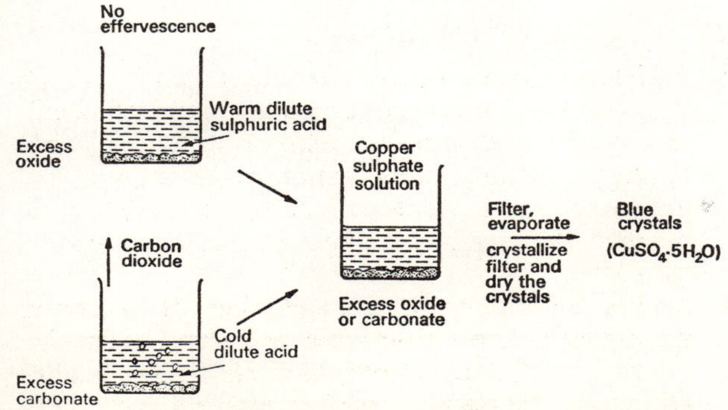

Fig. 19.2 Preparation of copper(II) sulphate from oxide or carbonate

To prepare calcium chloride from calcium carbonate

1. Add about 50 cm^3 of cold dilute hydrochloric acid to a beaker.
2. Add limestone or marble chips, a few at a time, until no more effervescence occurs. The gas evolved is carbon dioxide.
3. Filter off the excess carbonate.
4. Crystals cannot be obtained in the usual manner because the chloride is deliquescent. Therefore, evaporate the colourless solution to dryness; the chloride deposits as a white powder.

$$CaCO_3(s) + 2HCl(aq) = CO_2(g) + H_2O(l) + CaCl_2(aq)$$

Calcium nitrate is prepared similarly by using dilute nitric acid.

To prepare sodium nitrate from sodium hydroxide

1. Clamp a burette in its stand. Place a small funnel in the top and add dilute nitric acid until the acid level is near the zero mark on the burette. Note the reading.
2. Use a measuring jar or pipette to add 50 cm^3 of dilute sodium hydroxide to each of three beakers or flasks. Add two or three drops of indicator to two of the beakers or flasks.

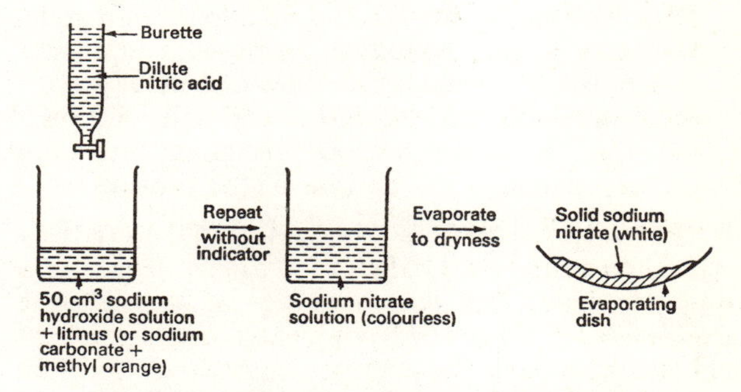

Fig. 19.3 Preparation of sodium nitrate, $NaNO_3$

3. Run the acid, about 1 cm^3 at a time, into the first beaker or flask and shake well until the indicator changes colour. Note the final burette reading. Calculate how much acid has been added; this volume is approximate only.
4. Fill the burette with acid again (if necessary). Note the reading.
5. Run the acid into the second alkaline solution until 1 cm^3 less than the approximate volume has been added. Then add the acid one drop at a time until the indicator just changes colour. Note the burette reading. The volume of acid added is the exact volume required to neutralize the alkaline solution.
6. Now add this volume of acid to the third alkaline solution, which contains no indicator. The product is sodium nitrate solution:

$$NaOH + HNO_3 = H_2O + NaNO_3$$

7. Evaporate the solution to dryness (sodium nitrate is deliquescent and therefore it is not easy to obtain crystals).

Alternative method Use two conical flasks each containing 50 cm^3 of sodium hydroxide solution, and do steps 1–5 as above. The product is sodium nitrate solution with indicator. Boil the mixture with animal charcoal powder, which absorbs the indicator. Filter off the nitrate solution and evaporate to dryness.

Sodium carbonate solution can be used instead of sodium hydroxide in both of the above methods. Methyl orange must be the indicator as litmus or phenolphthalein are unsuitable.

$$Na_2CO_3 + 2HNO_3 = CO_2 + H_2O + 2NaNO_3$$

To prepare lead(II) chloride and sulphate by precipitation

The two salts are insoluble in cold water. They are precipitated when a soluble lead salt reacts with a soluble chloride or sulphate in aqueous solution. Lead ethanoate or nitrate is used:

$$Pb^{2+}(aq) + 2Cl^-(aq) = PbCl_2(s)$$

$$Pb^{2+}(aq) + SO_4^{2-}(aq) = PbSO_4(s)$$

Lead(II) chloride, $PbCl_2$

1. Dissolve lead nitrate or lead ethanoate in distilled water. A milky precipitate forms with tap-water, which usually contains chlorides and other salts.
2. Add excess dilute hydrochloric acid or a solution of any chloride. A dense white precipitate forms and quickly settles to the bottom of the beaker.

$$Pb(NO_3)_2 + 2HCl = PbCl_2(s) + 2HNO_3(aq)$$

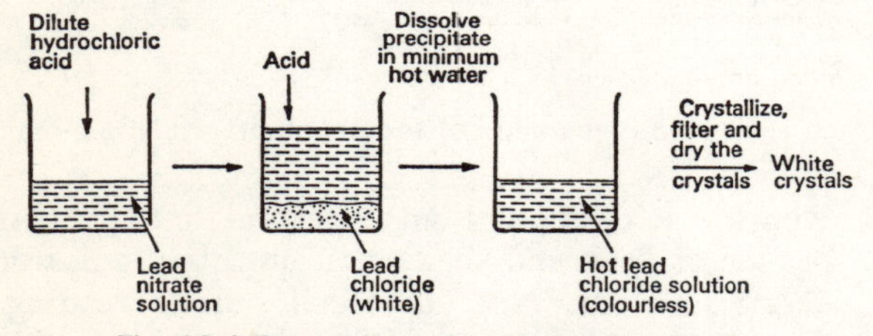

Fig. 19.4 Preparation of lead chloride, $PbCl_2$

3. The clear solution contains excess chloride and a nitrate. Pour it down a sink and retain the precipitate.
4. Add water to the precipitate and boil the mixture. Continue adding small quantities of water and keep the mixture boiling until the lead chloride just dissolves; it is soluble in hot water.
5. Allow the solution to crystallize slowly in the usual way. Filter off the crystals. Wash twice with a small volume of cold water to remove traces of soluble impurities that may be on it. Allow the white needle-shaped crystals to dry.

Lead(II) sulphate, $PbSO_4$

1. Add dilute sulphuric acid or a solution of any sulphate to lead nitrate or lead ethanoate solution. A dense white precipitate forms.

$$Pb(NO_3)_2 + H_2SO_4 = PbSO_4(s) + 2HNO_3(aq)$$

2. Filter the mixture and discard the filtrate. Wash the precipitate twice with cold water to remove excess soluble sulphate and nitrate on it. Dry in an oven or in air. Lead sulphate is insoluble in hot water and crystals cannot be obtained.

Other insoluble salts prepared by precipitation are:
silver chloride from silver nitrate and a soluble chloride;
barium sulphate from barium nitrate or chloride and a soluble sulphate;
calcium sulphate from calcium chloride and a soluble sulphate;
carbonates from a soluble metallic salt (such as copper sulphate) and sodium or potassium carbonate.

Equations for these reactions are given below.

$$Ag^+(aq) + Cl^-(aq) = AgCl(s) \quad \text{white}$$

$$Ba^{2+}(aq) + SO_4^{2-}(aq) = BaSO_4(s) \quad \text{white}$$

$$Ca^{2+}(aq) + SO_4^{2-}(aq) = CaSO_4(s) \quad \text{white}$$

$$Cu^{2+}(aq) + CO_3^{2-}(aq) = CuCO_3(s) \quad \text{green}$$

To prepare lead(II) sulphate from lead or lead(II) carbonate

Both lead and lead carbonate are insoluble and so is lead sulphate. Two reactions are needed. The first reaction produces a soluble lead salt and then lead sulphate is precipitated as in the previous experiment.

1. Dissolve the lead in concentrated nitric acid or the lead carbonate in dilute nitric acid. Hydrochloric acid or sulphuric acid are unsuitable because they react to form insoluble salts which quickly stop further reaction.

$$Pb(s) + HNO_3 = Pb^{2+}(aq) + \text{other products}$$

2. Use the lead nitrate solution in the manner already described to obtain lead sulphate or lead chloride.

To prepare copper(II) chloride from copper(II) sulphate

Both of the salts are soluble. Two reactions are needed. The first reaction produces an insoluble copper compound, the hydroxide or carbonate, and then the soluble copper chloride is prepared in the usual way.

1. Add sodium hydroxide solution or sodium carbonate solution to a solution of the copper sulphate:

$$Cu^{2+}(aq) + 2OH^-(aq) = Cu(OH)_2(s) \quad \text{pale blue}$$

$$Cu^{2+}(aq) + CO_3^{2-}(aq) = CuCO_3(s) \quad \text{green}$$

2. Filter off the precipitate. Add excess of the solid to dilute hydrochloric acid in the manner already described for the preparation of lead nitrate or calcium chloride.

SUMMARY

Ionic theory

Acid: a compound which forms hydrogen ions as the only cations when dissolved in water. $\mathrm{pH} < 7$.
Alkali: a compound which forms hydroxide ions as the only anions when dissolved in water; a soluble base. $\mathrm{pH} > 7$.
Base: contains hydroxide or oxide ions; reacts with hydrogen ions to form a salt and water (neutralization).
Strong acids or bases are completely or almost completely ionized in dilute aqueous solutions. Weak acids or bases are only slightly ionized; their pH is nearer that of neutral water ($\mathrm{pH} = 7$).
Basicity: the number of hydrogen ions formed by one molecule of acid. Hydrogen ions in water $H^+(aq)$ are always hydrated and form oxonium ions (H_3O^+).

Brönsted–Lowry theory

Acid: a proton donor; loses a proton (H^+) to become a base.
Base: a proton acceptor; gains a proton (H^+) to become an acid.
Conjugate pair: two compounds that together act as acid and base.
Weak acids form strong conjugate bases; *strong* acids form weak conjugate bases.

Salts: compounds in which all or part of ionizable hydrogen is replaced by another cation.
Soluble salts: all salts of Na^+, K^+, and NH_4^+; all nitrates; all chlorides except $PbCl_2$ (dissolves in hot water) and $AgCl$; all sulphates except $PbSO_4$, $BaSO_4$, and $CaSO_4$.
Insoluble salts: all carbonates except Na_2CO_3, K_2CO_3, and $(NH_4)_2CO_3$.
Preparation of salts: action of acid on metal, insoluble base, insoluble carbonate, soluble base, or soluble carbonate; precipitation of salt; synthesis of salt. To obtain insoluble salt from another insoluble salt, first prepare a soluble salt.

For example:

$$\underset{\text{insoluble}}{PbCO_3} \rightarrow \underset{\text{soluble}}{Pb(NO_3)_2} \rightarrow \underset{\text{insoluble}}{PbSO_4}$$

QUESTIONS

1. Explain the meaning of the following terms when they are applied to an acid or its aqueous solution: concentration, strength, and basicity.
2. Explain how water, the ammonium ion and the oxonium ion, H_3O^+, may be regarded as acids. Show that water is an acid and a base.
3. The hydrogen ion, H^+, or a proton is the most important acid. Discuss this statement and point out any limitations to its truth.
4. What is a salt? Indicate three general methods of preparing salts, and write suitable equations.
5. Starting from lead(II) oxide and concentrated nitric acid, describe carefully how to prepare crystals of lead nitrate. Starting from lead nitrate, state how you would obtain specimens of lead(II) oxide and lead. (L.)
6. You are given a white powder which contains only sodium carbonate and lead carbonate. Describe fully how you would prepare crystals of sodium sulphate and of one lead compound from this powder.
7. Describe in detail how you would prepare samples of potassium nitrate and lead chloride. How is copper carbonate prepared from copper sulphate? (C.)
8. Starting from concentrated sulphuric acid and solid sodium hydroxide, describe fully how you would obtain crystals of sodium hydrogensulphate. Why is this compound sometimes called an acid salt?

20 Acid-alkali Titrations

The object of volumetric analysis is to determine the mass of an acid or alkali or carbonate present in a solid or solution. Aqueous solutions of the acid and the alkali are prepared. One of these, usually the acid, is placed in a burette and 25 cm^3 of the other is added to each of four conical flasks. The first solution is then added from the burette to the second solution in one flask until neutralization occurs. An indicator shows that this neutral-point or end-point has been reached. The process of adding one solution to the other in this manner is called titration. The whole process, including the titration, is called volumetric analysis.

STANDARD AND 1·00 M SOLUTIONS

A *standard solution* is a solution whose concentration is known. A 1·00 M or M solution contains one mole of a solute in 1 dm^3 of solution. Note that a 1·00 M aqueous solution contains less than 1000 cm^3 of water.

For example, a solution which contains 4·72 g of a solute in 199 cm^3 of solution is a standard solution; any other figures could be used. One mole of hydrochloric acid is 1 + 35·5 = 36·5 g. Therefore molar or M hydrochloric acid contains 36·5 g of acid in 1000 cm^3 of solution. However, 36·5 g of the acid dissolved in 1000 cm^3 of water is not a 1·00 M solution.

One mole of sodium hydroxide is 23 + 16 + 1 = 40 g. Therefore 4 g of the hydroxide dissolved in 100 cm^3 of solution is a 1·00 M solution because the concentration is 40 g per dm^3 (g dm^{-3}).

Decimolar (0·1 M or M/10) solutions are often used in volumetric analysis because 1·00 M solutions are too concentrated. A decimolar solution contains one-tenth of a mole of solute in 1 dm^3 of solution, that is, it is ten times less concentrated than a 1·00 M solution. Decimolar hydrochloric acid contains 3·65 g of acid per dm^3 and decimolar sodium hydroxide contains 4 g of hydroxide per dm^3.

The concentration of a solution expressed in moles per dm^3 is called the *molarity* of the solution.

$$\text{Molarity} = \frac{\text{Concentration in grams per 1000 cm}^3}{\text{One mole of solute}}$$

If a solution of known molarity is diluted:

$$\underset{\text{(original solution)}}{\text{Molarity} \times \text{volume}} = \underset{\text{(diluted solution)}}{\text{molarity} \times \text{volume}}$$

The following is a list showing 1 mole of compounds used in volumetric analysis.

Hydrochloric acid, HCl	36·5 g
Sulphuric acid, H_2SO_4	98 g
Nitric acid, HNO_3	63 g
Ethanoic (acetic) acid, CH_3COOH	60 g
Ethanedioic acid, $(COOH)_2 . 2H_2O$	126 g
Sodium hydroxide, $NaOH$	40 g
Potassium hydroxide, KOH	56 g
Sodium carbonate, Na_2CO_3	106 g
Sodium carbonate, $Na_2CO_3 . 10H_2O$	286 g
Sodium hydrogencarbonate, $NaHCO_3$	84 g
Potassium hydrogencarbonate, $KHCO_3$	100 g

SPECIMEN CALCULATIONS USING MOLARITIES

(*a*) Consider a reaction in which 1 mole of acid reacts with 1 mole of alkali or carbonate; for example:

$$HCl + NaOH = NaCl + H_2O$$

$$\underset{\text{1000 cm}^3\text{ M acid}}{H_2SO_4} + \underset{\text{1000 cm}^3\text{ M solution}}{Na_2CO_3} = Na_2SO_4 + H_2O + CO_2$$

$$\therefore \frac{\text{(Molarity} \times \text{volume) of acid}}{\text{(Molarity} \times \text{volume) of alkali}} = \frac{1 \times 1000}{1 \times 1000} = \frac{1}{1}$$

(*b*) Consider a reaction in which 1 mole of acid reacts with 2 moles of alkali or carbonate; for example,

$$\underset{\text{1000 cm}^3\text{ M acid}}{H_2SO_4} + \underset{\text{2000 cm}^3\text{ M solution}}{2NaOH} = Na_2SO_4 + 2H_2O$$

$$\therefore \frac{\text{(Molarity} \times \text{volume) of acid}}{\text{(Molarity} \times \text{volume) of alkali}} = \frac{1000}{2000} = \frac{1}{2}$$

(*c*) Consider a reaction in which 2 moles of acid react with 1 mole of alkali or carbonate; for example,

$$\underset{\text{2000 cm}^3\text{ M acid}}{2HCl} + \underset{\text{1000 cm}^3\text{ M solution}}{Na_2CO_3} = 2NaCl + CO_2 + H_2O$$

$$\therefore \frac{\text{(Molarity} \times \text{volume) of acid}}{\text{(Molarity} \times \text{volume) of carbonate}} = \frac{2000}{1000} = \frac{2}{1}$$

If the left side of the general equation is:

a (acid)	+	b (alkali or carbonate)
a dm^3 of M acid		b dm^3 of M solution
or $2a$ dm^3 of 0·5 M acid		or $4b$ dm^3 of 0·25 M solution
or $\frac{1}{2}a$ dm^3 of 2 M acid,		or $\frac{1}{4}b$ dm^3 of 4 M solution
and so on		and so on

then

$$\frac{\text{(Molarity} \times \text{volume) of acid}}{\text{(Molarity} \times \text{volume) of alkali or carbonate}} = \frac{a}{b}$$

EXAMPLE 1

What is the molarity of a solution which contains 21·2 g of anhydrous sodium carbonate in 400 cm^3 of solution?

One mole of the carbonate, Na_2CO_3, is

$$(2 \times 23) + 12 + (3 \times 16)\ \text{g} = 106\ \text{g}$$

106 g in 1000 cm^3 of solution is a 1·00 M solution.

$$\therefore \quad 21{\cdot}2\ \text{g in 1000 cm}^3\text{ of solution is } \frac{21{\cdot}2}{106}\ \text{M}$$

$$\therefore \quad 21{\cdot}2\ \text{g in 400 cm}^3\text{ of solution is } \frac{21{\cdot}2 \times 1000}{106 \times 400}\ \text{M} = 0{\cdot}5\ \text{M}$$

The molarity of the solution is 0·5.

EXAMPLE 2

What mass of alkali is present in 250 cm^3 of 2 M sodium hydroxide solution?

$$1 \text{ mole of NaOH} = 23 + 16 + 1 \text{ g} = 40 \text{ g}$$

$\therefore$ 1000 cm^3 of M solution contain 40 g of alkali

i.e. 250 cm^3 of M solution contain 10 g of alkali

i.e. 250 cm^3 of 2 M solution contain 20 g of alkali

The mass of alkali present is 20 g.

EXAMPLE 3

What is the maximum volume of 0·1 M or M/10 sulphuric acid that can be made from 250 cm^3 of 0·5 M or M/2 acid?

Using the relationship:

$$\underset{\text{of diluted solution}}{(\text{Molarity} \times \text{volume})} = \underset{\text{of original solution}}{(\text{Molarity} \times \text{volume})}$$

$$0{\cdot}1 \times \text{volume} = 0{\cdot}5 \times 250 \text{ cm}^3$$

$$\text{i.e. volume of diluted solution} = \frac{0{\cdot}5 \times 250}{0{\cdot}1} \text{ cm}^3$$

$$= 1250 \text{ cm}^3.$$

Note It is wrong to assert that 1000 cm^3 of distilled water added to the original solution will produce a 0·1 M solution because 250 cm^3 of 0·5 M solution plus 1000 cm^3 of water may not form 1250 cm^3 of solution —contraction or expansion in volume may result from the mixing.

EXAMPLE 4

What volume of 0·1 M sodium hydroxide would exactly neutralize 25 cm^3 of 0·08 M hydrochloric acid? The equation is:

$$\underset{1 \text{ mole}}{NaOH} + \underset{1 \text{ mole}}{HCl} = NaCl + H_2O$$

$$\therefore \frac{(\text{Molarity} \times \text{volume}) \text{ of alkali}}{(\text{Molarity} \times \text{volume}) \text{ of acid}} = \frac{1}{1}$$

$$\text{i.e. } 0{\cdot}1 \times \text{volume of alkali} = 0{\cdot}08 \times 25 \text{ cm}^3$$

$$\text{i.e. Volume of alkali} = \frac{0{\cdot}08 \times 25}{0{\cdot}1} = 20 \text{ cm}^3$$

EXAMPLE 5

25 cm^3 of 0·04 M sodium carbonate solution exactly neutralize 24 cm^3 of hydrochloric acid. What is the molarity of the acid?

The equation is:

$$\underset{1 \text{ mole}}{Na_2CO_3} + \underset{2 \text{ moles}}{2HCl} = 2NaCl + CO_2 + H_2O$$

$$\therefore \frac{(\text{Molarity} \times \text{volume}) \text{ of acid}}{(\text{Molarity} \times \text{volume}) \text{ of carbonate}} = \frac{2}{1}$$

$$\text{i.e. Molarity of acid} \times 24 \text{ cm}^3 = 2 \times 0{\cdot}04 \times 25 \text{ cm}^3$$

$$\text{i.e. Molarity of acid} = \frac{2 \times 0{\cdot}04 \times 25}{24}$$

$$= 0{\cdot}083 \text{ M}$$

CALCULATIONS USING EQUIVALENTS OF ACIDS AND BASES AND NORMALITIES

Calculations arising from acid–alkali titrations have so far been done using the mole and molarities. They can be done by an alternative method which uses the equivalents of acids and bases.

The *equivalent weight of an acid* is its mass in grams which can supply, in aqueous solution, 1·008 g of hydrogen ions. The *equivalent weight of an alkali or carbonate* is the mass in grams which reacts with 1·008 g of hydrogen ions (i.e. with the equivalent weight of an acid).

Equivalent weights of the common acids are:

Hydrochloric acid, HCl ($=H^+$)	36·5 g
Nitric acid, HNO_3 ($=H^+$)	63·0 g
Sulphuric acid, H_2SO_4 ($=2H^+$)	49·0 g (i.e. 98/2)
Ethanoic (acetic) acid, CH_3COOH ($=H^+$)	60·0 g
Oxalic acid, $(COOH)_2.2H_2O$ ($=2H^+$)	63·0 g (i.e. 126/2)

Note that the equivalents of sulphuric acid and oxalic acid are half of one mole, because one molecule of these acids produces two hydrogen ions.

The reactions of some alkalis and carbonates with hydrogen ions may be written:

$$\underset{40 \text{ g}}{NaOH} + H^+ = H_2O + Na^+$$

$$\underset{106 \text{ g}}{Na_2CO_3} + 2H^+ = H_2O + 2Na^+ + CO_2$$

The equivalent weight of sodium hydroxide is 40 g, and that of sodium carbonate is 106/2 g = 53 g (because 106 g of the carbonate react with 2·016 g of hydrogen ions).

Other equivalent weights of alkalis and carbonates are:

Potassium hydroxide, KOH	56 g
Ammonium hydroxide, NH_4OH	35 g
Potassium carbonate, K_2CO_3	69 g
Hydrated sodium carbonate, $Na_2CO_3 . 10H_2O$	143 g
Sodium hydrogencarbonate, $NaHCO_3$	84 g
Potassium hydrogencarbonate, $KHCO_3$	100 g

STANDARD AND NORMAL SOLUTIONS

A *standard solution* is a solution whose concentration is known. A *normal* or *N solution* contains the equivalent weight of solute in 1000 cm^3 of solution (not of water).

The equivalent weight of sodium hydroxide is 40 g. Therefore solutions of the hydroxide which contain 40 g in 1000 cm^3 of solution are normal or N solutions.

Decinormal (0·1 N or N/10) solutions are often used in volumetric analysis because normal solutions are usually too concentrated. They contain one-tenth of the equivalent weight of solute in 1000 cm^3 of solution, i.e. they are ten times more dilute or ten times less concentrated than normal solutions.

The concentration of a solution expressed in gram-equivalents per dm^3 is called its *normality*.

$$\text{Normality} = \frac{\text{Concentration in grams per 1000 cm}^3}{\text{Equivalent weight of solute}}$$

If a solution of known normality is diluted:

$$\underset{\text{(original solution)}}{\text{Normality} \times \text{Volume}} = \underset{\text{(diluted solution)}}{\text{Normality} \times \text{Volume}}$$

If a solution of an acid just neutralizes a solution of an alkali or carbonate:

$$\underset{\text{(of acid)}}{\text{Normality} \times \text{Volume}} = \underset{\text{(of alkali or carbonate)}}{\text{Normality} \times \text{Volume}}$$

This equation can be used in almost all calculations connected with acid-alkali titrations.

Preparing standard solutions

For various reasons, the common acids and alkalis cannot be used to prepare standard acidic or alkaline solutions. It is impossible to weigh a known mass of any one of them because:

hydrochloric acid is too volatile;
sulphuric acid is hygroscopic;
nitric acid is volatile and cannot be obtained pure because it readily decomposes. (All three concentrated acids contain an unknown percentage of water. Even if the concentrated acid is weighed exactly the actual mass of acid present is still unknown.);
glacial acetic (ethanoic) acid is hygroscopic;
hydroxides of sodium and potassium absorb both water and carbon dioxide whilst they are being weighed;
ammonia solution loses ammonia during the weighing.

It is easy to obtain pure, dry, anhydrous sodium carbonate and ethanedioic acid and they are used to prepare standard alkaline and acidic solutions. The molarities of other acids and alkalis are then determined by titration.

Methods of weighing a substance

1. *Using a watch-glass* Weigh exactly a clean, dry watch-glass. The approximate mass should be marked on it and then it is easy in future to determine the exact mass in less than a minute.
2. Add the solute to the watch-glass and determine the new mass. The increase is the mass of the solute.
3. *Using a weighing bottle* If necessary, heat the bottle in a steam oven and allow to cool in a desiccator to ensure that the bottle is perfectly dry.
4. Weigh the bottle exactly; this can be done quickly if its approximate mass has been marked on it. Handle the bottle with a clean, dry cloth to prevent grease from the fingers affecting its mass.
5. Add solute to the bottle and determine the new mass. The increase is the mass of the solute. All of this solute must be used to make the standard solution.

Alternatively, weigh the bottle and solute, transfer some of the solute to a beaker and use it to make the standard solution, and then weigh the bottle and residual solute. The decrease in mass is the mass of solute used.

USE OF A BURETTE

A burette can deliver up to 50 cm^3 of liquid and is graduated to 0·1 cm^3. Its glass tap is smeared with a thin coat of vaseline so that it turns smoothly; too much grease may block the hole in the tap.

Place a small funnel in the top of the burette. Fill the burette with the solution it is to contain and then pour away this solution. Repeat if necessary. This procedure ensures that the inside of the burette is wet only with the solution. Then fill the burette above the zero mark with fresh solution and open the tap so that the level falls to or below the zero mark. Ensure that the jet is filled with solution and contains no air bubbles.

Hold white paper behind the meniscus and place your eyes level with it; read the bottom level of the meniscus. Estimate the reading to 0·05 cm^3.

USE OF A PIPETTE

The pipette delivers a known volume, usually 25 cm^3, of a liquid.

Place the tip of the pipette well below the surface of the liquid being used. Suck up the liquid until its level is above the graduation mark. If any liquid enters your mouth, spit it out at once and wash out your mouth with water. Let the liquid run away into a sink. Fill the pipette once again above the mark. Use your finger—never your thumb—to hold it there. Place your eyes level with the mark. Let the liquid run out slowly and steadily into a sink until the bottom of the meniscus is exactly at the graduation mark. Place the tip of the pipette inside a conical flask, which must be dry or contain only distilled water. Let the liquid run into the flask. Allow the pipette to drain for fifteen seconds and then touch the liquid with the tip of the pipette. A little liquid remains in the tip. Do not blow it into the flask. The pipette is made to deliver 25 cm^3 when used exactly as described.

INDICATORS

	Alkaline	*Acidic*	*NOT suitable for*
Litmus	Blue	Red	Carbonates
Universal Indicator	Various colours		
Phenolphthalein	Pink	No colour	Weak alkalis, e.g. ammonia solution
Methyl orange	Yellow	Red	Weak acids, e.g. ethanoic and ethanedioic acids
Screened methyl orange	Green	Violet	

It is easier to observe the colour change with screened methyl orange than with ordinary methyl orange;

these two indicators are used with sodium carbonate solution which is weakly alkaline. Phenolphthalein is used with ethanoic acid and other weak acids.

To prepare standard sodium carbonate (***exactly*** 0·05 M ***or*** M/20)

One mole of sodium carbonate, Na_2CO_3, is $(2 \times 23) + 12 + (3 \times 16)$ g = 106 g.

1000 cm³ of M sodium carbonate contains 106 g

$$\therefore \; 250 \text{ cm}^3 \text{ of M/20 carbonate contains } \frac{106}{4 \times 20} \text{ g}$$

$$= 1{\cdot}325 \text{ g}$$

The method given below describes how a pupil can prepare his own standard solution. However, the method is not easy and it is usual for a teacher to prepare for a whole class a solution containing 5·3 g per dm³.

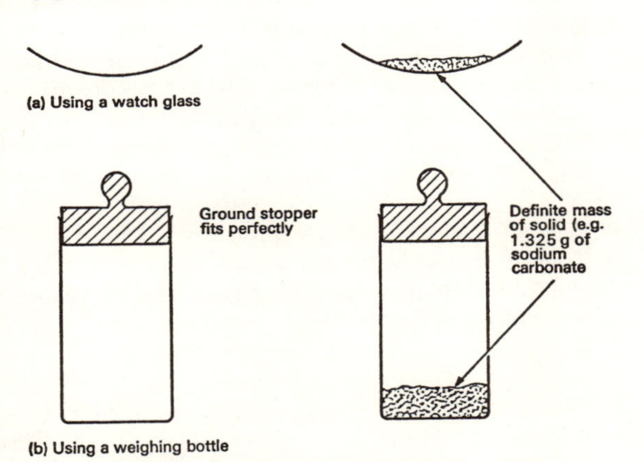

Fig. 20.1 Weighing a definite mass of a solid

Method

1. Dry some pure, anhydrous sodium carbonate by heating it in an evaporating dish for ten to fifteen minutes. Allow the dry carbonate to cool in a desiccator because it absorbs moisture from the atmosphere.
2. Weigh a watch-glass or a weighing bottle on a chemical balance. Leave the weights on the right-hand pan of the balance.
3. Add additional weights of 1·325 g to the scale-pan. Use a spatula to add the dry carbonate to the glass or bottle until it exactly balances the additional weights.
4. Transfer the carbonate to a beaker containing about 50 cm³ of hot distilled water—the carbonate dissolves very slowly in cold water. Wash every trace of carbonate into the beaker by using cold, distilled water from a wash bottle.
5. Place a small funnel in the neck of a 250 cm³ graduated flask that has been rinsed twice with distilled water. Pour the cool carbonate solution into the funnel.
6. Add distilled water to the beaker several times, rinse well, and then add the water to the graduated flask. Every trace of carbonate solution must be added to the flask.
7. Add more water to the flask until the meniscus is at the 250 cm³ graduation mark; your eye must be level with the mark when you observe this.
8. Stopper the flask and shake vigorously for two minutes or so to ensure that the whole solution is uniform. The meniscus is usually below the mark after shaking because some solution is on the stopper and on the neck above the mark. However, do not add any more water because the total volume of solution is still the required 250 cm³.

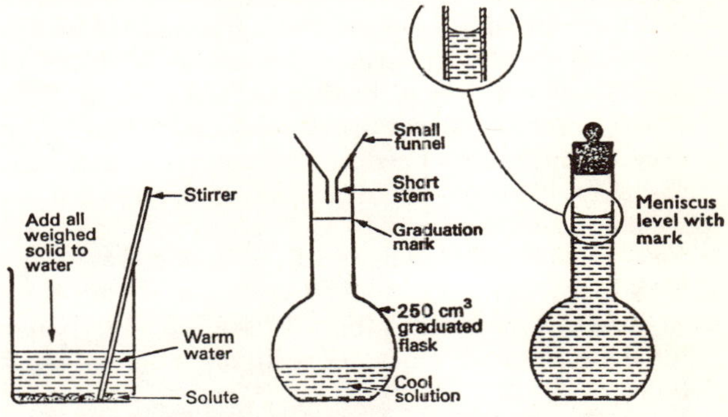

Fig. 20.2 Preparing 250 cm³ of standard solution

The solution prepared as above is a standard solution and is exactly 0·05 M. The method below describes how to prepare a standard solution of known molarity but not exactly 0·05 M.

To prepare standard sodium carbonate (***about*** 0·05 M ***or*** M/20)

1. Dry some pure anhydrous sodium carbonate by heating it in an evaporating dish for ten to fifteen minutes. Allow to cool in a desiccator.
2. Add about 1·35 g (i.e. slightly more than 1·325 g) of the carbonate to a watch-glass or a weighing bottle. Weigh the container and carbonate exactly.
3. Transfer most of the carbonate to a beaker containing hot distilled water. Stir gently until it dissolves.
4. Weigh the watch-glass or weighing bottle and the small amount of residual carbonate. The difference between the two weighings is the mass of the carbonate used.

5–8. Proceed as in paragraphs 5–8 above.

SPECIMEN WEIGHINGS AND CALCULATION OF MOLARITY

Mass of glass or bottle + carbonate =

Mass of glass or bottle + residual carbonate =

Mass of carbonate in the solution = 1·20 g

250 cm³ of solution contain 1·20 g of carbonate,

∴ 1000 cm³ of solution contain 4·80 g of carbonate.

$$\text{Molarity} = \frac{\text{Concentration in grams per 1000 cm}^3}{\text{One mole of carbonate (106 g)}}$$

$$= \frac{4{\cdot}80}{106} = 0{\cdot}045 \text{ M}$$

The carbonate solution is 0·045 M.

***To prepare hydrochloric acid (approximately* 0·1 M *or* M/10)**

One mole of hydrochloric acid, HCl, is

$$1 + 35{\cdot}5 \text{ g} = 36{\cdot}5 \text{ g}$$

Concentrated hydrochloric acid is 10 to 12 M. If it is diluted about one hundred times, the diluted acid is about 0·1 M or M/10.

Measure about 10 cm³ of the concentrated acid in a measuring cylinder. Add it to about 1000 cm³ of distilled water in a large beaker or flask. Stir or shake for two to three minutes to obtain a uniform solution.

The volumes used need not be known or measured accurately because the exact concentration of the pure acid is not known.

Sulphuric acid The density of the concentrated acid is 1·84 g per cm³ and the acid contains about 98 per cent acid by mass. Its molarity is about 18. If it is diluted about 360 times, the diluted acid is about 0·05 M or M/20.

Nitric acid The density of the concentrated acid is about 1·4 g per cm³ and the acid contains about 68 per cent acid by mass. Its molarity is about 22. If it is diluted 220 times the diluted acid is about 0·1 M or M/10.

To standardize hydrochloric acid, i.e. to find its exact molarity or concentration

The acid, in a burette, is added to exactly 25 cm³ of standard sodium carbonate in a conical flask. The exact volume of acid which reacts with the carbonate is determined by making four titrations.

1. Fill the burette with the diluted hydrochloric acid, about 0·1 M. Pour this acid down a sink. The inside of the burette is now wet only with acid.
2. Clamp the burette vertically in its stand. Place a small funnel in the top. Pour acid from a beaker into the funnel until the burette is filled above the zero mark. Open the tap and let acid run out into a second beaker until the acid level is on or a little below the zero mark. This ensures that the part below the tap is filled with acid and contains no air bubbles.
3. Read the burette accurately, move your eyes level with the meniscus and hold white paper behind the meniscus to make it clear.
4. *Measuring the carbonate solution with a pipette* Wash out four conical flasks with distilled water; they need not be dried because this water does not interfere with the titrations to be done.
5. Suck the standard sodium carbonate solution into a pipette until it is filled well above the graduation mark. Run out this solution into a sink.
6. Again fill the pipette above the mark with the carbonate solution, and then let the meniscus fall smoothly to the mark. Run out the 25 cm³ of solution from the pipette into one conical flask.
7. Add 25 cm³ of carbonate solution to each of the other conical flasks. Refer to the use of a pipette described earlier in this chapter on p. 108.
8. *Titration* Add two to three drops of methyl orange or screened methyl orange to each conical flask.
9. Add the acid, 1 cm³ at a time, to the first flask. The flask should be on a white tile or white paper so that a colour change can be seen easily. Add acid until the indicator changes colour. Read the burette. Calculate the volume of acid used; this value is approximate.
10. Repeat the titration with the carbonate in a second flask. Add acid of volume about 1 cm³ less than that used in the first titration, and then add more acid one drop at a time until the indicator just changes colour. Read the burette.
11. Repeat the accurate titration with the carbonate in the third and fourth flasks.

SPECIMEN CALCULATION

Hydrochloric acid is in the burette. 25 cm³ of 0·045 M carbonate are placed in each flask. The indicator is methyl orange.

	Trial	*1*	*2*	*3*
Burette reading (finish)	22·0	43·25	23·5	44·7
Burette reading (start)	0·5	22·0	2·3	23·5
Acid used =	>21	21·25	21·2	21·2
Average =		21·2 cm³		

i.e. 25 cm³ 0·045 M carbonate ≡ 21·2 cm³ acid.

The equation is:

$$\underset{\text{2 moles}}{2HCl} + \underset{\text{1 mole}}{Na_2CO_3} = 2NaCl + CO_2 + H_2O$$

$$\therefore \quad \frac{\text{(Molarity × volume) of acid}}{\text{(Molarity × volume) of carbonate}} = \frac{2}{1}$$

i.e. molarity of acid × 21·2 = 2 × 0·045 × 25

$$\text{Molarity of acid} = \frac{2 \times 0{\cdot}045 \times 25}{21{\cdot}2} = 0{\cdot}106$$

Concentration in grams per 1000 cm³ (cubic decimetre)

M acid contains 36·5 g of acid in 1000 cm³

∴ 0·106 M acid contains 0·106 × 36·5 g of acid in 1000 cm³

= 3·87 g per 1000 cm³

The concentration of the hydrochloric acid is 3·87 g per 1000 cm³ (or per dm³).

To prepare exactly* 0·1 M *hydrochloric acid

The standard acid is 0·106 M. Using the formula:

$$(\text{Molarity} \times \text{volume})_{\text{of diluted acid}} = (\text{Molarity} \times \text{volume})_{\text{of original acid}}$$

$$0{\cdot}1 \times \text{volume of diluted acid} = 0{\cdot}106 \times 500 \text{ cm}^3$$

i.e. $$\text{Volume of dilute acid} = \frac{0{\cdot}106 \times 500}{0{\cdot}1} \text{ cm}^3$$

$$= 530 \text{ cm}^3$$

Therefore add distilled water to 500 cm^3 of the 0·106 M acid until the volume of solution is 530 cm^3; this may not be the same as adding 30 cm^3 of water, because dilution may cause a slight contraction or expansion.

A similar calculation shows that 471 cm^3 of the original 0·106 M solution forms exactly 500 cm^3 of 0·1 M acid.

***To prepare sodium hydroxide solution* (*about* 0·1 M)**

One mole of sodium hydroxide, NaOH, is 23 + 16 + 1 g = 40 g.

1000 cm^3 of 0·1 M (M/10) hydroxide contains 4 g

∴ 250 cm^3 of 0·1 M (M/10) hydroxide contains 1 g

The solid hydroxide absorbs both water and carbon dioxide from the air. Therefore it is not possible to weigh 1 g exactly.

1. Weigh about 1 g of solid hydroxide on a watch-glass or in a weighing bottle.
2. Use a wash bottle to wash the hydroxide into a beaker of cold distilled water. Stir well until the solute dissolves. Make the volume of solution up to about 250 cm^3.

Neither the volume nor the weighing need be exact because the molarity of the solution cannot be known exactly.

To standardize sodium hydroxide solution

1. Take four clean conical flasks, which need not be dry. Use a pipette to add 25 cm^3 of the sodium hydroxide solution to each one. Add two to three drops of indicator (litmus or methyl orange) to each solution. The alkali may contain some sodium carbonate, formed by carbon dioxide in the air, and therefore phenolphthalein is not a suitable indicator.
2. Fill a burette with standard hydrochloric acid in the usual way.
3. Titrate the alkali against the acid. The first reading is a trial, and the other three readings are accurate. Record the results in a table as before.

SPECIMEN CALCULATION

Suppose that:

$$25 \text{ cm}^3 \text{ alkali} \equiv 27{\cdot}5 \text{ cm}^3\ 0{\cdot}106 \text{ M acid}$$

$$(\text{Molarity} \times \text{volume})_{\text{(of alkali)}} = (\text{molarity} \times \text{volume})_{\text{(of acid)}}$$

$$\therefore \text{Molarity of alkali} \times 25 \text{ cm}^3 = 0{\cdot}106 \times 27{\cdot}5 \text{ cm}^3$$

i.e. $$\text{Molarity of alkali} = \frac{0{\cdot}106 \times 27{\cdot}5}{25}$$

$$= 0{\cdot}117$$

Concentration in grams per 1000 cm^3 (dm^3)

M sodium hydroxide contains 40 g per dm^3

∴ 0·117 M hydroxide contains 0·117 × 40 g per dm^3

$$= 4{\cdot}68 \text{ g per dm}^3$$

The sodium hydroxide solution is 0·117 M and contains 4·68 g of sodium hydroxide in 1000 cm^3 of solution.

Calculate the volume of exactly 0·1 M sodium hydroxide that can be prepared from 100 cm^3 of the 0·117 M hydroxide (it is 117 cm^3).

Also calculate the volume of 0·117 M hydroxide required to prepare 100 cm^3 of 0·1 M solution (it is 100/1·17 cm^3, i.e. 85 cm^3).

***To prepare ethanoic* (*acetic*) *acid solution* (*about* 0·1 M)**

The density of pure ethanoic acid is 1·055 g per cm^3 at 15 °C. One mole of the acid, CH_3COOH, is 12 + 3 + 12 + 32 + 1 g = 60 g, which is about 57 cm^3. 1000 cm^3 of 0·1 M acid contains 6 g or about 5·7 cm^3 of the pure acid.

Add about 5·7 cm^3 of the acid to about 1000 cm^3 of distilled water in a beaker. Stir well to make a uniform solution which is about 0·1 M.

To standardize ethanoic* (*acetic*) *acid solution

1. Fill a burette with the acid, in the usual way.
2. Add 25 cm^3 of standard sodium hydroxide solution to each of four conical flasks. Add two to three drops of phenolphthalein.
3. Titrate the acid and alkali.

Calculation The equation is:

$$\underset{1 \text{ mole}}{CH_3COOH} + \underset{1 \text{ mole}}{NaOH} = CH_3COONa + H_2O$$

$$\therefore \frac{\text{Molarity} \times \text{volume (of acid)}}{\text{Molarity} \times \text{volume (of alkali)}} = \frac{1}{1}$$

The two volumes are determined by the titration and the molarity of the alkali is known. Therefore the molarity of the acid can be calculated.

***To determine x in the formula* $Na_2CO_3 . xH_2O$**

Hydrated sodium carbonate, washing soda crystals, has the above formula, and x is a whole number.

	Na_2CO_3	xH_2O
1 mole	$= (2 \times 23) + 12 + (3 \times 16)$	$+ x(2 + 16)$
Composition by mass	$= 106$	$18x$

i.e. $$\frac{\text{Mass of water}}{\text{Mass of anhydrous carbonate}} = \frac{18x}{106}$$

A standard solution of the hydrated carbonate is prepared. The mass of the anhydrous carbonate in the solution is determined by titration. The difference between the two masses is the mass of water. x can then be calculated from the above equation.

1. Weigh accurately about 3·5 g of pure hydrated carbonate.
2. Dissolve the carbonate in distilled water and make the volume of solution up to 250 cm³; the solution is about 0·05 M.
3. Fill a burette with standard acid—about 0·1 M hydrochloric or nitric acid or 0·05 M sulphuric acid is suitable.
4. Titrate 25 cm³ of the alkaline solution against the standard acid, using methyl orange as indicator.

Calculate the mass of anhydrous carbonate and the mass of water in the crystals. Substitute the masses in the equation given above and calculate x. The nearest whole number is the true value because x must be a whole number.

SUMMARY

Volumetric analysis

1·00 M *solution:* contains 1 mole of solute dissolved in 1000 cm³ of solution.

Standard solution: solution of known concentration.

$$\textit{Molarity} = \frac{\text{Concentration of solute (g per 1000 cm}^3\text{)}}{\text{1 mole of solute}}$$

Dilution: $\underset{\text{original solution}}{\text{Molarity} \times \text{Volume}} = \underset{\text{diluted solution}}{\text{Molarity} \times \text{Volume}}$

Titration: If x moles of acid react with y moles of alkali:

$$\frac{\text{(Molarity} \times \text{Volume) of acid}}{\text{(Molarity} \times \text{Volume) of alkali}} = \frac{x}{y}$$

Standard solutions

PREPARATION BY WEIGHING

Alkaline solution: 1·325 g pure dry anhydrous sodium carbonate in warm water; make up volume to 250 cm³ to make 0·05 M Na_2CO_3.

Acid solution: from pure ethanedioic acid, $(COOH)_2.2H_2O$.

PREPARATION BY DILUTION: make up all other acid and alkali solutions to approximate concentration; then titrate to find exact concentration.

0·1 M HCl: dilute concentrated acid (about 10 M) 100 times.

0·05 M H_2SO_4: dilute concentrated acid (about 18 M) 360 times.

0·1 M HNO_3: dilute concentrated acid (about 22 M) 220 times.

0·1 M NaOH: dissolve about 1 g solid NaOH in 250 cm³ of solution.

0·1 M KOH: dissolve about 1·4 g solid KOH in 250 cm³ of solution.

INDICATORS

Weak acids: use phenolphthalein.

Weak alkalis: use methyl orange or screened methyl orange.

Equivalent weights

Of an acid: weight in grams which supplies 1·008 g of hydrogen ions in aqueous solution. HCl, 36·5 g; H_2SO_4, 98/2 = 49 g.

Of an alkali: weight in grams which reacts with 1·008 g of hydrogen ions in aqueous solution. NaOH, 40 g; KOH, 56 g; Na_2CO_3, 106/2 = 53 g.

Normal solution contains equivalent weight of solute in 1000 cm³ of solution.

$$\textit{Normality} = \frac{\text{Concentration of solute (grams per 1000 cm}^3\text{)}}{\text{Equivalent weight of solute}}$$

Titration: $\underset{\text{acid}}{\text{Normality} \times \text{Volume}} = \underset{\text{alkali}}{\text{Normality} \times \text{Volume}}$

QUESTIONS

1. What is the mass of acid present in: (*a*) 500 cm³ of 0·1 M sulphuric acid, (*b*) 4000 cm³ of 0·5 M hydrochloric acid? (H = 1, S = 32, O = 16, Cl = 35·5.)
2. What is the molarity of each of the following solutions: (*a*) sulphuric acid containing 4·9 g of acid in 500 cm³; (*b*) nitric acid containing 6·3 g of acid in 250 cm³; (*c*) sodium hydroxide containing 20 g of the alkali in 2000 cm³; (*d*) sodium carbonate containing 53 g of the anhydrous carbonate in 4 dm³? (N = 14, Na = 23, C = 12.)
3. What volume of (*a*) M acid can be made from 1 dm³ of 1·1 M acid; (*b*) 0·1 M acid from 100 cm³ of 0·12 M acid; (*c*) decimolar alkali from 250 cm³ of M/5 alkali?
4. How many cm³ of M HCl are required to neutralize: (*a*) 25 cm³ of 0·8 M NaOH; (*b*) 20 cm³ of 1·25 M KOH; (*c*) 50 cm³ of M/10 Na_2CO_3; (*d*) 50 cm³ of 0·1 M $Na_2CO_3 . 10H_2O$?
5. How many cm³ of 0·1 M NaOH are required to neutralize: (*a*) 25 cm³ of 0·2 M HCl; (*b*) 25 cm³ of 0·2 M H_2SO_4; (*c*) 500 cm³ of decimolar nitric acid; (*d*) 1 dm³ of M/10 acetic acid?
6. 125 cm³ of sulphuric acid containing 9·8 g of acid in 1000 cm³ exactly neutralized 100 cm³ of sodium carbonate solution. Calculate the concentration of the carbonate solution in grams of anhydrous carbonate per 1000 cm³. (Na = 23, C = 12.)
7. 25 cm³ of 0·1 M sulphuric acid exactly neutralized 20 cm³ of sodium hydroxide solution. What is the concentration of the alkali? What volume of the alkali would you add to 100 cm³ of this acid in order to prepare sodium hydrogen sulphate?
8. When 20·0 g of a mixture of sodium hydrogen carbonate and sodium chloride were dissolved in water and made up to 1000 cm³, 25 cm³ of this solution were neutralized by 20·0 cm³ of 0·125 M sulphuric acid. Calculate the percentage of sodium chloride in the original mixture.
9. 50 cm³ of a solution of an acid containing 30·0 g in 1000 cm³ were found to need for complete neutralization 32·6 cm³ of molar sodium hydroxide. The

molecular mass of the acid is 92·0; calculate its basicity. (L.)

10. 10·00 g of solid sodium hydroxide was added to ammonium chloride solution. The solution was boiled until no more ammonia was evolved. The remaining solution then required 6·125 g of sulphuric acid for complete neutralization. (*a*) What test would you use to determine when ammonia is no longer evolved? (*b*) Calculate the mass of ammonium chloride originally present. (*c*) What substances are present in the residue if the final solution is evaporated to dryness? (Na = 23, O = 16, H = 1, N = 14, Cl = 35·5.)

11. In the previous question, what is the mass of ammonia evolved and its volume measured at s.t.p.?

12. A crystal of calcium carbonate weighing 4·37 g is placed in 100 cm^3 of hydrochloric acid. When no more reaction takes place, the crystal is washed and dried. It now weighs 1·87 g. Calculate the molarity of the acid and the volume at s.t.p. of the carbon dioxide liberated. (Ca = 40, C = 12, O = 16.) (C.)

21 Carbon and its Compounds. Silicates

VARIOUS FORMS OF CARBON

Diamond and graphite are two crystalline forms of carbon. They were proved to be pure carbon about two centuries ago by combustion to give carbon dioxide. Coke, charcoal, lampblack or soot, and sugar carbon are usually called *amorphous* or non-crystalline carbon. However, it is probable that they are merely forms of graphite.

The existence of an element or compound in more than one solid form is called *polymorphism*, and the forms are *polymorphs*. The existence of an element in more than one form in the same state of matter is called *allotropy*, and the forms are *allotropes*. Carbon is both polymorphic and allotropic, and the cause is the different arrangement of the atoms in diamond and graphite; refer to p. 78 in chapter 14. The element phosphorus exists as red and white solid allotropes or polymorphs.

Properties and uses of diamond

Diamond is the densest form of carbon because its atoms are closer than in the other forms. It is the hardest natural substance known. It is a colourless transparent solid and specially cut diamonds sparkle because of their high refractive index (that is, their great bending effect on light). It is a non-conductor of electricity because its electrons are not mobile.

It burns in oxygen or air when heated above 800 °C, but otherwise it is chemically inert and does not react with chlorine, acids, or alkalis.

Diamond is used for cutting and grinding hard substances such as glass, pottery, and rock, and as a die for drawing metal into fine wire used as lamp filaments, watch springs and so on.

Properties and uses of graphite

Graphite consists of flat layers of carbon atoms, which are arranged in hexagonal rings. The distance between the layers is more than double that between atoms in a layer or in diamond, and therefore graphite is less dense than diamond. It is a good conductor because the valency electrons of the atoms are mobile—they are not attached to any particular atom and can move readily.

Graphite is an opaque, shiny-black solid. It is soft and feels greasy. It starts to burn in air at about 700 °C —more readily than diamond burns. It has been changed into tiny diamonds at high temperatures and at a pressure greater than 50 000 atmospheres.

It is used for the brushes of dynamos and motors, for electrodes during commercial electrolysis because it is not attacked by chlorine and only slowly by oxygen, and for pencils in which the 'lead' is a mixture of graphite and clay. Graphite rods are used in atomic piles because they slow down neutrons emitted by uranium, so preventing the reaction becoming violent or explosive. Heavy machinery is lubricated by a suspension of graphite in oil; the layers of carbon atoms readily slide over each other.

Properties and uses of amorphous carbon

Coke is the residue when coal is heated in the absence of air; wood charcoal is the residue when wood is heated similarly. Animal charcoal is left when bones are heated without air; it contains 10 per cent of charcoal mixed with calcium and magnesium phosphates. The processes are called the *destructive distillation* of coal, wood, and bones. Lampblack, which is very fine soot, is obtained by burning petroleum or ethyne in a limited supply of air.

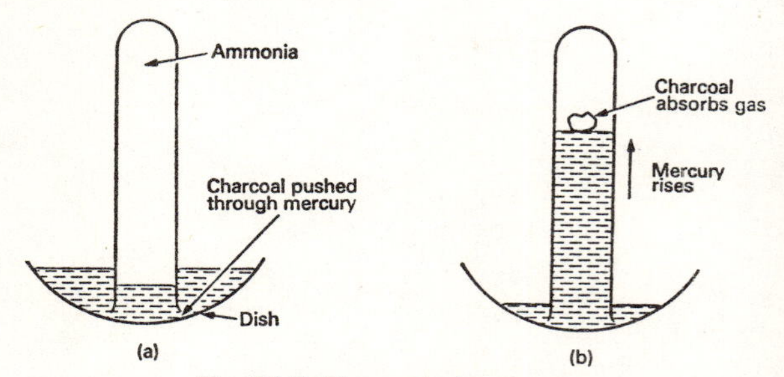

Fig. 21.1 Charcoal absorbs gas

Charcoal is very porous and so full of air that it floats in water. Its relative density when air-free is 1·5; that is, less than that of graphite and less than half that of diamond. Its surface area is great and surface forces make it extremely reactive. Wood charcoal absorbs gases because molecules of the gas form a layer on its large surface, and it is used in gas masks. Animal charcoal absorbs indicators from their warm aqueous solutions and it is used in the refining of brown sugar.

Charcoal burns in air when heated to about 500 °C; it thus burns more readily than diamond or graphite. It is a good reducing agent and reduces copper oxide and lead oxide to the metals:

$$2CuO + C = 2Cu + CO_2; \quad 2PbO + C = 2Pb + CO_2$$

Zinc, iron, and lead are extracted from their oxides by reducing them with carbon (usually coke, because it is cheapest) or with carbon monoxide obtained from coke. Oxides of metals above zinc in the reactivity series, such as aluminium and magnesium, are not reduced by carbon. Carbon reduces hot nitric acid and hot sulphuric acid:

$$C + 4HNO_3 = CO_2 + 2H_2O + 4NO_2\text{, brown}$$

$$C + 2H_2SO_4 = CO_2 + 2H_2O + 2SO_2\text{, colourless}$$

and it also reduces steam to hydrogen.

Lampblack is used as a black pigment in paint, printers' ink, shoe polish, and on carbon paper. Car tyres contain lampblack in order to increase their useful life.

Sugar charcoal is a pure form of carbon. It is prepared by heating sugar in the absence of air:

$$\text{(sucrose) } C_{12}H_{22}O_{11} = 11H_2O(g) + 12C(s)$$

COMPARISON OF THE ALLOTROPES OF CARBON

Diamond	*Graphite*	*Amorphous carbon*
Transparent, colourless, brilliant appearance	Opaque, greyish-black, metallic lustre	Opaque, black, and dull appearance
$1\frac{1}{2}$ times denser than graphite	Density $2{\cdot}3$ g/cm^3	Floats on water. Density $1{\cdot}5$ g/cm^3 when air-free
Giant structure in three dimensions	Giant layer structure. Weak forces between layers	Probably minute crystals of impure graphite
Hardest naturally-occurring substance	Flaky, soft, and greasy	Soft and porous
Burns above 800 °C	Burns above 700 °C	Burns above 500 °C
Poor conductor	Good conductor	Moderate conductor

To study the action of heat on carbonates

1. Heat copper carbonate in a hard-glass test-tube, and pass any gas evolved through lime water, or heat the carbonate on asbestos paper.
2. Repeat with the carbonates of lead, zinc, and magnesium.

The carbonates of sodium, potassium, and lithium were studied in chapter 12. Calcium carbonate exists in three different physical forms: *limestone*, *marble*, and *chalk*; it is discussed in chapter 24.

PROPERTIES OF CARBONATES AND HYDROGENCARBONATES

All carbonates exist as solids; all these solids are insoluble except the carbonates of sodium, potassium, and ammonium. Lithium carbonate is sparingly soluble.

Only the hydrogencarbonates of potassium, sodium, and ammonium ($KHCO_3$, $NaHCO_3$, NH_4HCO_3) exist in the solid state, and those of calcium, magnesium, and lithium exist only in solution.

Action of heat Carbonates of potassium and sodium do not decompose, but the others form the oxide and carbon dioxide:

$$\text{(white) } Li_2CO_3 = CO_2 + Li_2O \text{ (white)}$$

$$\text{(white) } ZnCO_3 = CO_2 + ZnO \text{ (white when cold)}$$

$$\text{(green) } CuCO_3 = CO_2 + CuO \text{ (black)}$$

The reactivity series indicates the ease of decomposition. Copper carbonate (copper is bottom of the series) decomposes most readily; carbonates of the alkali metals (which are at the top of the series) are stable. Lithium carbonate is an exception. Aluminium carbonate does not exist. Refer to p. 73.

Heat decomposes both solid and aqueous hydrogen carbonates:

$$2HCO_3^- \text{ (s or aq)} = CO_2(g) + H_2O + CO_3^{2-}$$

Action of dilute acids Refer to p. 32 in chapter 6.

TESTS FOR CARBONATES AND HYDROGENCARBONATES

1. Dilute nitric acid releases carbon dioxide: test with lime water which turns milky. Do not use dilute hydrochloric acid or sulphuric acid because they sometimes form insoluble compounds, such as $CaSO_4$, $PbSO_4$, and $PbCl_2$, which quickly prevent further evolution of carbon dioxide.
2. Aqueous magnesium sulphate forms a white precipitate with soluble carbonates, but not with hydrogencarbonates.

$$Mg^{2+}(aq) + CO_3^{2-}(aq) = MgCO_3(s) \text{ white}$$

To prepare carbon dioxide

Add dilute hydrochloric or nitric acid (not sulphuric acid, which forms an insoluble sulphate) to marble or limestone chips. Collect the gas over water or by downward delivery. If necessary it can be dried by concentrated sulphuric acid, calcium chloride or silica gel.

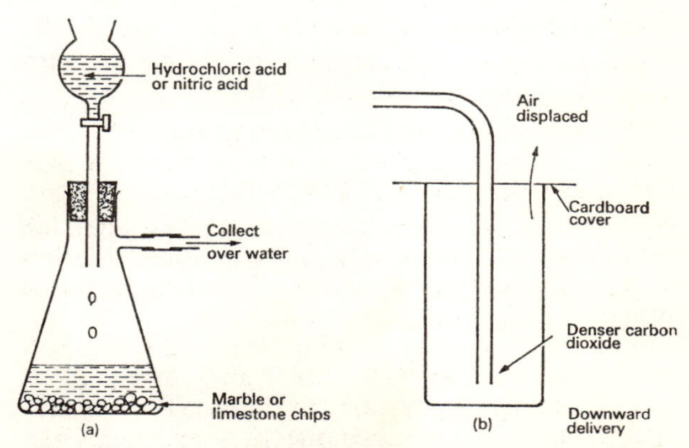

Fig. 21.2 Preparation and collection of carbon dioxide

Kipp's apparatus (see p. 28) is used to obtain a readily available supply of the gas, which should be bubbled through water to remove acid spray.

PROPERTIES OF CARBON DIOXIDE

It is a colourless gas with practically no smell or taste, but its solution has a pleasant sharp taste. Water dissolves its own volume of the gas at room temperature. The solution contains about one per cent of the gas as carbonic acid, which is weakly acidic to indicators and turns litmus pink:

$$H_2O + CO_2 \rightleftharpoons H_2CO_3 \rightleftharpoons H^+ + HCO_3^- \rightleftharpoons 2H^+ + CO_3^{2-}$$

The gas does not burn and does not support combustion or respiration, but it is not poisonous. Its density, relative to air, is 1·53; therefore it can be 'poured' downwards in air. It extinguishes a burning candle when 'poured' into a gas jar and a beaker becomes heavier when the gas is poured into it. Solid carbon dioxide is a white solid, called 'dry ice' because it sublimes at −78 °C and does not melt.

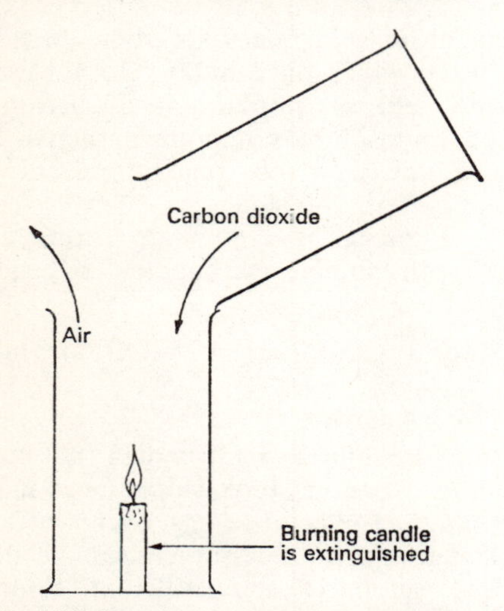

Fig. 21.3 'Pouring' carbon dioxide downwards

Chemical properties It is not very reactive chemically. The alkali metals and magnesium continue to burn in the gas if they are sufficiently hot:

$$4Na + 3CO_2 = 2Na_2CO_3 + C$$

$$2Mg + CO_2 = 2MgO + C$$

The gas turns lime water milky owing to precipitation of fine calcium carbonate; excess gas reacts with this precipitate and the solution becomes clear again as the soluble hydrogencarbonate forms:

$$Ca(OH)_2 + CO_2 = CaCO_3(s) + H_2O$$

$$\text{(insoluble) } CaCO_3 + CO_2 + H_2O \rightleftharpoons Ca(HCO_3)_2 \text{ (soluble)}$$

Solutions of the caustic alkalis react in a similar way chemically, but the carbonate is not precipitated:

$$2OH^- + CO_2 = CO_3^{2-} + H_2O$$

$$CO_3^{2-} + CO_2 + H_2O \rightleftharpoons 2HCO_3^-$$

Potassium hydroxide is used to absorb the gas as sodium carbonate may be precipitated from concentrated sodium hydroxide solution.

USES OF CARBON DIOXIDE

Some fire extinguishers contain an acid and sodium carbonate or hydrogencarbonate. The carbon dioxide produced when the reactants mix smothers a fire by keeping away air. The gas is very useful in extinguishing petrol and petroleum fires, which are made worse by water on which the burning liquids float. Some small extinguishers contain the gas under pressure and are used on electrical fires; the use of carbon dioxide foam or water on these would be dangerous because the liquids are conductors.

Aerated drinks and mineral waters contain the gas under slight pressure, and it produces the pleasant 'acid' taste.

Soda water is a solution of the gas in water, but other aerated drinks contain flavouring and colouring substances.

Dry ice or *carbon dioxide snow* is used for cooling ice-cream, packing fish, and to 'deep freeze' some foods. It is much colder than ice and leaves no residue because it sublimes.

The gas is used in the ammonia-soda process for the manufacture of sodium carbonate; see p. 66.

To find the volume (or mass) of carbon dioxide formed when one gramme of carbon burns

Oxygen is passed over a known mass of carbon heated in a silica tube. The volume of carbon dioxide produced is measured after collecting the gas in a syringe.

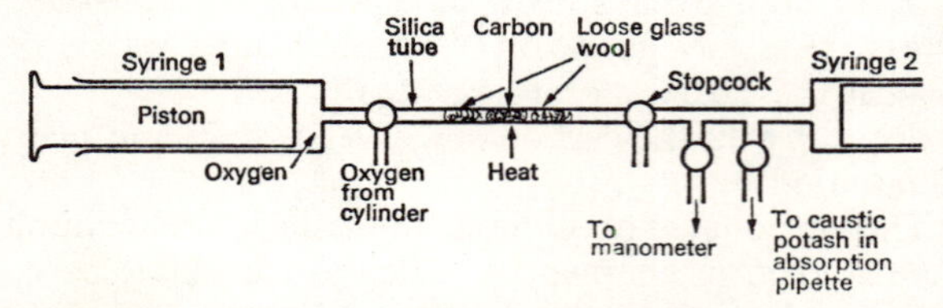

Fig. 21.4 Measuring the carbon dioxide formed by combustion of carbon

1. Place about 0·024 g of carbon, which may be charcoal, graphite or industrial diamond, in the apparatus shown in Fig. 21.4. Weigh the silica tube with the glass wool and the carbon. The glass wool keeps the carbon in position. Syringe 2 should be well greased to prevent tiny pieces of industrial diamond entering and damaging it.
2. Fill syringe 1 with 50 cm³ of oxygen. Also add 50 cm³ of air if charcoal is being used as hot charcoal sometimes explodes slightly in pure oxygen.
3. Heat the carbon strongly. Pass the oxygen slowly over the hot element until no more reaction occurs. There may be a little ash left as a solid residue.
4. Cool the hot tube with a damp cloth.

5. Measure the volume of gas in the syringes.
6. Move all the gas into syringe 2, that is, near to the absorption pipette containing caustic potash. Move the gas into and out of the pipette until no more is absorbed. Open the tap to the manometer at the end to ensure that the gas is at atmospheric pressure.
7. Measure the volume of gas in the syringes.
8. Weigh the silica tube and contents. The reduction in mass equals that of the carbon used.

SPECIMEN RESULTS AND CALCULATION

1.	Mass of tube, wool, and carbon at start	=
2.	Mass of tube, wool, and ash at finish	=
1 − 2.	Mass of carbon burnt	= w g
3.	Volume of gas at start	=
4.	Volume of gas at finish	=
5.	Volume of gas minus carbon dioxide	=
4 − 5.	Volume of carbon dioxide formed	= v cm³

w g of carbon form v cm³ of carbon dioxide

∴ 1 g of carbon forms v/w cm³ of carbon dioxide

The different forms of carbon (charcoal, graphite, and diamond) give the same result and do not form any other product. This is evidence that they are allotropes.

Readings 3 and 4 are the same, which means that 1 volume of oxygen produces 1 volume of carbon dioxide.

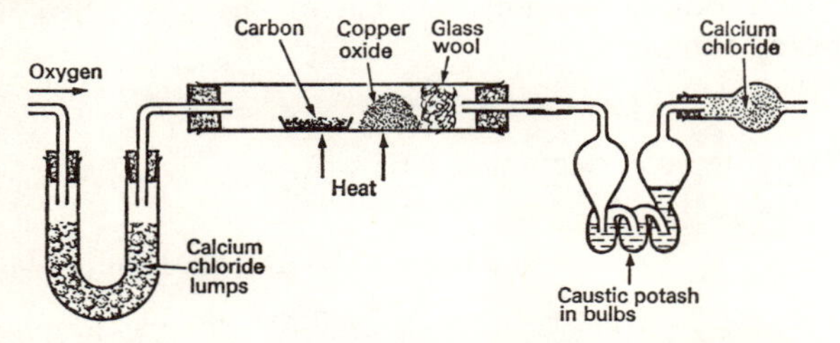

Fig. 21.5 Measuring the mass of carbon dioxide formed by combustion of carbon

Figure 21.5 shows a different apparatus for doing the above experiment. The dry oxygen passes over the hot carbon. The copper oxide oxidizes any carbon monoxide to dioxide. The dioxide is absorbed in caustic potash bulbs. The calcium chloride tube on the right prevents entry of water vapour.

The porcelain boat and carbon are weighed at the start and at the finish, and so are the potash bulbs. The results are calculated from the four weighings.

FORMULA OF CARBON DIOXIDE

By using the density of carbon dioxide and the temperature and pressure of the dioxide produced in either of the above experiments, the mass of the gas can be calculated. Accurate experiments show that:

1 g of carbon combines with 2·66 g of oxygen,

∴ 12 g of carbon combine with 32 g of oxygen,

i.e. 1 mole of carbon combines with 2 moles of oxygen atoms.

The formula of the gas is therefore CO_2.

Formula by using Avogadro's Principle If readings 3 and 4 are the same, carbon dioxide contains its own volume of oxygen, and

Carbon + 1 volume of oxygen
= 1 volume of carbon dioxide

i.e. Carbon + 1 molecule of oxygen
= 1 molecule of carbon dioxide

i.e. Carbon + 2 atoms of oxygen
= 1 molecule of carbon dioxide

The formula of the dioxide must be C_xO_2.

To study the action of heat on coal and on wood

1. Arrange the apparatus as shown in Fig. 21.6. Heat the coal strongly. The coal does not burn because there is so little air in the test-tube.
2. Some of the brownish fumes which are evolved do not condense in the cooled test-tube. Either collect over water the gas which does not condense or burn it at a jet.

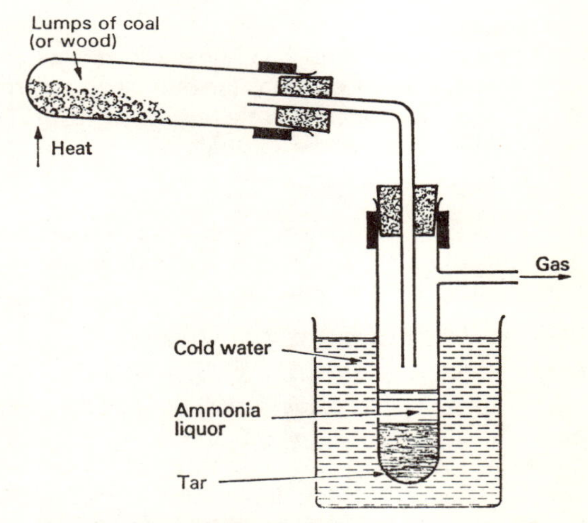

Fig. 21.6 Destructive distillation of coal or wood

3. At the end of the experiment, test the water distillate and the black tarry distillate with litmus paper. Examine the solid residue in the test-tube.
4. Repeat the experiment with lumps of wood or wood shavings.

The solid residue left by coal resembles coke, but the maximum temperature possible in a glass apparatus is not high enough for the production of good coke.

The solid residue from wood is charcoal. Coal forms an alkaline watery distillate, called ammonia liquor, and an acidic coal tar. Wood forms an acidic distillate. The gases which do not condense are called coal gas and wood gas respectively.

COAL, COKE, AND COAL GAS

Origins of coal Coal was probably formed by bacterial decomposition of plants some two hundred million years or so ago. The plants were large tree-ferns, giant reeds, and the undergrowth of smaller plants. Heat and pressure also played some part in the chemical change. The approximate composition by mass of coal is:

Carbon	80 per cent,	Hydrogen	5 per cent,
Oxygen	8 per cent,	Sulphur	1 per cent

together with nitrogen and phosphorus compounds.

Destructive distillation of coal Coal produces a mixture of liquids and gases and a solid residue when it is heated strongly in the absence of air:

$$\text{Coal} = \text{coke(s)} + \text{ammonia liquor(aq)} + \text{coal tar(l)} + \text{coal gas(g)}$$

The approximate composition by volume of coal gas is:

Hydrogen	50 per cent
Methane, CH_4	30 per cent
Carbon monoxide	10 per cent

and small amounts of other gases.

Hydrogen sulphide must be removed from coal gas, otherwise it burns to sulphur dioxide, which forms corrosive acid in the atmosphere. Moist hydrated iron(III) oxide removes the sulphide:

$$Fe_2O_3 + 3H_2S = 3H_2O + Fe_2S_3$$

The iron sulphide may be converted back to the oxide by periodically passing air over it. In this way the sulphide can be used several times.

$$2Fe_2S_3 + 3O_2 = 2Fe_2O_3 + 6S$$

The final solid is a mixture of oxide, sulphide, and sulphur called 'spent oxide'; this is used to form sulphur dioxide for the manufacture of sulphuric acid—see p. 139.

Coal tar This is a black liquid mixture of up to three hundred organic compounds. It is fractionally distilled; oils are collected over various temperature ranges and a non-volatile residue called *pitch* remains. The conditions of fractionation are arranged so that the production of pitch is a minimum and that of the more volatile fractions is a maximum.

Products obtained from coal tar include benzene, toluene (methylbenzene), naphthalene, and phenol. The action of chemicals such as sulphuric acid, nitric acid, halogens, and oxidizing agents on these primary products produces a very large number of coal-tar derivatives which are used in the manufacture of dyes, drugs, insecticides, plastics, and other organic chemicals.

OTHER FUEL GASES

Producer gas is formed by blowing air through hot coke in a 'gas-producer'. It has a lower calorific value than coal gas but is much cheaper, and it is easier to use in many processes than the solid coke. The gas is a mixture of carbon monoxide and nitrogen in the ratio by volume of 1:2; the nitrogen in the mixture accounts for the low *calorific value* of the gas—the heat produced by unit volume of the gas when it burns completely.

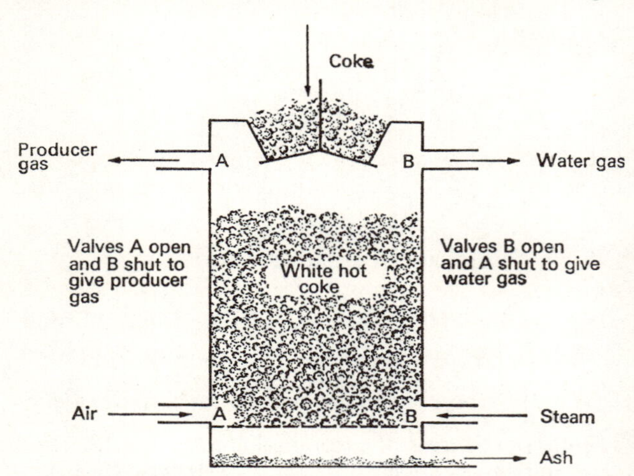

Fig. 21.7 Manufacture of producer gas (A) or water gas (B)

$$C + O_2 = CO_2 + \text{heat}; \quad CO_2 + C = 2CO - \text{heat}$$

The complete reaction can be written:

$$2C + \text{air } (O_2 + 4N_2) = 2CO \text{ (2 vol)} + 4N_2 \text{ (4 vol)} + \text{heat}$$

The reaction is exothermic, and the coke gradually becomes white hot. If the gas is used at once, most of the heat value of the coke is utilized.

Water gas is formed when steam instead of air is passed through white-hot coke. It is a mixture of about equal volumes of hydrogen and carbon monoxide, with small amounts of carbon dioxide and nitrogen. Its calorific value is very high because more than 90 per cent of the constituents are combustible. The reactions which take place are:

$$C + H_2O(g) = CO \text{ (1 vol)} + H_2 \text{ (1 vol)} - \text{heat}$$
$$CO + H_2O \rightleftharpoons CO_2 + H_2$$

Heat is absorbed because the reaction is endothermic, and the hot coke cools rapidly. Therefore producer gas and water gas are usually made alternately in the same gas producer. Production of the producer gas makes the coke white hot, and production of water gas cools the coke to about 1000 °C. Sometimes, both gases are made together and the product is *semi-water gas* which contains carbon monoxide and hydrogen, with about 50 per cent nitrogen.

Water gas is added to coal gas in order to increase its calorific value. It is a cheap source of commercial hydrogen. Methanol is manufactured from water gas and hydrogen by the reaction:

$$CO + 2H_2 = CH_3OH, \text{ methanol}$$

A high pressure and a catalyst are required.

Town gas is a mixture of coal gas, water gas, natural gas which consists mainly of methane, and gases made from light oils produced in petroleum refineries (p. 162). The natural gas greatly increases the calorific value of town gas. Much of the gas now supplied in many towns does not contain any coal gas and often it consists largely of natural gas.

REACTIONS IN A COKE FIRE

Figure 21.8 shows where the three reactions take place in a coke fire. The excess air at the bottom oxidizes the carbon to carbon dioxide, which is reduced to

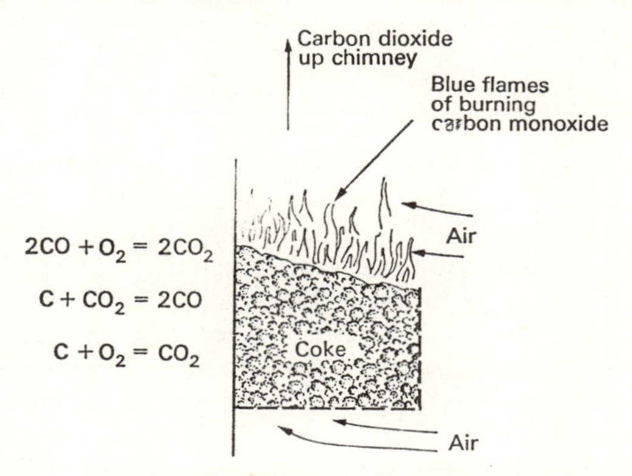

Fig. 21.8 Reactions in a coke fire

monoxide in the middle of the coke where oxygen is absent. Above the coke the monoxide burns in excess air with a blue flame.

To reduce carbon dioxide to carbon monoxide

1. Pass carbon dioxide from a Kipp's apparatus over very hot charcoal heated in a silica combustion tube.

$$CO_2 + C = 2CO$$

Fig. 21.9 Preparation of carbon monoxide from carbon dioxide

To prepare carbon monoxide from methanoic acid

Arrange the apparatus of Fig. 11.5, with water in the trough. Carefully warm concentrated sulphuric acid in a beaker and add it to the empty flask. Place methanoic acid (a corrosive liquid) in the funnel; add it drop by drop to the flask. The methanoic acid is dehydrated, p. 140:

$$HCOOH(l) = CO(g) + H_2O(l)$$

PROPERTIES OF CARBON MONOXIDE

It is a colourless, tasteless, odourless gas. It is very poisonous because the haemoglobin in blood reacts with it in preference to oxygen, forming carboxy-haemoglobin. The reduced amount of haemoglobin in the blood is then unable to carry enough oxygen round the body.

Chemical properties It burns in air with a blue flame, forming the dioxide. This is a test for the gas. The very dry gas does not burn because water catalyses the reaction. Like hydrogen, carbon monoxide is a reducing agent, and it reduces the oxides of zinc and metals below it in the reactivity or electrochemical series:

$$Fe_2O_3 + 3CO = 2Fe + 3CO_2$$

$$PbO + CO = Pb + CO_2$$

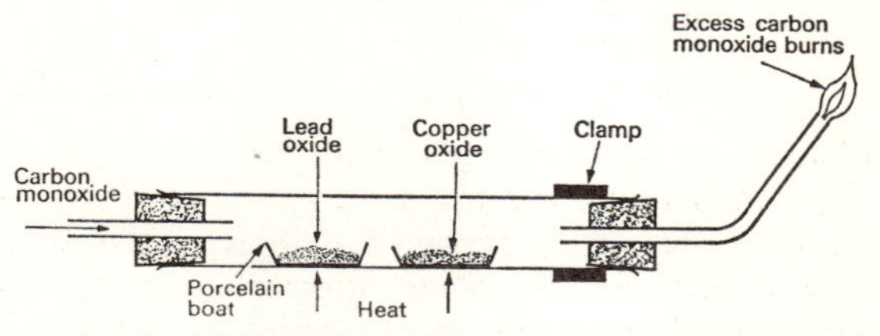

Fig. 21.10 Reducing action of carbon monoxide

Uses It is a fuel in producer gas and water gas, and is the reducing agent in the extraction of iron in the blast furnace—see p. 149. A mixture of the gas and hydrogen is used to manufacture methanol—see the uses of water gas opposite.

DANGER FROM CARBON MONOXIDE

Carbon monoxide is a product of the incomplete combustion of carbon and its compounds. The exhaust gases from internal combustion engines burning paraffin, petrol or diesel oil contain about 10 per cent of the gas and therefore they are very poisonous. Engines should never be run in a confined or badly-ventilated space such as a closed garage. Coal, coke, and coal-gas fires can produce carbon monoxide if they receive insufficient air supply. Gas water-heaters in bathrooms are dangerous if they are used when the windows are closed. Methane and other hydrocarbon gases occur in coal mines. They form explosive mixtures with air and a flame can ignite them. Usually carbon monoxide is one of the main products and it frequently kills more miners than the original explosion.

$$2CH_4 + 3O_2 = 2CO + 4H_2O$$

Coal gas contains about 30 per cent of carbon monoxide and therefore leaking pipes, ovens or heaters can be very dangerous.

THE CARBON CYCLE

The element carbon undergoes a series of changes into various compounds. The changes are called the carbon cycle because they are continuously repeated.

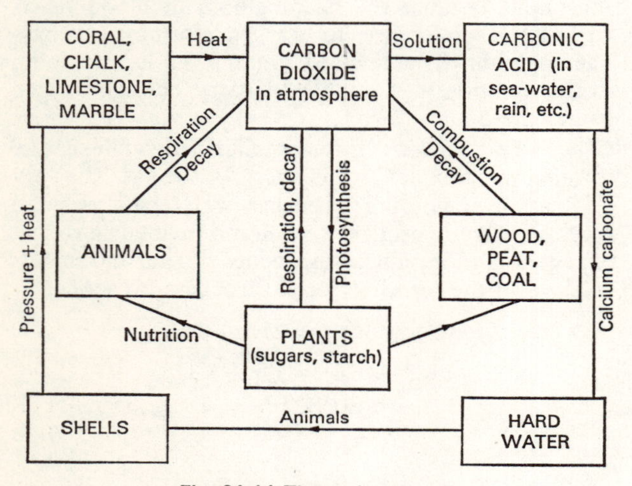

Fig. 21.11 The carbon cycle

Processes by which carbon dioxide is added to the atmosphere include the combustion of carbon or fuels containing carbon compounds, respiration of animals and plants, the fermentation of sugar to produce ethanol, and the thermal decomposition of calcium carbonate in the manufacture of lime:

Combustion:

$$CH_4 + 2O_2 = CO_2 + 2H_2O$$

Respiration:

$$C_6H_{12}O_6 + 6O_2 = 6CO_2 + 6H_2O + \text{energy}$$

Fermentation:

$$C_6H_{12}O_6 = 2CO_2 + 2C_2H_5OH$$

Manufacture of lime:

$$CaCO_3 = CO_2 + CaO$$

$C_6H_{12}O_6$ is the formula of glucose.

Processes which remove carbon dioxide from the atmosphere include solution of the gas in natural waters (such as rivers, lakes, and seas), photosynthesis, and the combination with lime of slaked lime in cement, plaster, etc. The reaction in photosynthesis is the reverse of that given for respiration above.

SILICATES

Silicon and carbon are both in group 4 of the Periodic Table. They have high melting points and their crystals are giant structures. Silicon has a great affinity for oxygen; the silicates contain strong silicon-oxygen bonds and silicon(IV) oxide, SiO_2, silica, is an acidic oxide like carbon dioxide.

Glass Ordinary glass or soda-glass is a mixture of sodium and calcium silicates. Hard glass is a mixture of potassium and calcium silicates, which has a highe melting point. A glass is a transparent substance formed by fusing white sand (SiO_2, an acidic oxide) with the carbonates or oxides of calcium, sodium, potassium, and lead.

$$SiO_2 + Na_2CO_3 = CO_2 + Na_2SiO_3, \text{ sodium silicate}$$

$$SiO_2 + CaO = CaSiO_3, \text{ calcium silicate}$$

Glasses are non-crystalline and do not have definite melting points; they gradually soften on heating. Addition of the oxides of aluminium and boron produces glasses with low coefficients of expansion which are suitable for vessels that have to be heated such as cooking vessels. Ordinary bottle glass contains iron(II) compounds which make it green. Cobalt and copper compounds are in blue and red glasses respectively.

Ceramics The common ceramics are pottery, stoneware, and the bricks, tiles, and pipes used as building materials. They are made from clay, which is a mixture of silica and a hydrated aluminium silicate of formula $Al_2O_3 . 2SiO_2 . 2H_2O$. They are made by first shaping the starting materials and then heating them to form the final product. They cannot be melted and then cast into the required shape like metals and glasses. Usually they are glazed with a thin coat of sodium silicate or lead silicate (i.e. a glass) to improve their appearance and make them non-porous. A ceramic consists of a myriad of tiny crystals.

Modern ceramics include materials used for electrical, electronic, and nuclear engineering. They are inorganic compounds such as oxides, carbides, nitrides, and borides. They are used in unbreakable tableware and can replace the metal parts of jet engine blades and of reactors, making possible higher temperatures and better performance. Television sets contain magnetic ceramics and all spark-plugs contain aluminium oxide ceramic. There could be little industry without refractory bricks made from ceramics, as they line the furnaces in which metals, glasses, cement, town gas, coke, and even ceramics themselves are made.

SUMMARY

Different forms of carbon

Polymorphism: the existence of an element or compound in two or more different solid forms or *polymorphs.*

Carbon, ammonium nitrate, and phosphorus all have polymorphs.
Allotropy: the existence of an element in two or more different forms or *allotropes* in the same state of matter: for example, carbon, solid sulphur, liquid sulphur, gaseous sulphur.

Fuel gases
Producer gas. Pass air over heated coke; contains CO and N_2.
Water gas: pass steam over white hot coke; contains CO and H_2; produced alternately with producer gas because reaction is endothermic.

Allotropes of carbon

Diamond	*Graphite*	*Amorphous carbon*
Transparent, colourless	Opaque, metallic lustre	Opaque, dull black
Density 3·45 g/cm³	Density 2·3 g/cm³	Floats on water; density 1·5 g/cm³ when air-free
Hardest natural substance	Flaky, soft, and greasy	Soft and porous
Burns above 800 °C	Burns above 700 °C	Burns above 500 °C
Poor conductor	Good conductor	Moderate conductor

Amorphous carbon includes wood charcoal, animal charcoal, coke, carbon black, and sugar charcoal.

Properties of carbon: burns in air; reduces oxides of zinc, iron, lead, and copper, nitric acid to NO_2 and sulphuric acid to SO_2; forms water gas with steam at 1000 °C.

Carbonates and hydrogencarbonates
Carbonates are all insoluble except those of Na^+, K^+, and NH_4^+; they are all white except green $CuCO_3$. They all decompose on heating except Na^+ and K^+ to give the oxide and CO_2.
Hydrogencarbonates: those of Li^+, Ca^{2+}, and Mg^{2+} exist in solution only; those of Na^+, K^+, and NH_4^+ exist as solids. They all decompose on heating to give the carbonate, CO_2 and water.
Test: dilute HNO_3 releases CO_2 which turns lime water milky. Carbonates form a precipitate with $MgSO_4$ solution, but hydrogen carbonates do not.

Carbon dioxide: colourless, odourless gas, denser than air, moderately soluble in water forming carbonic acid. Does not support combustion, but burning alkali metals and magnesium continue to burn in it. Turns lime water milky; excess gas turns the mixture clear again. Reacts readily with caustic alkalis.
Preparation: add cold dilute hydrochloric or nitric acid to limestone chips. Collect over water or by downward displacement.
Uses: in fire extinguishers, aerated drinks and as solid dry ice for cooling foods; in ammonia-soda process for manufacture of Na_2CO_3 and in cooking from baking powder to make mixture rise when heated.

Coal: formed from decomposing plants 200 million years ago. Destructive distillation gives the following:
Ammonia liquor: a watery alkaline distillate containing NH_3;
Coal tar: a thick black liquid mixture of up to 300 organic compounds which can be separated and used in the manufacture of dyes, drugs, insecticides, and plastics;
Coal gas: which contains mainly hydrogen, methane, and carbon monoxide with some hydrogen sulphide which must be removed with moist iron(III) oxide;
Coke: the solid residue, used as a reducing agent in many metallurgical processes, in particular the removal of oxygen in the manufacture of iron and steel. Burns to carbon monoxide at high temperatures or in limited oxygen, and to carbon dioxide in excess air at moderate temperatures.

Town gas: a mixture of fixed calorific value made from natural gas, and sometimes coal gas and water gas.

Carbon monoxide: colourless, odourless gas; very poisonous. A neutral oxide: does not react with acids or alkalis. Burns with a blue flame to carbon dioxide. Reducing agent which releases many metals from their oxides.
Uses: as a fuel in producer gas and water gas, as a reducing agent in the extraction of metals, and with hydrogen in the manufacture of methanol.

Carbon cycle
Release to atmosphere as carbon dioxide by combustion, respiration, fermentation, and in manufacture of lime.
Removed from atmosphere as carbon dioxide by solution in rivers and seas, by photosynthesis in leaves of plants and by combination with lime or slaked lime in cement and plaster.

Glass: a mixture of metal silicates (mainly calcium and sodium) made by melting white sand (SiO_2, an acid oxide) with carbonates of metals. Colour of glass results from metal impurities which can also change its properties: cobalt (blue), copper (red), iron (green).
Ceramics include pottery, stoneware, and bricks. Made from clay which contains aluminium and silicon oxides. Many ceramics have very high melting points and are used to contain melted metals.

QUESTIONS

1. Make a labelled diagram to show the preparation and collection of dry carbon dioxide. Describe and explain what happens when carbon dioxide is passed into a solution of calcium hydroxide until no further change occurs.
2. State and explain what happens when: (*a*) burning magnesium is plunged into a jar of carbon dioxide, (*b*) a jar of carbon dioxide is placed with the mouth of the jar below the surface of a solution of sodium hydroxide, (*c*) carbon dioxide is passed over red hot charcoal.
3. Describe an experiment to show the destructive distillation of coal. Draw a diagram of the apparatus you would use. Indicate on the diagram where four named products would be found.

4. How could you prove that chalk contained carbon, without assuming the composition of carbon dioxide? How can it be shown that diamond and graphite are forms of carbon? (C.)

5. Describe and explain how you would distinguish between carbon monoxide and hydrogen. How would you obtain a sample of silicon(IV) oxide from a mixture of silicon(IV) oxide (SiO_2) and calcium carbonate? (C.)

6. Discuss the origins and the thermal decomposition of coal, and the uses of coal as a source of energy (include the fuel gases obtained from coal).

7. Draw a candle flame and state simply what is happening in each zone. State and explain differences in appearance of a coke fire and a coal fire. Account for the drops of liquid that form on the bottom of a beaker of cold water when it is first placed over a Bunsen flame burning coal gas. (C.)

8. Combustion of a fuel is an exothermic reaction. Explain the meaning of the words *combustion* and *exothermic* in this statement. Explain what is meant by saying that the calorific value of water gas is higher than that of producer gas. Suggest reasons (both economic and chemical) why natural gas is often added to coal gas.

22 Nitrogen and its Compounds

The atomic number of nitrogen is 7 and its electronic structure is 2, 5. The atom can share three pairs of electrons with one or more atoms of another element and complete its octet. Three electrons from each atom are shared and the covalency of nitrogen is 3; for example, NH_3. Occasionally a nitrogen atom can receive three electrons by transfer from a more electropositive element and it then forms the nitride ion, N^{3-}, as in Li_3N and Mg_3N_2.

The ammonia molecule can accept a proton, H^+, and form the ammonium ion, $NH_4{}^+$. The nitrogen has a covalency of 4 and the ion has an electrovalency of 1:

Ammonium ion
$$\left[\begin{array}{ccc} & H & \\ & \bullet\times & \\ H\,\overset{\bullet}{\times} & N & \overset{\times}{\times}\,H \\ & \bullet\times & \\ & H & \end{array}\right]^+$$

INDUSTRIAL PREPARATION AND USES OF NITROGEN

Nitrogen is prepared by fractional distillation of liquid air, see p. 40 in chapter 7. A mixture of nitrogen, hydrogen, and carbon monoxide is made by passing air and steam over hot coke, and the mixture is used in the Haber process for making ammonia. Liquid nitrogen is used as a cooling agent in preference to liquid air, which sometimes causes explosions. Gaseous nitrogen is used as an inert gas in some industrial processes in order to exclude oxygen; for example, in the manufacture of nylon 6 and nylon 66.

To prepare ammonia gas

Slaked lime $Ca(OH)_2$ or quicklime CaO and an ammonium salt such as the chloride or sulphate are mixed well and then heated in a flask, arranged as in Fig. 22.1. The gas is dried by quicklime because both sulphuric acid and calcium chloride react with ammonia, and it is collected by upward delivery because it is about half as dense as air.

$$NH_4{}^+(s) + OH^-(s) = NH_3(g) + H_2O$$
$$2NH_4{}^+(s) + O^{2-}(s) = 2NH_3(g) + H_2O$$

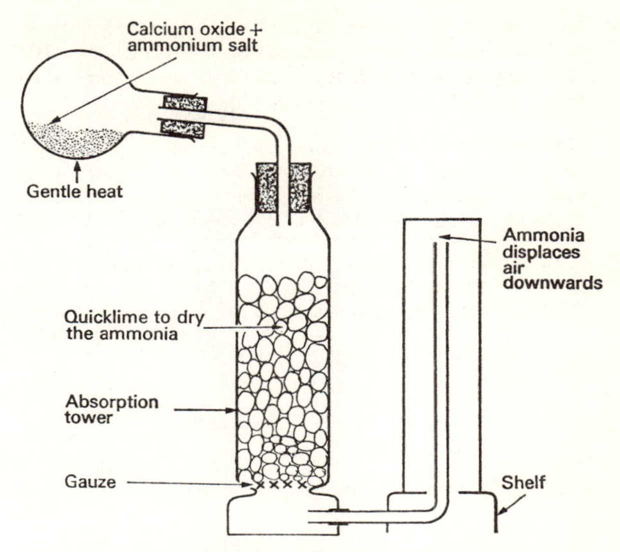

Fig. 22.1 Preparation of ammonia

PROPERTIES OF AMMONIA

It is a colourless, poisonous gas with a characteristic pungent smell. It is easily liquefied either by cooling (b.p. −33 °C) or by a pressure of about 10 atmospheres at ordinary temperatures, and it has a high molar latent heat. Both of these facts suggest that the liquid consists of $(NH_3)_x$ molecules rather than the single NH_3 molecules. The joining together of two or more like molecules to form a single molecule is called *association* or *polymerization*.

It is very soluble in water and forms an alkaline solution. 1 cm^3 of water dissolves 1300 cm^3 of gas at 0 °C and about 700 cm^3 at 20 °C. The solution softens water (p. 146) and removes stains from clothes. The concentrated solution contains about 35 per cent by mass of the gas and is called eight-eighty ammonia because its relative density is 0·880. The formula NH_4OH, for ammonium hydroxide, is frequently written for the solution but it is incorrect because nitrogen cannot have a covalency of 5; its outer shell would then have ten electrons. The reaction is:

$$NH_3(g) + H_2O(l) \rightleftharpoons NH_3(aq)$$
$$\rightleftharpoons NH_4^+(aq) + OH^-(aq)$$

The ammonia hydrate is a weak base.

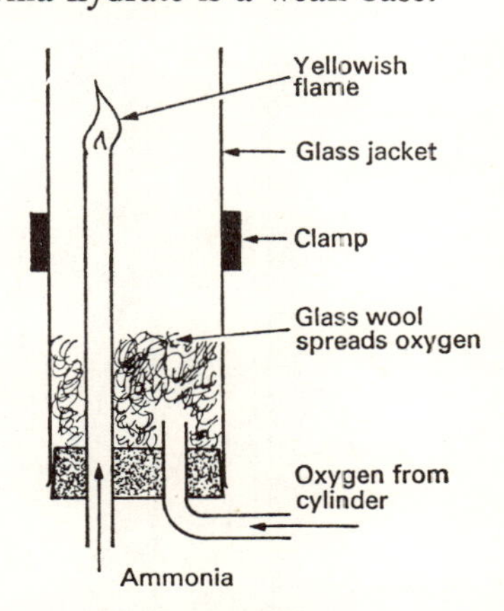

Fig. 22.2 Ammonia burns in oxygen

Ammonia extinguishes a burning splint and does not burn in air. It burns in oxygen and the flame is yellowish:

$$4NH_3 + 3O_2 = 2N_2 + 6H_2O$$

In the presence of heated platinum, oxygen oxidizes ammonia to oxides of nitrogen, which react with water to form nitric acid:

$$4NH_3 + 5O_2 = 4NO + 6H_2O; \quad 2NO + O_2 \rightleftharpoons 2NO_2$$

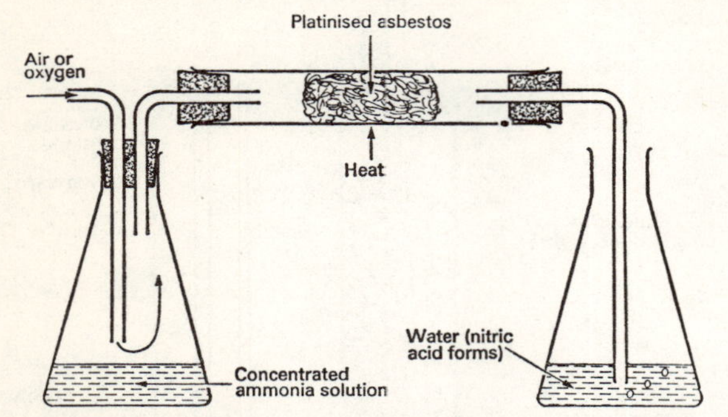

Fig. 22.3 Catalytic oxidation of ammonia to oxides of nitrogen

Ammonia reacts with hydrogen chloride to form white fumes of ammonium chloride:

$$NH_3(g) + HCl(g) = NH_4Cl(s)$$

and the two compounds also react in solution. Ammonia reacts with chlorine and is oxidized to nitrogen:

$$2NH_3 + 3Cl_2 = N_2 + 6HCl,$$

then $$NH_3 + HCl = NH_4Cl$$

i.e.

$$8NH_3 + 3Cl_2 = N_2 + 6NH_4Cl \text{ (white fumes)}$$

Copper(II) oxide or lead(II) oxide oxidizes ammonia to nitrogen and water. The diagram (Fig. 22.4) shows a method of condensing the water and collecting the nitrogen. The experiment shows that ammonia is a

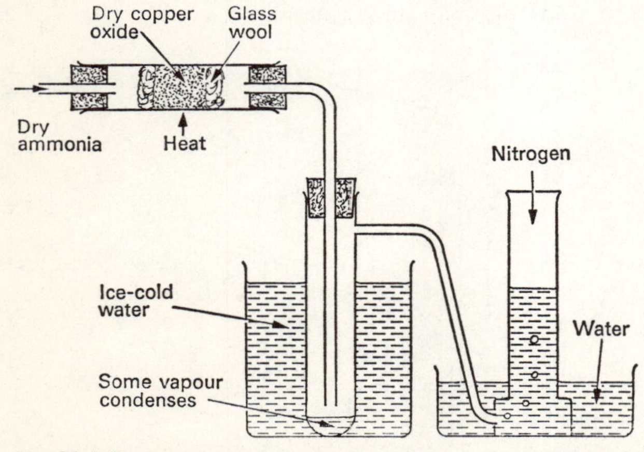

Fig. 22.4 The reaction between ammonia and copper(II) oxide

hydride of nitrogen because water contains hydrogen and this element cannot have come from the dry oxide. The glass wool prevents pieces of oxide passing into the water.

$$3CuO + 2NH_3 = N_2(g) + 3Cu(s) + 3H_2O(l)$$

Ammonia decomposes reversibly when it is heated or sparked:

$$2NH_3 \rightleftharpoons N_2 + 3H_2 - \text{heat}$$

At equilibrium the product contains about 6 per cent of ammonia. Hydrogen is prepared commercially by catalytic decomposition of ammonia, which is more easily transported as a liquid than hydrogen gas.

To demonstrate the solubility of ammonia

1. Fill a large dry flask with dry ammonia. Close the flask with a stopper through which passes a glass tube having a narrow jet at one end and a clip at the other. Open the clip.

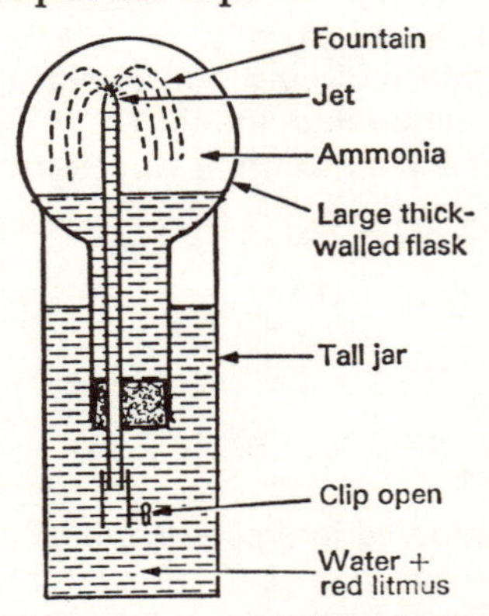

Fig. 22.5 Fountain experiment to show that ammonia is very soluble in water

2. Invert the flask in a tall jar of water, coloured with red litmus solution, as shown in Fig. 22.5. The ammonia in the glass tube slowly dissolves and the water rises up the tube. When a drop of water reaches the jet and passes into the flask, most of the ammonia dissolves rapidly and leaves a partial vacuum in the flask. Atmospheric pressure forces water from the jar up the glass tube and out at the jet as a kind of fountain of water. The litmus, which turns blue in the flask, makes the water easily visible.

Alternative method Figure 22.6 shows a second apparatus which can be used to demonstrate that ammonia is very soluble in water.

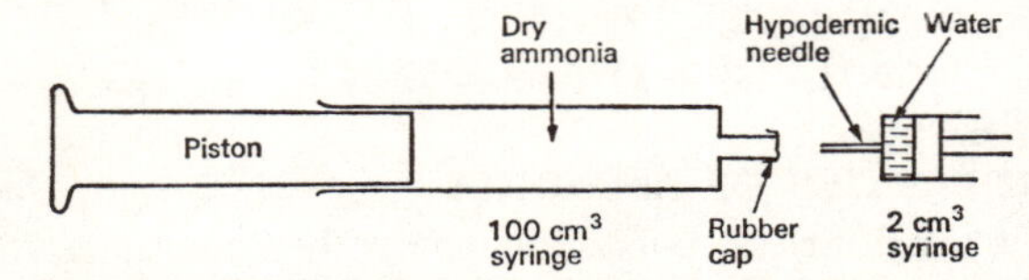

Fig. 22.6 To demonstrate that ammonia is very soluble in water

1. Fill a 100 cm³ syringe with dry ammonia and close it with a rubber cap.
2. Fill a 2 cm³ syringe with water and fit it with a hypodermic needle.
3. Inject water from the small syringe through the rubber cap into the large syringe. The water

quickly dissolves the ammonia and leaves a partial vacuum in the large syringe. Atmospheric pressure forces the piston into the barrel of the large syringe.

Both of the above tests can also be done with hydrogen chloride or with sulphur dioxide, which are also very soluble in water.

The great solubility of ammonia in water can also be demonstrated by inverting a gas-jar of the gas in a trough of water. The jar quickly fills with water as the ammonia dissolves.

To obtain ammonia by synthesis

1. Connect a combustion tube, filled with iron wool, between two dry gas syringes as in Fig. 22.7.
2. Add about 60 cm^3 of dry hydrogen from a cylinder to one syringe and about 20 cm^3 of dry nitrogen to the second syringe.

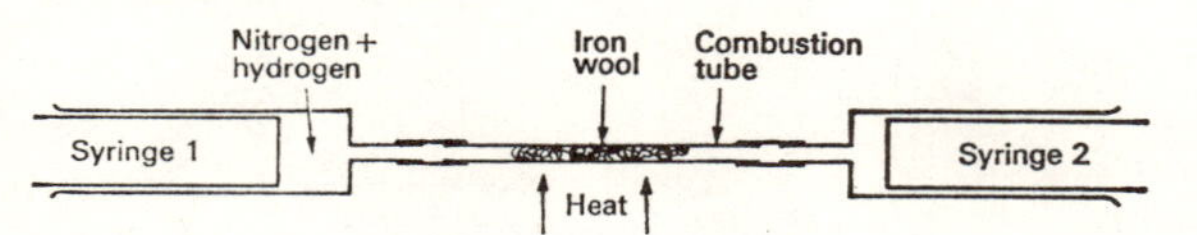

Fig. 22.7 Synthesis of ammonia

3. Mix the two gases well by moving them backwards and forwards from one syringe to the other several times.
4. Heat the iron strongly. Pass the gas mixture over it slowly several times.
5. Remove the flame, allow the apparatus to cool, and push damp litmus or indicator paper into the gas in one syringe. The colour change should indicate the presence of alkaline ammonia.

To demonstrate the thermal decomposition of ammonia

1. Use the apparatus of Fig. 22.7 once again. Add about 40 cm^3 of dry ammonia to one syringe and leave the other syringe empty.
2. Heat the iron strongly. Pass the ammonia several times over the hot iron.
3. Cool the tube with a damp cloth and observe the volume of the products, which consist mainly of hydrogen and nitrogen. The volume is about 80 cm^3.

$$\underset{\text{(2 volumes)}}{2NH_3} = \underset{\text{(1 volume)}}{N_2} + \underset{\text{(3 volumes)}}{3H_2}$$

MANUFACTURE OF AMMONIA

FROM COAL

The nitrogen present in coal comes off as ammonia during destructive distillation; see pp. 117–8. The gas dissolves in water, which is also formed, and forms ammonia liquor. The liquor is treated with a little lime, CaO, to decompose any ammonium salts present and is heated to expel the ammonia, which is absorbed in dilute sulphuric acid. The ammonium sulphate is used as a fertilizer:

$$2NH_3 + H_2SO_4 = (NH_4)_2SO_4$$

HABER PROCESS from nitrogen and hydrogen

This process depends on passing synthesis gas, a mixture of nitrogen and hydrogen in the ratio 1:3 by volume, over a heated iron catalyst at about 500 °C and a high pressure of 250 atmospheres.

$$N_2 + 3H_2 \rightleftharpoons 2NH_3; \ \Delta H = -92 \text{ kJ g-equation}^{-1}$$

Obtaining the synthesis gas Nitrogen is obtained either by fractionation of liquid air or from producer gas. Hydrogen is obtained either from water gas or by the action of steam on hydrocarbons; see p. 30 in chapter 5. If a mixture of producer gas and water gas is used, the carbon monoxide in it is first converted to carbon dioxide by the action of steam and a catalyst:

$$CO + H_2O \rightleftharpoons CO_2 + H_2$$

and the carbon dioxide is removed by washing with potassium carbonate under pressure.

The catalyst chamber The catalyst for the synthesis of ammonia is iron; this is mixed with aluminium oxide which helps to maintain a large active surface of iron. The synthesis gas is dried, heated, and passed over the iron at a pressure of 250 atmospheres. Ten to fifteen per cent of ammonia by volume is present after reaction.

Since the reaction is exothermic, the gases leaving the catalyst chamber are hot. They pass through pipes in a *heat-interchanger* and warm the synthesis gas before it enters the catalyst chamber.

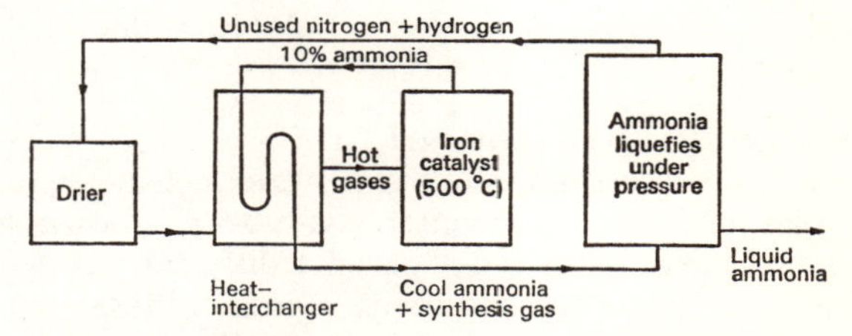

Fig. 22.8 Synthesis of ammonia by Haber process

The ammonia produced is liquefied in condensers cooled by water, or it is dissolved in water. The unchanged nitrogen and hydrogen are mixed with more synthesis gas and circulated again.

Effect of pressure and temperature Four volumes of synthesis gas form two volumes of ammonia. Therefore a high pressure increases the yield of ammonia. The rate of reaction is also greater because the four molecules of reactants collide more often at high pressures, since they are closer.

The rate of reaction is fast at high temperatures of 1000 °C or so. However, the yield of ammonia is very small because the heat decomposes the gas. The equilibrium yield at low temperatures is great, but the time taken to attain equilibrium is far too long.

The following approximate figures show the percentage of ammonia at equilibrium and the time taken to reach equilibrium under various conditions:

Temperature (°C)	*Pressure* (atm)	*Percentage ammonia*	*Time taken*
1000	1000	1	Seconds
1000	1	0·01	Seconds
250	1000	96	Minutes
250	200	75	Minutes

Clearly, the yields at 1000 °C are too low and the rates of reaction at 250 °C are too slow. At about 500 °C, the optimum or best temperature, there is a reasonable yield in a fairly short time provided a catalyst speeds the reaction. Only the surface of the iron is active catalytically and the metal must therefore be finely-divided.

AMMONIUM COMPOUNDS

AMMONIA SOLUTION $NH_3(aq)$

Aqueous ammonia contains the hydrate $NH_3 \cdot H_2O$ and small quantities of ammonium and hydroxide ions. It behaves as a weak base and like the caustic alkalis, precipitates insoluble hydroxides from solutions of metal salts:

$$Fe^{2+}(aq) + 2OH^-(aq) = Fe(OH)_2\text{, green}$$

$$Fe^{3+}(aq) + 3OH^-(aq) = Fe(OH)_3\text{, reddish-brown}$$

Copper salts form a blue precipate; zinc and lead salts form white precipitates. Refer to p. 33 of chapter 6.

An inverted funnel as in Fig. 11.3 on p. 57 must be used when dissolving the very soluble ammonia in water.

AMMONIUM CHLORIDE NH_4Cl

This is a white solid, soluble in water. Heat a little of the solid in a horizontal test-tube as shown in Fig. 22.9. The glass wool permits different gases to diffuse at different rates (denser gases diffuse more

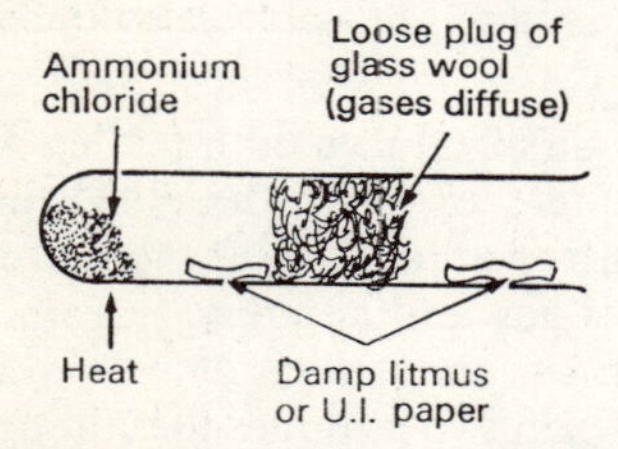

Fig. 22.9 Thermal dissociation of ammonium chloride

slowly). One litmus paper turns blue and the other turns red, showing that ammonium chloride produces both alkaline and acidic gases:

$$NH_4Cl(s) \rightleftharpoons NH_3(g) + HCl(g)$$

The hydrogen chloride diffuses more slowly than the less dense ammonia. If the two gases are not separated, they recombine on cooling to form the ammonium chloride again. This change is called *thermal dissociation,* which means the decomposition of a substance into simpler substances which recombine on cooling. The ammonium chloride *dissociates* into ammonia and hydrogen chloride on heating. Substances formed as a result of thermal decomposition do not recombine on cooling.

Ammonium chloride does not melt when it is warmed. It *sublimes,* that is, it changes directly from a solid to a gas. The chloride is formed as a *sublimate* when the gas is cooled. Figure 22.10 shows the apparatus used to demonstrate the sublimation of ammonium chloride and the formation of a sublimate. Ammonium chloride is readily separated from salt, sand, and other impurities by sublimation.

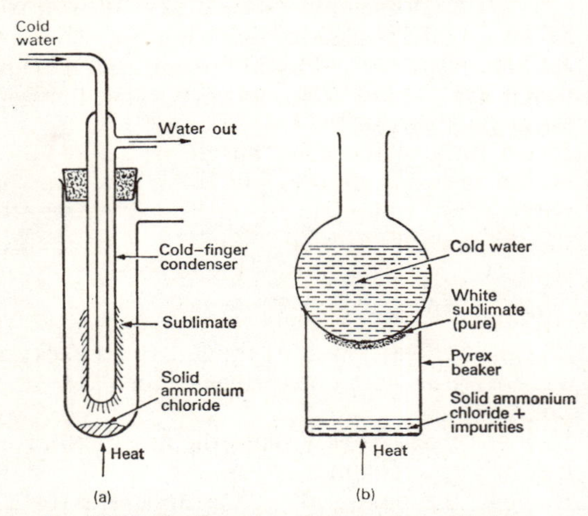

Fig. 22.10 Sublimation of ammonium chloride

Ammonium chloride sublimes because it dissociates. Iodine, naphthalene, and solid carbon dioxide sublime because their vapour pressure is 760 mmHg before they melt; they vapourize completely at a temperature lower than their melting points.

AMMONIUM SULPHATE $(NH_4)_2SO_4$

This can be prepared by adding dilute ammonia solution to dilute sulphuric acid in an evaporating basin until the mixture smells of ammonia. Evaporate the solution until crystals can be obtained on cooling.

It is manufactured from the ammonia liquor of gasworks, and also by passing synthetic ammonia and carbon dioxide into a suspension of gypsum (calcium sulphate):-

$$CaSO_4 + 2NH_3 + CO_2 + H_2O = CaCO_3(s) + (NH_4)_2SO_4(aq)$$

The sulphate decomposes on heating but does not readily sublime because sulphuric acid is less volatile than hydrogen chloride. Ammonium hydrogensulphate and possibly some sulphuric acid remain and ammonia is evolved:

$$(NH_4)_2SO_4 \rightleftharpoons NH_4HSO_4 + NH_3(g) \rightleftharpoons H_2SO_4(l) + 2NH_3(g$$

AMMONIUM CARBONATE $(NH_4)_2CO_3$
This is obtained as a sublimate when calcium carbonate is heated with ammonium sulphate. It always contains the hydrogencarbonate, NH_4HCO_3. The mixture is called sal volatile or smelling salts; it readily decomposes to form ammonia:

$$(NH_4)_2CO_3 \rightleftharpoons 2NH_3 + CO_2 + H_2O$$
$$NH_4HCO_3 \rightleftharpoons NH_3 + CO_2 + H_2O$$

AMMONIUM NITRATE NH_4NO_3
This is prepared by neutralization of dilute nitric acid with slight excess of ammonia and then obtaining crystals. It is a white crystalline solid, readily soluble in water. The process of solution is endothermic and a mixture of ammonium nitrate and water can be used as a freezing mixture. The solid exists in different forms or polymorphs (p. 114).

Heat a little of the solid nitrate in a test-tube. It melts and then decomposes into colourless gases and also into some white fumes which are the result of a little thermal dissociation:

$$NH_4NO_3 = 2H_2O + N_2O, \text{ dinitrogen monoxide}$$
$$NH_4NO_3 \rightleftharpoons HNO_3 + NH_3, \text{ white fumes when cool}$$

The last trace of nitrate always explodes; brown nitrogen dioxide is one of the products.

Ammonium nitrate itself is not an explosive; a mixture with trinitrotoluene (TNT) or aluminium is a common explosive used in bombs.

Ammonium nitrate is a fertilizer. By itself it 'cakes' or forms lumps; powdered chalk is added to it to prevent this. The mixture is called *nitro-chalk.*

AMMONIUM NITRITE NH_4NO_2
This is very unstable and readily decomposes into nitrogen and water:

$$NH_4NO_2 = N_2 + 2H_2O$$

TEST FOR AMMONIUM SALTS Add caustic alkali to the solid or solution, and warm gently if necessary.

$$NH_4^+ + OH^- = NH_3(g) + H_2O$$

Test for ammonia by its smell and by its action on moist red litmus or on a stopper moistened with concentrated hydrochloric acid.

NITRIC ACID

To prepare nitric acid HNO_3
Nitric acid is obtained when any nitrate is heated with concentrated sulphuric acid. The ionic equation for the reaction is:

$$H^+ + NO_3^- = HNO_3(g)$$

The reaction is reversible, but since nitric acid is much more volatile than sulphuric acid, it passes from the reaction mixture and therefore the reverse reaction cannot take place. Usually potassium nitrate, which is also called *nitre* or *saltpetre,* is the nitrate used.

1. Add solid potassium nitrate to a retort. Cover it with cold, concentrated sulphuric acid.
2. Arrange the apparatus as shown in Fig. 22.11. The sand tray is necessary so that the sand soaks up the hot sulphuric acid if the retort cracks during the experiment. The apparatus must be all glass because nitric acid fumes react with cork and rubber.

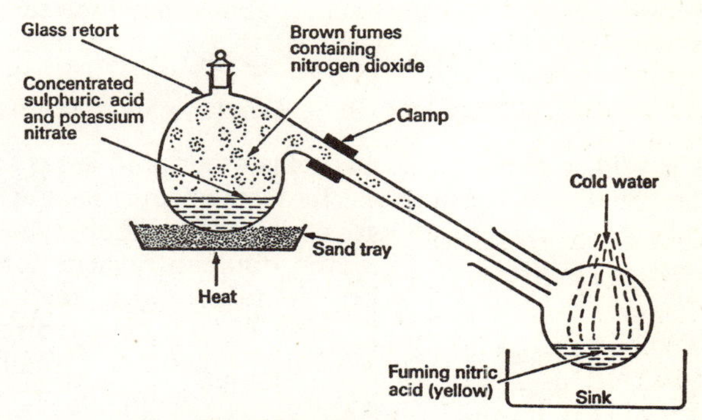

Fig. 22.11 Preparation of nitric acid

3. Heat the sand tray fairly strongly until the mixture in the retort appears to boil gently. Nitric acid is formed; it distils over and is collected in a water-cooled flask.

$$H.HSO_4 + KNO_3 = HNO_3(g) + KHSO_4(s)$$

The distillate is called fuming nitric acid because it fumes in air. It is yellow because it contains dissolved nitrogen dioxide, which is produced by thermal decomposition of some of the nitric acid:

$$4HNO_3 = 2H_2O + O_2 + 4NO_2 \text{ (brown)}$$

At the end of the reaction, solid potassium hydrogen sulphate remains in the flask when it is cold. The normal sulphate, K_2SO_4, is formed only at temperatures of about 1000 °C, which cannot be reached in this apparatus.

Note that the action of concentrated sulphuric acid on a chloride is similar to its action on a nitrate, as the following equations make clear:

$$H^+ + Cl^- = HCl(g)$$
$$H.HSO_4 + NaCl = HCl(g) + NaHSO_4$$

Properties of nitric acid
100 per cent nitric acid is a colourless fuming liquid of density 1·52 g/cm³. It decomposes slightly when it is exposed to light (the equation is given above) to form some nitrogen dioxide, which dissolves in the acid and turns it yellow. The fuming acid boils at 86 °C and some decomposition occurs. The boiling point does not remain constant and it gradually rises to 121 °C; the liquid which boils off then consists of 68 per cent acid and 32 per cent water. This mixture is called concentrated nitric acid—it is the concentrated acid usually used in laboratories. The concentrated

acid is usually yellowish because it contains some dissolved nitrogen dioxide. The acid becomes colourless when a little water is added, which reacts with the nitrogen dioxide, or when air is blown through the acid to drive out the nitrogen dioxide.

Acidic properties of dilute nitric acid The pure acid does not attack metals readily and does not react with carbonates because it contains no hydrogen ions or oxonium ions. The dilute acid is completely ionized:

$$HNO_3 = H^+ + NO_3^-$$

or $$HNO_3 + H_2O = H_3O^+ + NO_3^-$$

The acid system is nitric acid plus water, and it has all the usual acidic properties. Its taste is sour, it affects indicators, and it reacts with carbonates, sulphites, and bases (refer to p. 33). However, dilute nitric acid does not form hydrogen when it reacts with metals, although the very dilute acid (1 per cent) forms hydrogen when it reacts with magnesium. Usually the acid and metals produce oxides of nitrogen because the acid is a strong oxidizing agent.

Thermal decomposition of nitric acid Vapour from the fuming or concentrated acid readily decomposes on heating. Equations for the changes that occur are:

$$4HNO_3 = 2N_2O_5 + O_2$$

and then $$2N_2O_5 = O_2 + 4NO_2 \text{ (brown)}$$

i.e. $$4HNO_3 = 2H_2O + O_2 + 4NO_2$$

Dinitrogen pentoxide is the acid anhydride of nitric acid and it is unstable. The apparatus of Fig. 22.12 is used to demonstrate the action of heat on nitric acid.

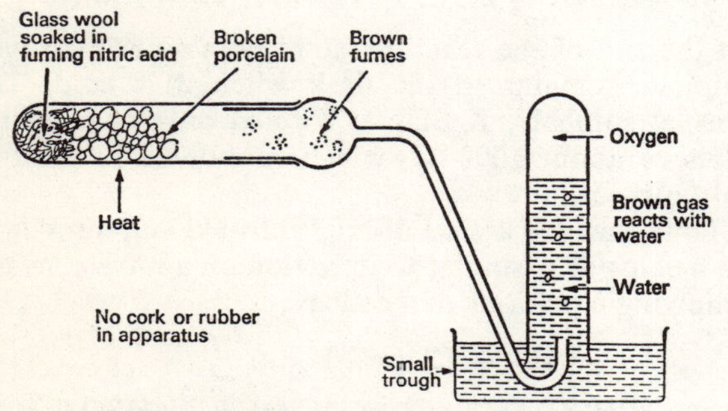

Fig. 22.12 Thermal decomposition of nitric acid

1. Push glass wool to the bottom of a hard-glass boiling-tube. Add fuming or concentrated nitric acid drop by drop to the glass wool until it can absorb no more. Arrange the apparatus as in Fig. 22.12. If possible, the apparatus should be all glass because nitric acid vapour reacts with cork and rubber.
2. Heat the broken porcelain strongly. Some heat passes by conduction down to the acid and boils it. The acid vapour must pass through the hot porcelain.
3. Note the colour of the fumes in the boiling-tube and the colour of the gas as it passes through the water in the small trough and when it collects above the water. Test the gas with a glowing splint.

The acid vapour decomposes to form a brown mixture of gases, which are nitrogen dioxide, oxygen, and steam. The reddish-brown dioxide reacts with the water in the trough and only oxygen collects in the test-tube.

Oxidizing properties of nitric acid The acid is a powerful oxidizing agent because it readily produces oxygen which reacts with reducing agents:

$$2HNO_3 = H_2O + 2NO_2 + [O]$$

and also $$2HNO_3 = H_2O + 2NO + 3[O]$$

The formation of reddish-brown nitrogen dioxide when nitric acid reacts is a sign that the acid is being reduced; nitrogen monoxide is colourless but it also forms the dioxide when it mixes with air.

Add *copper* to cold concentrated nitric acid. Reddish-brown fumes are evolved at once and green aqueous copper nitrate is formed. The essential change is that the acid removes electrons from the copper; see p. 129.

$$Cu(s) - 2e = Cu^{2+}(aq) \quad \text{copper oxidized}$$

$$4HNO_3 + 2e = 2NO_3^- + 2NO_2 + 2H_2O \quad \text{acid reduced}$$

i.e. $$4HNO_3 + Cu = Cu(NO_3)_2 + 2NO_2 + 2H_2O$$

Add *copper* to diluted nitric acid made by adding the concentrated acid to an equal volume of water—the diluted acid contains about 33 per cent nitric acid. Colourless nitrogen monoxide is formed and the gas turns brown in air. The copper is oxidized as above but the reduction of the acid by the copper proceeds further:

$$8HNO_3 + 6e = 6NO_3^- + 2NO + 4H_2O$$

i.e.

$$8HNO_3 + 3Cu = 3Cu(NO_3)_2 + 2NO + 4H_2O$$

then $$2NO + O_2 = 2NO_2$$

Nitric acid oxidizes *sulphur dioxide* to sulphuric acid, *sulphur* to sulphuric acid, and *hydrogen sulphide* to sulphur:

$$2HNO_3 + SO_2 = H_2SO_4(aq) + 2NO_2(g)$$

$$6HNO_3 + S = H_2SO_4(aq) + 6NO_2(g) + 2H_2O$$

$$2HNO_3 + H_2S = S(s) + 2NO_2(g) + 2H_2O$$

Iron(II) compounds are oxidized to iron(III) compounds:

$$Fe^{2+} - e = Fe^{3+}$$

Hot *sawdust* bursts into flames when a few drops of the fuming acid are dropped on to it. The acid supplies oxygen to start the combustion of the sawdust.

Concentrated nitric acid mixed with concentrated hydrochloric acid in the proportions by volume of 1:3 dissolves platinum and gold. The mixture is called *aqua regia*, which means 'royal water'. The hydrochloric acid is oxidized to chlorine, which reacts with the metals:

$$2Au + 3Cl_2 = 2AuCl_3, \text{ gold chloride}$$

MANUFACTURE OF NITRIC ACID

BY OXIDATION OF SYNTHETIC AMMONIA The ammonia produced by the Haber process is mixed with excess air and passed over a platinum-rhodium catalyst at about 900 °C and under a pressure of 10 atmospheres.

$$4NH_3 + 5O_2 = 6H_2O + 4NO \text{ (nitrogen monoxide)} + \text{heat}$$

The gases are cooled so that the excess air converts the nitrogen monoxide to dioxide at a temperature below 600 °C.

$$2NO + O_2 \rightleftharpoons 2NO_2$$

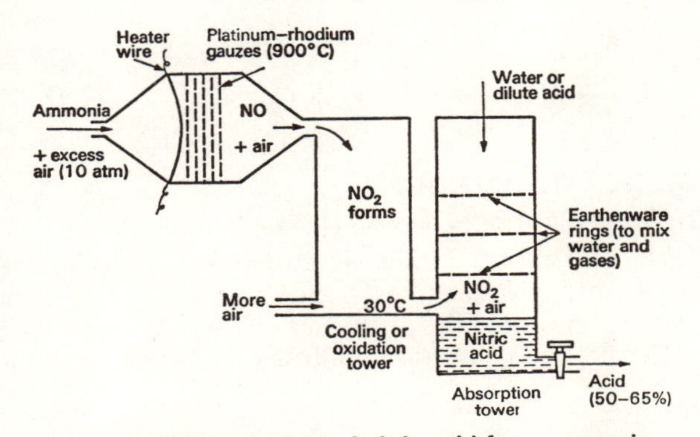

Fig. 22.13 Manufacture of nitric acid from ammonia

The dioxide, mixed with air, is absorbed in water or in dilute nitric acid.

$$4NO_2 + O_2 + 2H_2O = 4HNO_3$$

Catalytic furnace The layers of platinum-rhodium gauze are heated electrically by a heater wire to start the reaction. The oxidation of ammonia is an exothermic reaction and the heater is turned off as soon as the reaction commences. The high pressure increases the rate of oxidation.

Oxidation tower (cooling tower) The hot gases are cooled to about 30 °C in a large empty tower, to which more air is added. Nitrogen dioxide, mixed with excess air, is formed.

Absorption towers The cool gases are passed up several towers down which water or dilute nitric acid flows over earthenware rings. The high pressure assists the formation of nitric acid; the concentration of the acid is between 50 and 65 per cent.

Distillation of the acid produces 68 per cent acid which is called concentrated nitric acid. A more concentrated acid cannot be obtained by ordinary distillation. Distillation of the concentrated nitric acid with concentrated sulphuric acid produces fuming nitric acid, which is about 98 per cent acid. Colourless crystals of pure nitric acid separate when the fuming acid is cooled in a freezing mixture.

FROM CHILE SALTPETRE, impure sodium nitrate. This method is now very rarely used. The Chile saltpetre is heated with concentrated sulphuric acid in cast-iron retorts:

$$H^+ + NO_3^- = HNO_3(g)$$

The nitric acid vapour which is evolved is cooled and then condensed in silica or earthenware vessels and it forms about 95 per cent acid. A mixture of sodium sulphate and sodium hydrogensulphate remains in the retort.

ELECTRONIC THEORY OF OXIDATION AND REDUCTION

The original meaning of oxidation was the addition of oxygen to an element or compound and the removal of hydrogen from a compound. Reduction was the opposite process and the term meant the removal of oxygen from a compound or the addition of hydrogen to an element or compound. Oxidation and reduction must occur together, and oxidation-reduction reactions are called *redox reactions*. For example:

Oxidized		*Reduced*	
$2H_2$	+	O_2	$= 2H_2O$
H_2	+	CuO	$= H_2O + Cu$
$2CO$	+	O_2	$= 2CO_2$
Reducing agent		Oxidizing agent	

The definitions have now been extended to include transfer of electrons.

Oxidation is the removal or *loss* of electrons from a substance.

Reduction is the *gain* of electrons by a substance.

An oxidizing agent is an *electron acceptor* because it receives electrons from a substance; a reducing agent is an *electron donor* because it supplies electrons to a substance.

Oxidizing agent	+	*electrons*	=	*reducing agent*
Cl_2	+	2e	=	$2Cl^-$
Fe^{3+}	+	e	=	Fe^{2+}
S	+	2e	=	S^{2-}
$2HNO_3$	+	e	=	$NO_2 + NO_3^- + H_2O$
$2H_2SO_4$	+	2e	=	$SO_2 + SO_4^- + 2H_2O$

The following equations represent the oxidation of iron(II) compounds to iron(III) compounds:

$$4FeO + O_2 = 2Fe_2O_3; \quad 2FeCl_2 + Cl_2 = 2FeCl_3$$

The essential change in each of the reactions is:

$$Fe^{2+} - e = Fe^{3+}$$

Electrolysis The transfer of electrons occurs during electrolysis. The anode receives electrons and the cathode supplies electrons. Thus oxidation occurs at the anode and reduction occurs at the cathode, for example:

at anode:

$$2Cl^- - 2e = Cl_2;$$
$$4OH^- - 4e = 4OH = 2H_2O + O_2$$

at cathode:

$$Cu^{2+} + 2e = Cu(s);$$
$$2H^+ + 2e = H_2(g)$$

The following *oxidizing agents* have already been studied: oxygen, sulphuric acid, nitric acid, chlorine, and other halogens. The following are *reducing agents*: hydrogen, hydrogen chloride, carbon, sulphur, sulphur dioxide, and hydrogen sulphide.

NITRATES

These are prepared by the action of concentrated nitric acid on metals (copper and lead) or of dilute acid on metals (magnesium and zinc), oxides, hydroxides, and carbonates. All nitrates are soluble and all decompose on heating.

SODIUM AND POTASSIUM NITRATES

Chile saltpetre is crude sodium nitrate with sodium iodate, $NaIO_3$, as one of the other constituents. The sodium nitrate is purified by crystallization from hot water; the iodate remains in solution.

Potassium nitrate is prepared commercially by mixing hot saturated solutions of potassium chloride and sodium nitrate. Sodium chloride is the least soluble of the four possible salts and crystallizes out; potassium nitrate separates when the remaining solution is cooled:

$$KCl + NaNO_3 = NaCl(s) + KNO_3(aq)$$

Both nitrates melt to colourless liquids when heated, and then decompose to form oxygen and a nitrite; sodium nitrate decomposes more readily than potassium nitrate:

$$2NaNO_3 = O_2(g) + 2NaNO_2(l), \text{ pale yellow}$$

The nitrates are used as fertilizers. Potassium nitrate is used in explosives, but sodium nitrate is deliquescent and cannot be used as an explosive. *Gunpowder* is a mixture of potassium nitrate, carbon, and sulphur in the proportions by mass of 15:2:3. On explosion it forms nitrogen, carbon monoxide, and carbon dioxide, together with the carbonate, sulphate, and sulphide of potassium which are given off as smoke. The potassium nitrate serves as a source of oxygen.

OTHER NITRATES

All the other nitrates, except those of lead and ammonium, are hydrated. On heating, the hydrates lose their water of crystallization; some nitric acid is also formed.

$$Cu(NO_3)_2.3H_2O = Cu(NO_3)_2 + 3H_2O$$

All the anhydrous metal nitrates decompose to form an oxide, nitrogen dioxide, and oxygen:

In general,

$$2M(NO_3)_2 = 4NO_2 + O_2 + 2MO$$

For example,

$$2Zn(NO_3)_2 = 4NO_2 + O_2 + 2ZnO$$

Zinc oxide is yellow when hot and white when cold.

TEST FOR NITRATES

Mix the solid substance with copper turnings and cover with concentrated sulphuric acid in a test-tube. Heat gently. If brown fumes are evolved, the substance is a nitrate. The nitrate and acid first form nitric acid, which is reduced to brown nitrogen dioxide by the copper:

$$H^+ + NO_3^- = HNO_3$$
$$Cu + 4HNO_3 = 2NO_2 + Cu(NO_3)_2 + 2H_2O$$

OXIDES OF NITROGEN

To prepare dinitrogen tetraoxide (liquid), N_2O_4

Heat lead nitrate, which is anhydrous, on asbestos paper in a boiling tube as shown in Fig. 22.14. The asbestos prevents the residue of lead oxide combining with the glass. The lead nitrate crystals crackle at

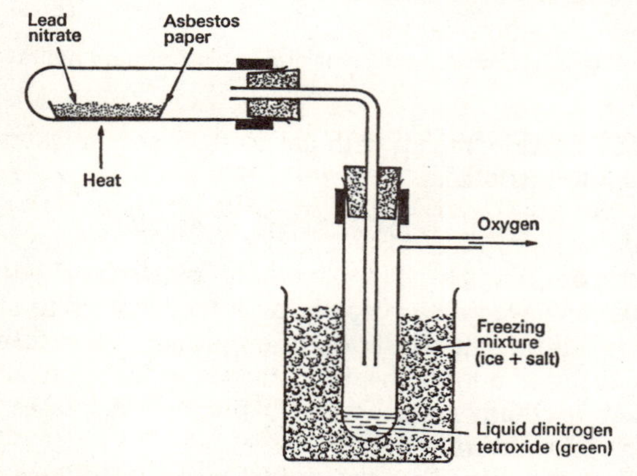

Fig. 22.14 Preparation of dinitrogen tetraoxide from a nitrate

first. This is called decrepitation; gas forms inside the crystals and bursts them open. A brown mixture of nitrogen dioxide (NO_2) and oxygen is evolved. The nitrogen dioxide condenses to a green liquid N_2O_4 and the oxygen passes on:

$$2Pb(NO_3)_2 = 4NO_2 + O_2 + 2PbO$$

The lead oxide is reddish-brown when hot. The pure dinitrogen tetraoxide is a yellow liquid, and the green colour is probably due to dissolved dinitrogen trioxide N_2O_3.

ACTION OF HEAT ON DINITROGEN TETRAOXIDE
The solid tetraoxide is colourless and melts at 9 °C to a pale yellow liquid. This boils at 22 °C to form a brown vapour. The colour gradually darkens on warming and at about 160 °C it is almost black. The colour then becomes paler and is colourless at about 620 °C. Thermal dissociation (p. 126) accounts for these colour changes. Slow cooling reverses all these colour changes.

$$\underset{\text{colourless}}{N_2O_4(s)} \rightleftharpoons \underset{\text{pale yellow}}{N_2O_4(\text{l or g})} \rightleftharpoons \underset{\text{dark brown}}{2NO_2(g)} \rightleftharpoons \underset{\text{both colourless}}{2NO + O_2}$$

COMPARISON OF NITROGEN DIOXIDE AND NITROGEN MONOXIDE

Nitrogen dioxide NO_2 or N_2O_4	*Nitrogen monoxide* NO
Reddish-brown gas	Colourless gas, turns brown in air
Reacts with water, forming nitric acid (and nitrous acid, HNO_2, if no air is present)	Almost insoluble
Acidic oxide	Neutral oxide
Denser than air	As dense as air

Both gases extinguish a burning splint and a burning candle, but support the combustion of substances such as phosphorus, carbon, and magnesium because the burning elements are hot enough to decompose the gases into nitrogen and oxygen and then they combine with the oxygen:

$$4Mg + 2NO_2 = N_2 + 4MgO$$
$$2Mg + 2NO = N_2 + 2MgO$$

To prepare dinitrogen monoxide N_2O

Mix ammonium sulphate and sodium nitrate well in a mortar. Add the mixture to a boiling tube or flask and heat it steadily. Collect the gas evolved over water, which should preferably be hot because the gas is fairly soluble in cold water.

$$NH_4^+ + NO_3^- = N_2O(g) + 2H_2O$$

Any ammonium salt and any nitrate can be used to obtain dinitrogen monoxide. The gas is evolved when ammonium nitrate itself is heated but the preparation in this way can be dangerous because ammonium nitrate sometimes explodes, especially if it is heated too strongly or if small quantities become very hot.

$$NH_4NO_3 = N_2O + 2H_2O$$

PROPERTIES OF DINITROGEN MONOXIDE
The gas is colourless and has a faint, pleasant smell. It is used as a mild anaesthetic in dental and other minor operations. It is moderately soluble in water and the solution is neutral.

Dinitrogen monoxide readily decomposes at temperatures above 600 °C and forms a mixture of nitrogen and oxygen containing one-third oxygen by volume. The gas re-lights a glowing splint and therefore it is sometimes confused with oxygen. Other burning substances, such as magnesium, sulphur, phosphorus, and a candle, burn brightly in the gas.

Dinitrogen monoxide differs from oxygen in four ways:

1. It has a faint but distinct smell.
2. It is more soluble in water. It is not absorbed by alkaline benzenetriol solution which readily absorbs oxygen.
3. It does not react with nitrogen monoxide to form brown fumes.
4. It leaves a residue of nitrogen gas when it reacts with hot copper whereas oxygen is completely absorbed by excess hot copper:

$$N_2O + Cu = CuO + N_2(g)$$

THE NITROGEN CYCLE

The element nitrogen undergoes a series of changes into various compounds. The changes are continuously repeated and they are therefore called the nitrogen cycle. Figure 22.15 summarizes the changes of the nitrogen cycle.

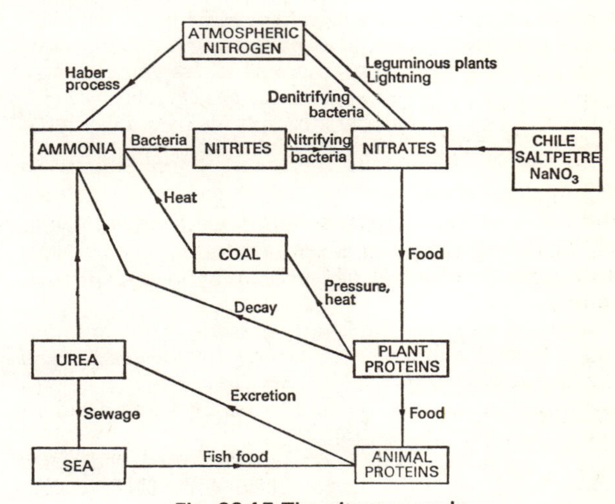

Fig. 22.15 The nitrogen cycle

All living things contain nitrogen which occurs in compounds called *proteins*. The proteins are made by plants from nitrogen in the soil where the element is present in nitrates; this nitrogen is known as *fixed nitrogen* because it is not the free element. Growing plants remove nitrogen from the soil; if more plants are to be grown in the same soil, this nitrogen should be replaced by adding ammonium salts and nitrates as *fertilizers*.

Some leguminous plants such as peas and beans can use atmospheric nitrogen and convert it to nitrates. Some bacteria called denitrifying bacteria can decompose nitrates in the soil and release their nitrogen as a gas. Plants form ammonia when they decay and nitrifying bacteria convert this ammonia to nitrates. Lightning in the atmosphere *fixes* nitrogen by changing it to nitrogen monoxide, which then forms nitrogen dioxide and finally nitric acid which comes down in rain.

SUMMARY

Nitrogen: unreactive gas prepared by fractional distillation of liquid air.

Ammonia: colourless gas, pungent smell, poisonous, easily liquefied. Very soluble; solution is alkaline, contains NH_3(aq) and a few NH_4^+ and OH^- ions.
Preparation: heat NH_4Cl or $(NH_4)_2SO_4$ with an alkali (CaO). Dry with CaO, collect by upward delivery.
Properties: does not support combustion, burns only in oxygen with yellowish flame to give nitrogen and steam. Heat or sparks decompose it. With platinum catalyst in air or oxygen, ammonia forms steam and NO; on cooling, this gives NO_2 and in water HNO_3.
With HCl: white fumes of NH_4Cl.
With Cl_2: N_2 and white fumes of NH_4Cl.
With hot CuO or PbO: N_2 with steam and metal.
With acids: ammonium salts.
Tests: characteristic smell; alkaline; white fumes with HCl.
MANUFACTURE: from *coal* by heating ammonia liquor with lime and absorbing ammonia in sulphuric acid; or from *Haber process* using synthesis gas ($H_2:N_2 = 3:1$ by volume) at 500 °C and 250 atmospheres over a catalyst of iron and aluminium oxide.

Ammonium compounds
Ammonia solution NH_3(aq); weak base which precipitates many metal hydroxides from salts.
Ammonium chloride NH_4Cl; sublimes by thermal dissociation.
Ammonium sulphate $(NH_4)_2SO_4$; made from ammonia liquor or from ammonia gas, carbon dioxide and aqueous suspension of gypsum $CaSO_4 . 2H_2O$.
Ammonium carbonate $(NH_4)_2CO_3$; readily forms ammonia gas, used as smelling salts.
Ammonium nitrate NH_4NO_3; fertilizer used with chalk (nitro-chalk), explosive when mixed with TNT; decomposes on heating to N_2O and steam.
Test: add NaOH solution and warm to give NH_3 gas.

Nitric acid: colourless liquid when pure; usually yellow from dissolved NO_2 formed by decomposition.
Preparation: heat a nitrate and concentrated sulphuric acid.
Acid properties only if dilute; hydrogen only released by magnesium.
Oxidizing properties: readily produces oxygen for reaction so releasing brown NO_2 fumes. Cu, HCl, H_2S, S, C, SO_2, KI, and sawdust all oxidized.
MANUFACTURE: from *Chile saltpetre* $NaNO_3$ by heating with sulphuric acid; or from *ammonia-air mixture* at 900 °C and 10 atmospheres over platinum-rhodium catalyst—NO forms NO_2 on cooling to give acid in water.

Redox reactions: simultaneous oxidation and reduction.
Oxidation: the loss of electrons from a substance.
Reduction: the gain of electrons by a substance.

Nitrates: all soluble and all decompose on heating.
Preparation from nitric acid and metals, oxides, hydroxides or carbonates.
Sodium and potassium nitrates decompose on heating to form O_2 and $NaNO_2$ or KNO_2.
Other nitrates decompose to O_2, NO_2, and oxide.
Test: heat solid with Cu and concentrated H_2SO_4 to give brown NO_2.

Oxides of nitrogen
Dinitrogen tetraoxide N_2O_4: green liquid from heating lead nitrate; NO_2 gas evolved which condenses to N_2O_4; forms HNO_3 in water with HNO_2 if air is present.
Nitrogen monoxide NO: colourless neutral gas that turns to brown NO_2 in air.
Dinitrogen monoxide N_2O: colourless neutral gas with pleasant smell, soluble in water; relights glowing splint but does not react with NO; used as anaesthetic.

Nitrogen cycle: conversion of atmospheric nitrogen to *fixed* nitrogen in soil for use by bacteria and by plants and animals in *proteins*. Nitrogen is released again by excretion and by decay.

Uses of nitrogen compounds
Ammonia: manufacture of nitric acid and ammonium salts; source of hydrogen by decomposition with catalyst and heat.
Ammonium salts: sulphate and nitrate as fertilizers; nitrate in explosives; carbonate and hydrogencarbonate as smelling salts.
Nitric acid: manufacture of nitrate fertilizers, explosives, dyes, and synthetic fibres and plastics.
Nitrates: K^+, Na^+, and NH_4^+ in fertilizers; KNO_3 with S and C in gunpowder.

QUESTIONS

1. Draw a fully labelled diagram to show the preparation and collection of some dry ammonia from ammonium chloride. State, with reasons, how you would alter the apparatus in order to prepare a concentrated solution of ammonia. Name the various substances which can be obtained by oxidizing ammonia. (C.)
2. Describe the preparation of nitric acid in the laboratory. Draw a diagram. How would you show that nitric acid (*a*) contains oxygen, (*b*) is an oxidizing agent? (C.)
3. In the laboratory preparation of nitric acid, reaction occurs between a nitrate and concentrated sulphuric acid. (*a*) Why is an apparatus, consisting of glass only, desirable for the preparation? (*b*) Nitric acid and sulphuric acid are both liquids. Why does nitric acid only distil off and not sulphuric acid? (*c*) Pure nitric acid is

colourless but the product in this preparation is usually pale yellow or even brown. Why is this? What precaution would you take in the preparation to reduce this coloration to a minimum? (*d*) If your product was yellow or brown, how would you obtain a colourless sample of nitric acid from it? (L.)

4. Write an essay on either ammonia or nitric acid. (In your answer include its manufacture, its physical and chemical properties when pure and in aqueous solution, and the economic importance of the substance.)
5. Give an account of the action of heat on three ammonium compounds. In each case make a comparison with the action of heat on the corresponding sodium salt. Discuss the use of ammonium salts. (L.)
6. How does nitrogen prepared from a compound differ from nitrogen prepared in the laboratory from the air? Mention two natural ways by which nitrogen is returned to the soil. State briefly how you would obtain nitrogen monoxide from nitric acid. (C.)
7. Describe reactions, and give equations where appropriate, in which oxides of nitrogen are produced (their collection is not required). Describe two similarities and two differences between the properties of two named oxides of nitrogen.
8. Describe either the Haber process for the manufacture of ammonia or the manufacture of nitric acid. Emphasize in your answer the raw materials used and the principles involved in the reactions.
9. Oxidation may be defined in terms of electron transfer. Discuss the application of this definition to the following: (*a*) dissolving a metal in acid, (*b*) burning lead in chlorine, (*c*) the change in oxidation state of a metal ion, (*d*) the rapid corrosion of a galvanized iron water tank when a copper water pipe is connected to it. (L.)

23 Sulphur and its Compounds

OCCURRENCE AND EXTRACTION

Frasch process In parts of the U.S.A., sulphur occurs 100 metres or more below the surface. It cannot be mined because above it there is loose wet sand called quicksand and also poisonous gases such as hydrogen sulphide. Water is heated under pressure and the superheated water is forced down the outer of three concentric pipes. It melts the sulphur. Compressed air is pumped down the narrow inner pipe. The air forces an emulsion of molten sulphur and water up the second pipe. The mixture is run off into vats to solidify, forming almost pure sulphur.

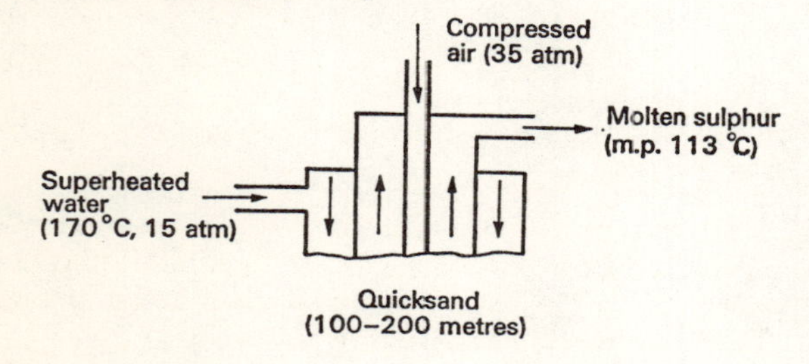

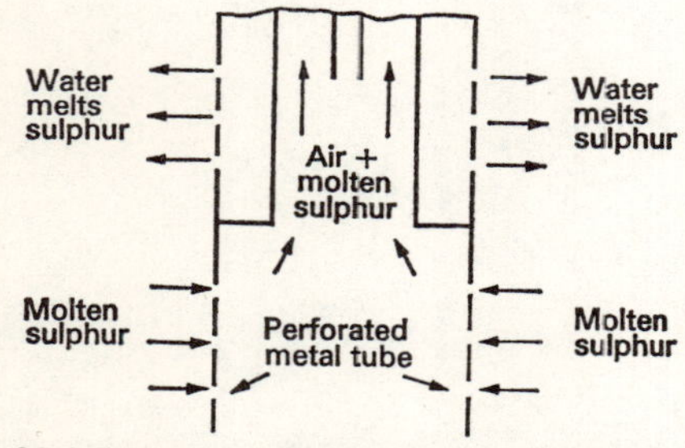

Fig. 23.1 Extraction of sulphur by Frasch process

From natural gas and petroleum Hydrogen sulphide is extracted from natural gas and from gases produced in oil refineries. Part of the sulphide is mixed with air and burnt to sulphur dioxide, which is then mixed with the rest of the hydrogen sulphide to precipitate the sulphur:

$$2H_2S + 3O_2 = 2H_2O + 2SO_2$$
$$2H_2S + SO_2 = 2H_2O + 3S(s)$$

Sicily process Rock containing up to 20 per cent of free sulphur occurs in volcanic regions of Sicily. The rock is stacked in two kilns or furnaces. In the first, the sulphur is burned and the hot gases pass through the rock in the second and melt the sulphur in it. The sulphur flows out at the bottom of the kiln. One-quarter of the sulphur is used as fuel, which is cheaper than imported coal or oil.

The crude sulphur is purified by distillation. It is boiled and the vapour is condensed in large brick chambers. At first the vapour forms a sublimate called *flowers of sulphur* on the cold walls. Later the vapour condenses to liquid sulphur, which then solidifies as *roll sulphur*.

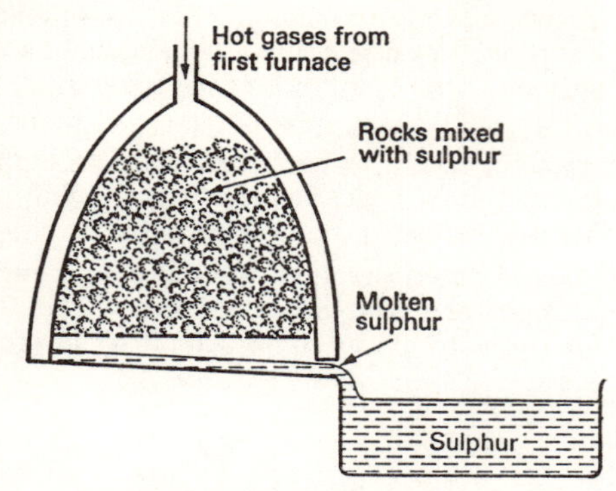

Fig. 23.2 Extraction of sulphur in Sicily

USES OF SULPHUR

Sulphuric acid, sulphur dioxide, and hydrogen sulphites are manufactured from sulphur; see p. 139. Sulphur is a fungicide for vines and hops, and sulphur ointment and drugs are widely used in medicine. The element is present in fireworks, matches, and gunpowder. Soft, pliable, natural rubber is converted into the tough rubber needed for motor car tyres and many other uses by treatment with sulphur; the process is called *vulcanizing* of rubber.

PROPERTIES OF SULPHUR

The yellow solid exists in several crystalline and amorphous allotropes or polymorphs. All are insoluble in water but the crystalline forms are soluble in carbon disulphide and hot benzene and toluene. The amorphous forms are insoluble in these liquids.

The element burns in air to form sulphur dioxide and a little sulphur trioxide, SO_3. Concentrated nitric and sulphuric acids oxidize sulphur to sulphuric acid and sulphur dioxide respectively:

$$S + 4HNO_3 = H_2SO_4 + 4NO_2 + H_2O$$
$$S + 2H_2SO_4 = 3SO_2 + 2H_2O$$

The element combines directly with copper, iron, and zinc and also with chlorine, forming S_2Cl_2, disulphur dichloride.

Action of heat The solid melts at 113 °C to an amber liquid. The molecules in both the solid and in this liquid are puckered rings of eight atoms, S_8 (Fig. 14.10 on p. 79). The clear liquid becomes dark red on further warming and suddenly becomes very viscous at about 160 °C because the rings break and form long chains with as many as 100 000 atoms in a chain.

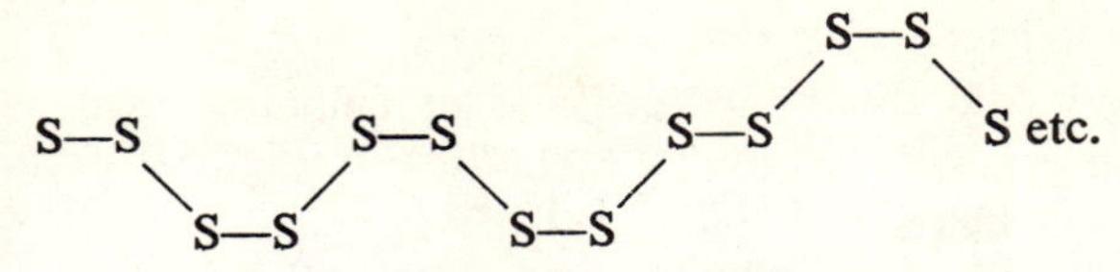

At higher temperatures the colour becomes almost black and the liquid is mobile again because short-chain molecules form and these can pass more readily over each other. Sulphur boils at 446 °C and the molecules are octatomic, S_8. At 1000 °C they seem to be diatomic, S_2, and at higher temperatures they are monatomic, S. The gaseous allotropes are probably in equilibrium:

$$S_8 \rightleftharpoons 4S_2 \rightleftharpoons 8S$$

The changes are examples of thermal dissociation (p. 126) and reversible reactions.

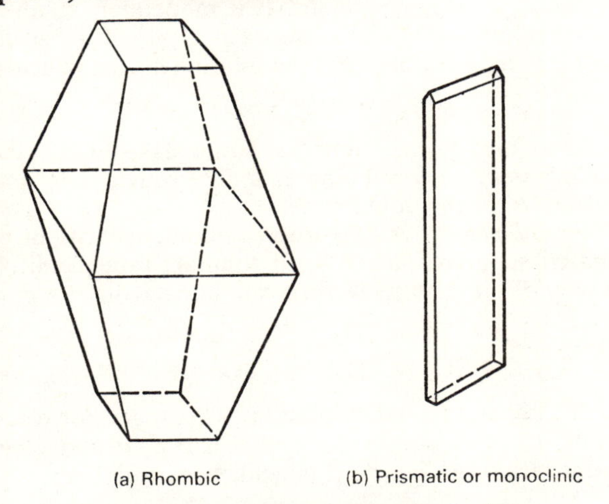

Fig. 23.3 Two crystalline allotropes of sulphur

The various allotropes of sulphur are:

1. Plastic sulphur, which is amorphous.
2. Rhombic or α-(alpha) sulphur, which is crystalline.
3. β-(beta) or monoclinic sulphur, which is crystalline.
4. Various liquid forms of sulphur, which consist of rings and chains of atoms.
5. S_8, S_2 and S in gaseous sulphur.

To prepare plastic sulphur

1. Fill a test-tube with powdered sulphur or small lumps of roll sulphur.
2. Heat the sulphur until it melts and then boils. Move the flame over all parts of the tube because the sulphur is a poor conductor and one part can boil while the rest is still cool.
3. Pour the boiling sulphur in a steady stream into a beaker of cold water.

Usually the plastic sulphur is black but the pure form is pale yellow. It is elastic and can be stretched like rubber, but it hardens in a day or two. It is insoluble in carbon disulphide. The sudden cooling of the boiling liquid preserves the equilibrium existing in the liquid, and the sulphur atoms in plastic sulphur are arranged in chains and not in rings. Plastic sulphur is therefore called a supercooled liquid.

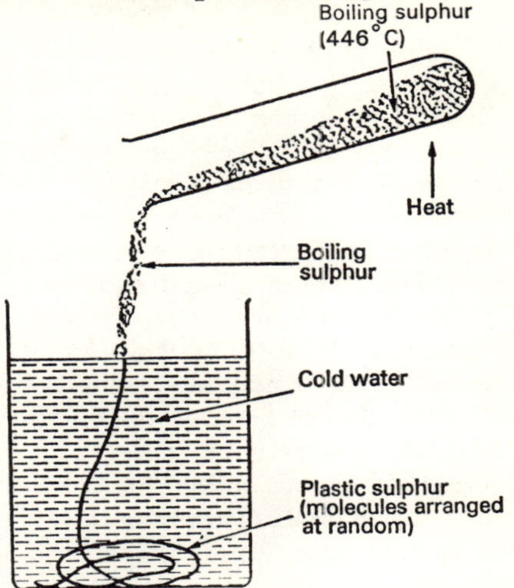

Fig. 23.4 Preparation of plastic sulphur

To prepare rhombic or α-sulphur

1. Dissolve powdered roll sulphur in cold carbon disulphide. Do not heat carbon disulphide; keep it away from flames because it is as inflammable as petrol and its vapour is poisonous.
2. Filter the solution through a dry filter paper; the disulphide will not pass through a wet filter paper because it does not mix with water.

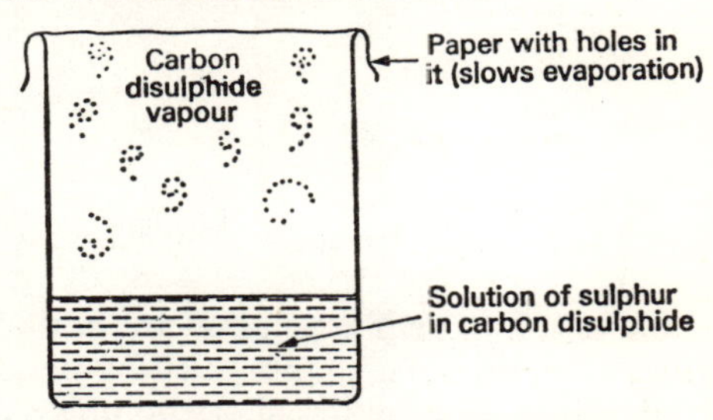

Fig. 23.5 Preparation of rhombic sulphur

3. Allow the solution to evaporate slowly in a fume cupboard or in the open air. Transparent, lemon-yellow crystals form.

To prepare monoclinic or β-sulphur

1. Half-fill a boiling-tube with dimethylbenzene (xylene) or methylbenzene (toluene) and add powdered roll sulphur.
2. Push a loose plug of glass wool or cotton wool into the mouth of the tube. The plug prevents vapour escaping freely when the liquid is warmed. Heat the tube gently with a medium flame. Do not boil the liquid because its vapour may catch fire.
3. If necessary, add more sulphur until the solvent is saturated near its boiling point. Keep the liquid away from flames when you add the sulphur.

4. Surround the tube with cotton wool so that it cools slowly. Allow it to stand for some time. Needle-shaped crystals form when the temperature is above 96 °C; if the temperature falls below this, rhombic crystals form.

Carbon disulphide cannot be used instead of xylene or toluene because its boiling point is below 96 °C.

Monoclinic sulphur can also be prepared when molten sulphur crystallizes because its temperature is much above 96 °C. The method is as follows.

1. Fill a test-tube with powdered roll sulphur.
2. Heat the tube slowly and move it continuously in the flame. The sulphur should melt to an amber liquid; if the liquid is red, it has been heated too quickly and it is too hot.
3. Take two dry filter papers together and fold them into one cone. Hold the cone in tongs. Pour the liquid sulphur into the cone.
4. Allow the liquid to cool until a solid crust forms on its surface, which cools faster than the other parts. Pour away into cold water the remaining liquid sulphur.
5. Open the filter paper cone and note the needle-shaped crystals.

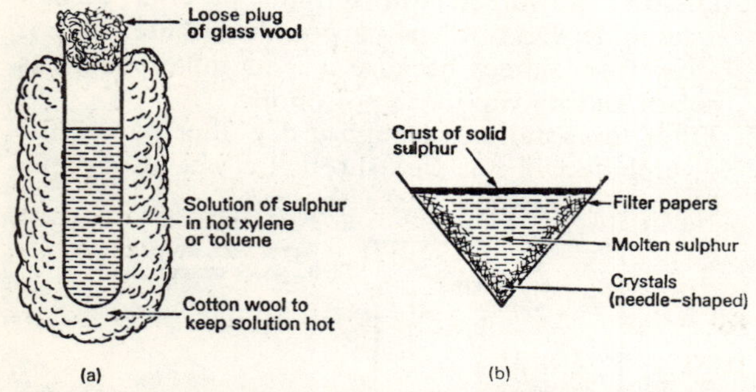

Fig. 23.6 Two methods of preparing monoclinic sulphur

Rhombic and monoclinic sulphur are both crystalline forms. Their colours, shapes, melting points, and relative densities are different. Rhombic sulphur changes into monoclinic sulphur at and above 96 °C; the reverse change occurs at and below 96 °C, which is called the *transition point* of the two allotropes:

$$\text{Rhombic sulphur} \xrightleftharpoons{96\ ^\circ\text{C}} \text{Monoclinic sulphur}$$

The two forms are known to be allotropes because one gram of one form changes into one gram of the other and nothing else.

HYDROGEN SULPHIDE, H_2S

When hydrogen is passed through liquid sulphur almost at its boiling point, some combination occurs:

$$H_2(g) + S(l) \rightleftharpoons H_2S(g)$$

Only about 1 per cent hydrogen sulphide is formed.

To prepare hydrogen sulphide

Add cold dilute hydrochloric or sulphuric acid to iron(II) sulphide in the apparatus of Fig. 5.9 or in Kipp's apparatus (Fig. 5.10); see p. 28:

$$FeS + 2HCl = H_2S + FeCl_2$$

or
$$S^{2-} + 2H^+ = H_2S(g)$$

The gas contains hydrogen, formed from iron in the sulphide, and acid spray. Collect the gas over water, which removes the acid spray.

Hydrogen sulphide gas can be dried by calcium chloride or phosphorus(V) oxide but not by concentrated sulphuric acid, which oxidizes it to sulphur.

PROPERTIES OF HYDROGEN SULPHIDE

Physical properties It is a colourless gas with a smell of bad eggs. It is poisonous. It is moderately soluble in water; about $2\frac{1}{2}$ cm³ dissolve in 1 cm³ of water at ordinary temperatures and the solution is a weak acid:

$$H_2S \rightleftharpoons H^+ + HS^- \rightleftharpoons 2H^+ + S^{2-}$$

It condenses to a colourless liquid at −61 °C. Its boiling point is low because, unlike water, it is not associated in the liquid state.

Chemical properties Hydrogen sulphide burns in a limited supply of air to form sulphur; in a plentiful supply of air it burns to form sulphur dioxide:

$$2H_2S + O_2 = 2H_2O + 2S \text{ (limited air)}$$
$$2H_2S + 3O_2 = 2H_2O + 2SO_2 \text{ (plentiful air)}$$

The gas is very ready to supply hydrogen (or electrons) to other substances and it is a good reducing agent (electron donor); it is oxidized to sulphur:

$$H_2S = S + 2[H] \quad \text{or} \quad S^{2-} = S + 2e$$

Sulphur is liberated from hydrogen sulphide by the action of air on its solution. The gas is also oxidized to sulphur by the following reactants:

Potassium manganate(VII) solution turns from purple to colourless.

Orange potassium dichromate(VI) solution is turned green.

Concentrated nitric acid, concentrated sulphuric acid, chlorine, bromine, and moist sulphur dioxide are all able to release sulphur from the gas:

$$H_2S + 2HNO_3 = S + 2NO_2 + 2H_2O$$
$$3H_2S + H_2SO_4 = 4S + 4H_2O$$
$$H_2S + Cl_2 = S + 2HCl$$
$$2H_2S + SO_2 = 3S + 2H_2O$$

Hydrogen sulphide is a dibasic acid and can form normal and acid salts with caustic alkalis:

$$2NaOH + H_2S = 2H_2O + Na_2S,$$
sodium sulphide

then
$$Na_2S + H_2S = 2NaHS,$$
sodium hydrogensulphide

Insoluble sulphides are precipitated when the gas is passed into solutions of copper, lead, and zinc salts:

$$Cu^{2+} + S^{2-} = CuS(s),\ \text{black copper sulphide}$$

$$Pb^{2+} + S^{2-} = PbS(s),\ \text{black lead sulphide}$$

$$Zn^{2+} + S^{2-} = ZnS(s),\ \text{white zinc sulphide}$$

SULPHUR DIOXIDE AND SULPHITES

To prepare sulphur dioxide (from a sulphite)

Add concentrated hydrochloric acid to sodium sulphite crystals in a flask arranged as in Fig. 11.1 (p. 57). Heat the mixture gently. Dry the gas with concentrated sulphuric acid and collect it by downward delivery:

$$2H^+ + SO_3^{2-} = SO_2(g) + H_2O$$

PROPERTIES OF SULPHUR DIOXIDE

Physical properties It is a colourless gas with a characteristic pungent odour of burning sulphur. It is very soluble in water (the apparatus of Fig. 22.5 or Fig. 22.6 on p. 124 can be used to show this) and the solution is *sulphurous acid*, a weak dibasic acid:

$$H_2O + SO_2 \rightleftharpoons H_2SO_3 \rightleftharpoons H^+ + HSO_3^- \rightleftharpoons 2H^+ + SO_3^{2-}$$

Sulphur dioxide is therefore an acidic oxide and an acid anhydride. It condenses to a colourless liquid (b.p. − 10 °C) and can be liquefied at room temperatures by a pressure of three atmospheres. The liquid is stored in thick-walled glass siphons.

Chemical properties The gas or its solution can act as an acid, a reducing agent, a bleaching agent, and an oxidizing agent.

It reacts as an *acid* with aqueous alkalis to form normal and acid salts:

$$2NaOH + SO_2 = 2H_2O + Na_2SO_3,\ \text{sodium sulphite}$$

then

$$Na_2SO_3 + H_2O + SO_2 = 2NaHSO_3,\ \text{sodium hydrogensulphite}$$

The moist gas and solution are good *reducing agents* because they can remove oxygen from substances or supply electrons to substances:

$$H_2SO_3 + [O] = H_2SO_4$$

$$SO_3^{2-} + H_2O = SO_4^{2-} + 2H^+ + 2e$$

The solution is slowly oxidized by air and is rapidly oxidized by potassium manganate(VII) solution (the purple solution is decolorized), potassium dichromate(VI) solution (orange solution is turned green), concentrated nitric acid, chlorine and other halogens, and iron(III) salts. No sulphur is precipitated and this distinguishes the changes from similar ones given by hydrogen sulphide:

$$SO_2 + 2HNO_3 = H_2SO_4 + 2NO_2$$

$$Cl_2 + H_2O + SO_3^{2-} = SO_4^{2-} + 2H^+ + 2Cl^-$$

$$2Fe^{3+} + H_2O + SO_3^{2-} = SO_4^{2-} + 2H^+ + 2Fe^{2+}$$

The moist gas *bleaches* by reduction some coloured compounds to colourless substances, and it is used to bleach straw and sponge, which are damaged by chlorine:

$$\text{Dye} + H_2SO_3 = H_2SO_4 + (\text{Dye} - O),\ \text{colourless}$$

or

$$\text{Dye} + H_2O + H_2SO_3 = H_2SO_4 + (\text{Dye} + 2H),\ \text{colourless}$$

The action of air and light restores the yellow colour to straw and newspapers.

Sulphur dioxide can sometimes act as an *oxidizing agent* by giving up its oxygen to a stronger reducing agent. For example, burning magnesium continues to burn in the gas (compare magnesium and carbon dioxide) and moist hydrogen sulphide is oxidized:

$$2Mg + SO_2 = 2MgO + S$$

$$SO_2 + 2H_2S = 2H_2O + 3S$$

USES OF SULPHUR DIOXIDE

Sulphuric acid is manufactured from sulphur dioxide; see p. 139. The yellow dye in wool and straw is bleached by it. Calcium hydrogensulphite is made from it and is used in the paper industry—the sulphite removes lignin from wood and leaves cellulose which is made into paper. Sulphur dioxide is used as a preservative in some foodstuffs and fruit juices and as a fumigant for grain. The liquid is used as a solvent in the petroleum industry.

To prepare sulphurous acid from sulphur

Burn sulphur in oxygen or air, as shown in Fig. 23.7. The sulphur dioxide formed reacts with water to form sulphurous acid; a little sulphuric acid is also formed from small amounts of sulphur(VI) oxide in the combustion gases.

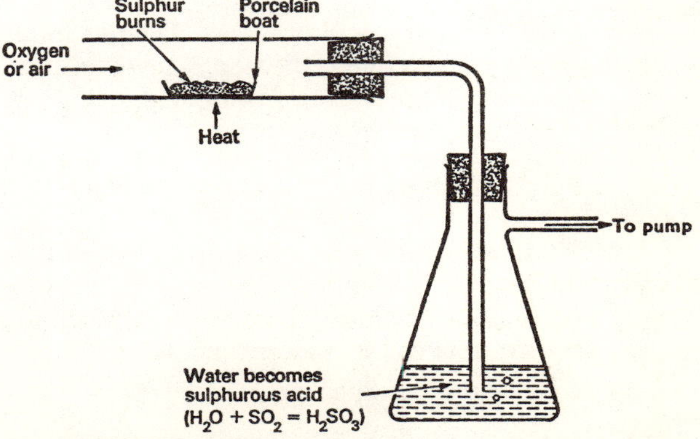

Fig. 23.7 Preparation of sulphurous acid from sulphur

Sulphurous acid exists only in solution. Sulphur dioxide is evolved when the solution is warmed.

Add aqueous barium nitrate or chloride to the solution. White barium sulphite is precipitated. Add excess dilute nitric acid to the precipitate; it dissolves and sulphur dioxide is evolved:

$$Ba^{2+} + SO_3^{2-} = BaSO_3(s)$$

$$BaSO_3 + 2H^+ = SO_2(g) + Ba^{2+} + H_2O$$

To prepare sodium hydrogensulphite and sodium sulphite

Divide sodium hydroxide solution into two equal volumes. Pass sulphur dioxide from a siphon, through one of them until the liquid smells strongly of the gas and is saturated with it. Sodium hydrogensulphite is present:

$$NaOH + SO_2 \text{ (excess)} = NaHSO_3(aq)$$

Solid hydrogensulphite cannot be obtained because crystals of $Na_2S_2O_5$ form if the solution is evaporated. *Sodium sulphite* To the aqueous hydrogensulphite add the other half of the sodium hydroxide:

$$NaHSO_3 + NaOH = H_2O + Na_2SO_3(aq)$$

Evaporate the solution until crystals ($Na_2SO_3.7H_2O$) can be obtained.

NaHS and Na_2S are prepared in the same way by using hydrogen sulphide.

Tests for sulphites and hydrogen sulphites

1. *Dilute acid* Add dilute acid to the solid or solution and warm gently. Identify sulphur dioxide by its smell and its action on filter papers dipped in potassium manganate(VII) solution (the purple colour is decolorized) and in potassium dichromate (the orange colour is turned green):

$$2H^+ + SO_3^{2-} = SO_2(g) + H_2O$$

2. *Barium salts* Add aqueous barium nitrate or chloride to a solution. A white precipitate forms:

$$Ba^{2+} + SO_3^{2-} = BaSO_3(s), \text{ white}$$

Now add dilute nitric acid in excess. The precipitate decomposes and sulphur dioxide is evolved; this distinguishes the precipitate from barium sulphate, which does not decompose.

$$BaSO_3 + 2H^+ = Ba^{2+} + SO_2(g) + H_2O$$

3. *Bromine water and chlorine water* Add one of these drop by drop to a solution. The reddish bromine and the yellowish chlorine are decolorized. The halogens are reduced to halogen acids:

$$SO_3^{2-} + Br_2 + H_2O = SO_4^{2-} + 2H^+ + 2Br^-$$

or $Br_2 + H_2O + Na_2SO_3 = 2HBr + Na_2SO_4$

SULPHURIC ACID AND SULPHATES

To oxidize sulphur dioxide to sulphur(VI) oxide

Sulphur dioxide and oxygen combine in the presence of hot platinum as a catalyst:

$$2SO_2 + O_2 \rightleftharpoons 2SO_3 + \text{heat}$$

Arrange the apparatus, which must be dry, as in Fig. 23.8. Pass sulphur dioxide from a siphon and oxygen from a cylinder through concentrated sulphuric acid to dry them. Heat the platinum, which is spread out on asbestos so that its surface area is a maximum, to about 500 °C with a medium flame. White fumes of sulphur(VI) oxide form and condense to white silky crystals in the freezing mixture. Sulphur(VI) oxide attacks rubber and cork, and therefore the apparatus must be all glass where sulphur(VI) oxide exists.

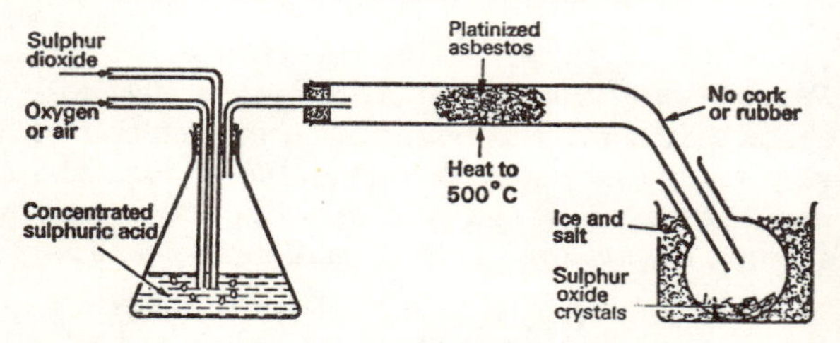

Fig. 23.8 Laboratory preparation of solid sulphur(VI) oxide

Add water, drop by drop, to some of the crystals. There is a hissing sound and sulphuric acid is formed. The oxide is an acidic oxide and an acid anhydride:

$$SO_3 + H_2O = H_2SO_4 + \text{heat}$$

The oxide fumes in moist air (compare hydrogen chloride) because its vapour combines with water vapour in the atmosphere to form droplets of sulphuric acid.

To study the action of heat on iron sulphates

Heat iron(II) sulphate in a test-tube. Observe the colour changes of the solid. Identify sulphur dioxide (from its smell and test with potassium manganate(VII) solution) and sulphur(VI) oxide in the vapours evolved:

$$\underset{\text{green}}{FeSO_4 . 7H_2O} = 7H_2O + \underset{\text{white}}{FeSO_4}$$

$$\underset{\text{white}}{2FeSO_4} = SO_2 + SO_3 + \underset{\text{red}}{Fe_2O_3}$$

Note that the residue is iron(III) oxide.

Heat iron(III) sulphate and identify the vapour evolved:

$$Fe_2(SO_4)_3 = 3SO_3(g) + \underset{\text{red}}{Fe_2O_3}$$

USES OF SULPHURIC ACID

The biggest use is in the production of superphosphate fertilizer. Phosphorus occurs as insoluble calcium phosphate, which is converted to the more soluble calcium hydrogenphosphate and calcium sulphate by the action of the acid. The mixture is called *superphosphate*:

$$Ca_3(PO_4)_2 + 2H_2SO_4 = Ca(H_2PO_4)_2 + 2CaSO_4$$

The second biggest use is in the manufacture of ammonium sulphate fertilizer from ammonia obtained from coal or by the Haber process.

The acid is also used to clean iron and steel sheets before coating them with zinc to make galvanized iron, or with tin to make tin-plate, or with enamel. The actions of the acids on common salt and sodium nitrate are used to manufacture hydrogen chloride and nitric acid. Rayon, transparent paper, drugs, dyes, pigments, plastics, and detergents are made by using sulphuric acid. Accumulators and car batteries contain the diluted acid. The acid is used to remove sulphur compounds during the refining of petroleum and its products.

MANUFACTURE OF SULPHURIC ACID

CONTACT PROCESS

Sulphur dioxide is oxidized to sulphur(VI) oxide by atmospheric oxygen, using vanadium(V) oxide, V_2O_5, as a catalyst at about 500 °C. The sulphur(VI) oxide is then combined with water:

$$2SO_2 + O_2 \rightleftharpoons 2SO_3 + \text{heat}; \quad SO_3 + H_2O = H_2SO_4$$

In the original process platinum was the catalyst, but it soon loses its catalytic power because compounds of arsenic 'poison' it, i.e. make it ineffective.

1. *Production of sulphur dioxide* The gas is obtained in four ways. Sulphur is burnt in air, or sulphide ores (ZnS, PbS, FeS_2) are roasted in air in the extraction of the metals:

$$\text{(zinc blende) } 2ZnS + 3O_2 = 2ZnO + 2SO_2(g)$$

$$\text{(galena) } 2PbS + 3O_2 = 2PbO + 2SO_2(g)$$

In the third method the spent oxide, which is a mixture of S, Fe_2O_3, and Fe_2S_3 from gas works, is roasted in air.

Anhydrite, $CaSO_4$, is the source of sulphur in the fourth method. It is heated strongly with coke and silicon(IV) oxide. The carbon reduces part of the anhydrite to calcium sulphide, which reacts with the rest of the anhydrite:

$$CaSO_4 + 2C = CaS + 2CO_2$$

$$CaS + 3CaSO_4 = 4CaO + 4SO_2$$

The basic calcium oxide combines with the acidic silicon(IV) oxide to form a silicate slag, $CaSiO_3$, which is used to make cement.

2. *Purification of the impure sulphur dioxide* The gas contains arsenic compounds, sulphuric acid droplets formed from sulphur(VI) oxide produced in the burners, and dust particles. They are removed by allowing some dust to settle, by passing through electrostatic precipitators which attract dust to the charged plates, and by washing with water. The pure gas is cooled and dried by concentrated sulphuric acid.

3. *Oxidation of dioxide to sulphur(VI) oxide.* The vanadium oxide catalyst is on perforated shelves in steel vessels. The sulphur dioxide mixed with excess air is passed over the catalyst at about 500 °C. The reaction is exothermic and no external heat is required once it has started. Ordinary pressure is used; high pressures would increase the yield of oxide but the extra yield does not justify the cost of providing the high pressure.

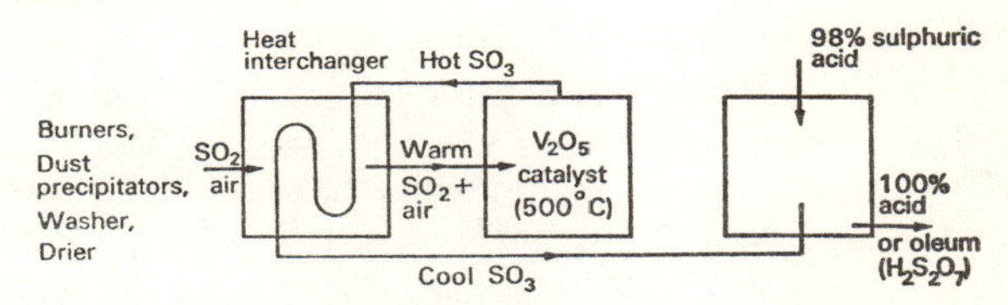

Fig. 23.9 Manufacture of sulphuric acid by the contact process

4. *Production of acid* Sulphur(VI) oxide cannot be absorbed in water because a mist of sulphuric acid droplets is formed. The cooled gases are passed up a tower down which 98 per cent sulphuric acid (i.e. 2 per cent water) flows. The oxide combines with this water to form 100 per cent acid. If necessary, further absorption of oxide produces *fuming sulphuric acid* or *oleum*:

$$H_2SO_4 + SO_3 = H_2S_2O_7, \text{ oleum}$$

The great advantage of the contact process is that it can produce the fuming acid and also acid of any desired concentration.

Effects of pressure and temperature Three volumes of reactants ($2SO_2 + O_2$) form two volumes of product ($2SO_3$). Therefore high pressures would increase the yield of the SO_3. The rate of reaction would also be greater because the three molecules of reactants would be much closer and would collide more often at high pressures. In practice, good yields are obtained without the expense of high pressures.

The rate of reaction is fast at temperatures of 1000 °C or so. However, the yield of oxide is very small because the heat decomposes it. The equilibrium yield at low temperatures is high but the time taken to attain equilibrium is far too long.

At about 500 °C (which is the optimum temperature) and in the presence of a catalyst, there is a reasonable yield in a fairly short time. Excess air is used so that as much as possible of the expensive sulphur dioxide is converted to sulphur(VI) oxide.

LEAD CHAMBER PROCESS

The acid is prepared in chambers lined with lead, on which the cold acid has little action. Sulphur dioxide and water (sulphurous acid) are oxidized to sulphuric acid by the action of air, and the catalyst is nitrogen monoxide, NO. The reactions are complicated but can be represented simply by:

$$2NO + O_2 \text{ (from air)} = 2NO_2$$
$$SO_2 + H_2O + NO_2 = H_2SO_4 + NO$$

The nitrogen monoxide then reacts with more air and the process is repeated. No oxide of nitrogen is used up in the reaction itself, but some escapes with waste gases and has to be replaced. The final reaction is:

$$2SO_2 + 2H_2O + O_2 \text{ (from air)} = 2H_2SO_4$$

PROPERTIES OF SULPHURIC ACID

Physical properties The pure acid is a colourless, viscous, dense liquid (1·84 g cm^{-3}). It boils at about 338 °C; some decomposition occurs and a little sulphur(VI) oxide is formed. The distillate is 98 per cent acid, called concentrated sulphuric acid (oil of vitriol). Its high boiling point and high viscosity indicate that it is associated $(H_2SO_4)_x$.

Chemical properties It has acidic, oxidizing, and dehydrating properties.

Great heat is evolved when the concentrated acid is diluted:

$$H_2SO_4 + aq = H_2SO_4(aq)$$
$$\Delta H = -70 \text{ kJ g-equation}^{-1}$$

When diluting, both acid and water must be cold. Always add the acid to water; add slowly and stir well. Never add water to the concentrated acid, otherwise the great heat evolved around drops of water convert it into steam, which sprays acid in all directions. When the acid is added to water in the correct manner, the heat cannot vaporize drops of acid because its boiling point is too high. The mixture must be stirred otherwise the dense acid settles to the bottom of the water.

Dilute sulphuric acid is a strong dibasic acid and has all the usual reactions on indicators, metals, bases, carbonates, and sulphites:

$$H_2SO_4 \rightleftharpoons H^+ + HSO_4^- \rightleftharpoons 2H^+ + SO_4^{2-}$$

or

$$H_2O + H_2SO_4 \rightleftharpoons H_3O^+ + HSO_4^-$$
$$H_2O + HSO_4^- \rightleftharpoons H_3O^+ + SO_4^{2-}$$

The acid system is therefore acid plus water.

Concentrated sulphuric acid displaces hydrogen chloride from chlorides, and nitric acid from nitrates respectively because its boiling point is much higher. The other two acids are volatile and therefore are removed from the reaction mixture and the reverse reaction cannot take place:

$$H^+ + Cl^- \rightleftharpoons HCl(g); \quad H^+ + NO_3^- \rightleftharpoons HNO_3(g)$$

e.g. $NaCl + H.HSO_4 \rightleftharpoons HCl(g) + NaHSO_4$

Oxidizing reactions The hot concentrated acid supplies oxygen to or removes electrons from substances because it is an electron acceptor:

$$H_2SO_4 = SO_2(g) + H_2O + [O]$$
(added to substance)

$$2H_2SO_4 + 2e = SO_2(g) + 2H_2O + SO_4^{2-}$$

Carbon is oxidized to carbon dioxide and sulphur to sulphur dioxide:

$$C + 2H_2SO_4 = 2H_2O + 2SO_2 + CO_2$$
$$S + 2H_2SO_4 = 2H_2O + 2SO_2 + SO_2 \text{ (i.e. } 3SO_2)$$

The cold concentrated acid oxidizes hydrogen sulphide to sulphur and therefore cannot be used to dry this gas:

$$3H_2S + H_2SO_4 = 4S(s) + 4H_2O$$

Metals reduce the hot concentrated acid to sulphur dioxide and the metals are oxidized to sulphates.

Dehydrating reactions Concentrated sulphuric acid is hygroscopic (p. 24 in chapter 5) and is used as a drying agent in desiccators and for gases. It dehydrates

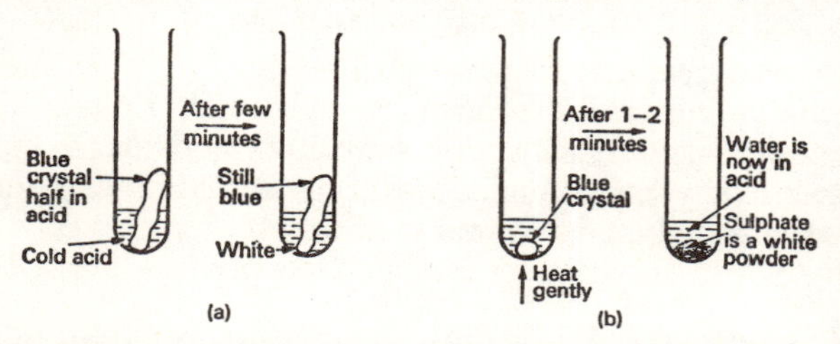

Fig. 23.10 Dehydrating action of concentrated sulphuric acid on hydrated copper sulphate

hydrated salts by removing their water and dehydrates organic compounds by removing hydrogen and oxygen, the elements of water. Blue hydrated copper sulphate is turned white, only slowly by the cold acid but quickly by the hot acid:

$$\underset{\text{blue}}{CuSO_4 . 5H_2O} = 5H_2O + \underset{\text{white powder}}{CuSO_4}$$

Sugar and starch are dehydrated to black porous carbon, ethanol to ethene (ethylene), and paper, cloth, and wood (cellulose compounds) to carbon. The 'burning' of flesh by the acid is a dehydration:

$$\text{(glucose) } C_6H_{12}O_6 = 6H_2O + 6C$$
$$\text{(ethanol) } C_2H_5OH = H_2O + C_2H_4 \text{ (ethene)}$$
$$\text{(cellulose) } C_6H_{10}O_5 = 5H_2O + 6C$$

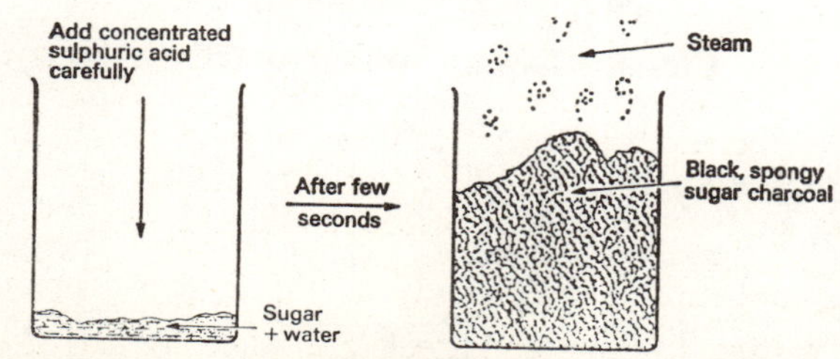

Fig. 23.11 Dehydrating action of concentrated sulphuric acid on sugar (or starch)

To prepare sodium sulphate and sodium hydrogensulphate

The equations for the two preparations are:

$$\underset{1\ \text{mole}}{H_2SO_4} + \underset{2\ \text{moles}}{2NaOH} = Na_2SO_4 + 2H_2O$$

$$\underset{1\ \text{mole}}{H_2SO_4} + \underset{1\ \text{mole}}{NaOH} = NaHSO_4 + H_2O$$

The normal salt is made by neutralizing the caustic alkali with sulphuric acid. The acid salt is then made by adding twice as much acid to the same volume of alkali.

Sodium sulphate Fill a burette with dilute sulphuric acid. Add 50 cm^3 of dilute sodium hydroxide to a beaker, add two to three drops of indicator, and titrate to find how much acid is required to neutralize the alkali. Throw away the product because the indicator makes it impure.

Now add this volume of acid to 50 cm^3 of the same sodium hydroxide without the indicator. Evaporate the product until crystals can be obtained. Alternatively, first boil the neutral solution with animal charcoal (p. 104) to remove the indicator; filter and then evaporate the solution.

Sodium hydrogensulphate Add 50 cm^3 of the same sodium hydroxide to a beaker. Now add twice the above volume of acid; no indicator should be added. Evaporate the product until crystals can be obtained.

Test for a sulphate

Add excess dilute nitric acid and then barium nitrate solution (or excess dilute hydrochloric acid and then barium chloride solution) to a solution of the substance under test. Sulphates form a white precipitate of barium sulphate:

$$Ba^{2+}(aq) + SO_4^{2-}(aq) = BaSO_4(s)\ \text{white}$$

The excess acid must be added in order to prevent precipitation of white barium carbonate or sulphite, which are insoluble in water but decomposed by acid.

ATMOSPHERIC POLLUTION

Hydrogen sulphide and sulphur dioxide are usually present in the atmosphere over and near industrial towns. The two gases are formed in fires and furnaces in which coal, coal gas, and oil are burnt. The gases have bad effects on people and certain objects and are said to pollute the atmosphere.

Hydrogen sulphide causes lead paints to turn black, owing to the formation of black lead sulphide; silver tarnishes because black silver sulphide, Ag_2S, forms.

Sulphur dioxide forms tiny drops of sulphurous acid in moist air, and some is oxidized to sulphuric acid. The acid droplets stay suspended in the air. They react with carbonates in building materials and increase the rate of weathering, they increase the speed of rusting of iron and steel, and they irritate the mucous membranes of people who breathe them. Many countries now restrict the burning of coal in open fires and ensure that coal gas and natural gas are free from sulphur compounds in an effort to reduce pollution of the atmosphere.

SUMMARY

Extraction of sulphur: In *Frasch process* sulphur is melted by steam and forced up by air under pressure; hydrogen sulphide from natural gas or from petroleum products is burnt in limited air; in *Sicily process* part of sulphur in volcanic rocks is burnt to melt the rest.
Uses of sulphur: manufacture of sulphuric acid and hydrogensulphites; fungicide; vulcanizing of rubber.
Preparation of sulphur allotropes
Plastic: pour boiling sulphur into cold water.
Rhombic: crystallize sulphur from its solution in carbon disulphide.
Monoclinic: crystallize from hot methylbenzene or dimethylbenzene or from molten sulphur.

Hydrogen sulphide: colourless gas, smell of bad eggs, moderately soluble; solution is weakly acidic.
Preparation Iron(II) sulphide and dilute HCl or H_2SO_4.
Properties Burns, forming S or SO_2. Reduces potassium manganate(VII), potassium dichromate, concentrated HNO_3 and H_2SO_4, Cl_2, SO_2, and iron(III) compounds. Sulphur is always precipitated. Forms normal and acid salts, Na_2S and NaHS. Precipitates sulphides from copper, lead and zinc salts.
Tests: smell; turns aqueous lead salts black.

Sulphur dioxide: colourless gas, pungent smell, easily liquefied. Very soluble; solution is acidic and contains H_2SO_3.
Properties: does not burn. Combines with oxygen in presence of Pt or V_2O_5. Reduces the same substances as H_2S but no sulphur is precipitated. Bleaches straw and sponge. Oxidizes Mg to MgO and H_2S to S.
Tests: smell; turns purple potassium manganate(VII) colourless and no sulphur forms.
Uses: manufacture of sulphuric acid and hydrogen sulphites; bleaching; food preservation.

Tests for sulphides, sulphites, and sulphates
Sulphides: add dilute acid and test for hydrogen sulphide.
Sulphites: add dilute acid, warm, and test for SO_2. Add aqueous $Ba(NO_3)_2$, which forms white $BaSO_3$, decomposed by excess dilute HNO_3.
Sulphates: add excess dilute HNO_3 and then aqueous $Ba(NO_3)_2$, which forms white $BaSO_4$.

Sulphuric acid: colourless, viscous, dense, hygroscopic liquid.
Preparation Pass SO_2 and O_2 over platinum catalyst at 500 °C, then add water to the SO_3 formed.
Properties Acidic properties when it contains water. Concentrated acid oxidizes H_2S to S when cold, C to CO_2, S to SO_2, and Cu to $CuSO_4$ when hot. Concentrated acid dehydrates hydrated salts, sugar, starch, paper, wood, cloth, ethanol and flesh.
Manufacture: in *Contact process*, SO_2 and excess air are passed over V_2O_5 catalyst at 500 °C and the SO_3 formed is dissolved in 98 per cent acid. In *lead chamber process*, SO_2 reacts with O_2 and water and NO is catalyst.

Uses: manufacture of superphosphate fertilizer and ammonium sulphate; cleans iron and steel; manufacture of HCl, HNO_3, rayon, paper, drugs, pigments, plastics, and detergents. Battery acid.

QUESTIONS

1. Describe how sulphur is obtained on a large scale. State how you would convert sulphur first into a solution of a hydrogensulphite, then into sulphur(VI) oxide, and finally into concentrated sulphuric acid. (C.)
2. Describe how you would obtain, starting from powdered roll sulphur: (*a*) plastic sulphur; (*b*) rhombic sulphur; (*c*) sodium sulphite crystals; (*d*) sodium sulphate crystals.
3. Describe and explain the changes which occur when iron(II) sulphate crystals are heated gently and then strongly in air. What chemical changes occur when hydrogen is passed over the solid product heated strongly in a combustion tube? Mention briefly three reactions in which iron(II) sulphate acts as a reducing agent, and give ionic equations for the reactions.
4. Describe how and under what conditions sulphuric acid reacts with (*a*) copper, (*b*) a named solid nitrate, (*c*) a named solid chloride, and (*d*) hydrated copper(II) crystals.
5. Give two ways in which sulphur dioxide is made on a large scale for the manufacture of sulphuric acid. State two other uses of sulphur dioxide. Explain the reaction by which sulphur dioxide bleaches wood pulp. Why does the colour slowly return to paper made from the pulp when it is exposed to the atmosphere? (C.)
6. Write ionic equations for (*a*) the precipitation of barium sulphite, (*b*) the neutralization of sulphuric acid with copper oxide, (*c*) the action of a dilute acid on sodium sulphite, and (*d*) the displacement of hydrogen from dilute sulphuric acid by a metal. Mention three reducing actions of hydrogen sulphide.
7. Describe the conditions under which the manufacture of sulphuric acid by the contact process is carried out. Explain why the reacting gases must be pure, why the temperature must not be too high, and why a catalyst is necessary.
8. What is the action of concentrated sulphuric acid on sugar, copper sulphate crystals, and flesh? From the equation

$$C + 2H_2SO_4 = CO_2 + 2H_2O + 2SO_2$$

calculate the mass of carbon oxidized by 19·6 g of sulphuric acid and the volume of sulphur dioxide, converted to s.t.p., liberated at the same time. (C.)

24 Metals and their Compounds

CLASSIFICATION OF ELEMENTS AS METALS AND NON-METALS

Iron and copper are typical metals; hydrogen, oxygen, chlorine, and sulphur are typical non-metals. Differences between the physical properties of metals and non-metals are listed below, but there are many exceptions and some of them will be mentioned later on this page.

Metals	*Non-metals*
Can be polished to shiny solids	Solids are not shiny
Malleable, that is, can be made into sheets	Not malleable
Ductile, that is, can be drawn into wire	Not ductile
Tough	Solids are soft
High melting points and high boiling points	Low melting points and low boiling points (many are gases at room temperature)
High relative densities	Low relative densities
Good conductors of heat and electricity	Poor conductors of heat and electricity

The essential difference between metals and non-metals is not in their physical properties but in their tendency to lose or gain electrons and become ions. All metallic elements readily lose electrons and form positively charged ions or cations. Such elements are called electropositive. Of the non-metals, the same is true only of hydrogen:

$$Na - e = Na^+; \quad Cu - 2e = Cu^{2+};$$
$$Fe - 3e = Fe^{3+}$$

All non-metallic elements, except hydrogen, readily gain electrons and form negatively charged ions or anions. Such elements are called electronegative:

$$Cl_2 + 2e = 2Cl^-; \quad S + 2e = S^{2-};$$
$$N_2 - 6e = 2N^{3-}$$

Chemical properties of metals and non-metals Every metal forms a basic oxide by donating electrons to oxygen and forming ionic oxides containing oxide ions O^{2-}. When the basic oxide is soluble in water, it produces an alkali; examples of such oxides are $Ca^{2+}O^{2-}$ and $Na_2{}^+O^{2-}$. Non-metals usually form acidic oxides which are covalent compounds; for example, carbon dioxide, sulphur dioxide, sulphur(VI) oxide, and some oxides of nitrogen. Some non-metallic oxides are called neutral oxides because they do not affect indicators, but this classification is not a good one. Water and carbon monoxide are sometimes called neutral oxides, but water can be regarded as an amphoteric oxide because it has both acidic and alkaline properties, p. 65, and carbon monoxide is an acidic oxide because it combines with hot, concentrated caustic alkalis to form sodium or potassium methanoate, H . COONa or H . COOK.

Many metals react with dilute acids and some even react with water; hydrogen is evolved as a result of donation of electrons by the metals to hydrogen or oxonium ions:

$$2Na + 2H^+ = 2Na^+ + H_2(g)$$

or $$2Na + 2H_3O^+ = 2Na^+ + 2H_2O + H_2(g)$$

Copper is less electropositive than hydrogen and never displaces hydrogen from acids. Non-metals do not displace hydrogen from water or dilute acids because they are not electron donors.

Many metallic halides are ionic compounds, for example, $(Na^+ + Cl^-)$ and $(Ca^{2+} + 2Cl^-)$. They are non-volatile, soluble in water, and their aqueous solutions are electrolytes (p. 79). Two exceptions are the chlorides of aluminium and iron, $AlCl_3$ and $FeCl_3$, which are fairly volatile and which readily react with water—anhydrous aluminium chloride fumes in moist air because it reacts to form hydrogen chloride with water vapour. Non-metallic halides are covalent and therefore are volatile and do not form conducting solutions; several of them react with water, for example, PCl_3 and PCl_5.

Metals rarely combine with hydrogen to form hydrides because they do not share electrons. Non-metals readily form covalent hydrides, such as H_2O, H_2S, NH_3, CH_4.

Metals form metallic crystals which contain mobile electrons. Non-metals form volatile diatomic molecules (O_2, H_2, N_2) or non-volatile giant structures such as diamond and graphite.

Metals are electron donors and therefore are reducing agents. Non-metals are electron acceptors and therefore are oxidizing agents.

Exceptional properties of metals and non-metals
Mercury is the only liquid metal and carbon is a non-metal with a very high melting point. The alkali metals are soft and have low relative densities and melting points. Iodine and graphite are lustrous non-metals. Graphite is a good conductor.

Iron(III) chloride and aluminium chloride are covalent and react with water. Tetrachloromethane, CCl_4, does not react with water. Strongly electropositive metals and hydrogen form ionic hydrides, e.g. NaH $(Na^+ + H^-)$ and CaH_2 $(Ca^{2+} + 2H^-)$.

Principles underlying the extraction of metals

Metals are extracted from their compounds by reduction processes in which the metallic ions are electron acceptors:

$$Na^+ + e = Na;$$
$$Zn^{2+} + O^{2-} = Zn + [O] \text{ (to reducing agent)}$$

Metals from lithium to aluminium (Li, K, Na, Ca, Mg, Al) in the electrochemical series are very electropositive; they are reluctant to accept electrons and therefore they are very difficult to reduce. They are extracted by electrolysis during which a cathode is the electron donor:

$$Al^{3+} + 3e = Al; \quad Na^+ + e = Na$$

Metals from zinc to lead in the series are less electropositive. They occur either as oxides (Fe_2O_3, Fe_3O_4) or as carbonates or sulphides (ZnS, $ZnCO_3$, FeS_2, SnO_2, PbS) which are easily converted to oxides. The oxides are reduced by carbon or carbon monoxide:

$$Zn^{2+}O^{2-} + C = Zn + CO$$
$$(2Fe^{3+} + 3O^{2-}) + 3CO = 2Fe + 3CO_2$$

Copper, mercury, silver, and gold at the bottom of the series occasionally occur as the free metals. Their compounds are easily reduced by various methods.

ALKALINE EARTH METALS

Magnesium and calcium

Group 2 of the periodic table contains magnesium, calcium, and four other alkaline earth metals. The electronic structures are: magnesium 2, 8, 2, and calcium 2, 8, 8, 2.

Magnesium and calcium are metals because their valency electrons, the two outer electrons in each atom, are mobile. They are greyish solids and fairly hard. They resemble the alkali metals in being 'light' metals but they are denser than water, and their melting points and boiling points are much higher than those of the alkali metals. The following table summarizes some of their properties:

	Magnesium	*Calcium*
Relative density	1·74	1·55
Melting point	651 °C	842 °C
Boiling point	1100 °C	1440 °C
Conduction of heat and electricity	Good	Good
Atomic radius	$1{\cdot}36 \times 10^{-8}$ cm	$1{\cdot}74 \times 10^{-8}$ cm
Ionic radius	$0{\cdot}78 \times 10^{-8}$ cm	$1{\cdot}06 \times 10^{-8}$ cm

Both metals are obtained by electrolysis of their fused chlorides, and most of the magnesium chloride used is obtained from sea water.

The actions of air, water, and acids on the metals have already been studied (pages 10, 26–7, 33 and 73). Calcium readily tarnishes in air owing to formation of its oxide and it decomposes cold water. Magnesium decomposes cold water very slowly and is stable enough in air to be used in many alloys. Magnesium is a constituent of light alloys which have great strength but low densities; the alloys are useful in engines and aircraft. The metal is used in fireworks, flares, and incendiary bombs because it burns readily.

MAGNESIUM OXIDE AND HYDROXIDE

MgO and $Mg(OH)_2$

The oxide is a white powder with a very high melting point. It is obtained by burning magnesium in air or by heating the hydroxide, carbonate or nitrate. Unlike calcium oxide, it is almost insoluble in water and it does not form the hydroxide when it is sprinkled with a limited amount of water, a process called 'slaking'. Since the oxide is chemically stable and has a high melting point it is used as a ceramic refractory in furnace linings.

The hydroxide is a white gelatinous precipitate formed when a caustic alkali or ammonia solution is added to a solution of a magnesium salt:

$$Mg^{2+}(aq) + 2OH^-(aq) = Mg(OH)_2(s)$$

Magnesium metal does not corrode in air because a film of oxide forms which prevents further reaction with the air.

CALCIUM OXIDE AND HYDROXIDE

CaO and $Ca(OH)_2$

Lime and quicklime are the common names of the oxide, which is formed by heating calcium carbonate to over 900 °C. This can be done by heating a lump of limestone or marble in a roaring Bunsen flame for about fifteen minutes or by using a crucible furnace, shown below.

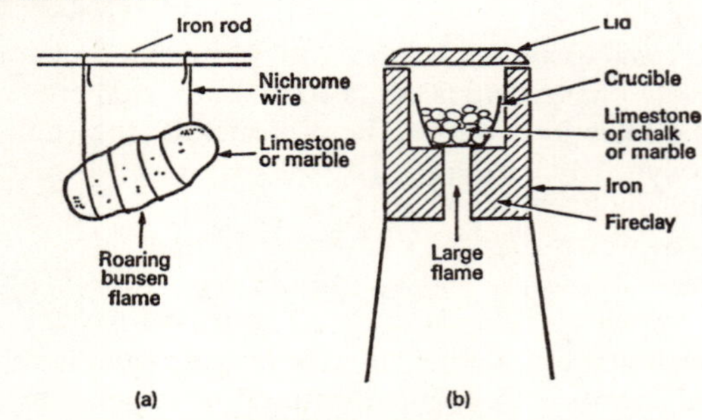

Fig. 24.1 Action of heat on calcium carbonate

Calcium oxide is a white, hygroscopic powder. Its melting point is almost 2600 °C because it has a giant structure. At high temperatures it becomes incandescent, that is it gives out a brilliant white light, called limelight.

Heat is evolved when water is added slowly to a lump of the oxide, a process called slaking of lime. Steam is given off, and the lime swells and forms a large volume of light powder; this is calcium hydroxide or slaked lime:

$$CaO(s) + H_2O(l) = Ca(OH)_2(s) + \text{heat}$$

Lime is a basic oxide which reacts with acids and acidic oxides to form salts:

$$CaO + SiO_2 = CaSiO_3, \text{ calcium silicate}$$

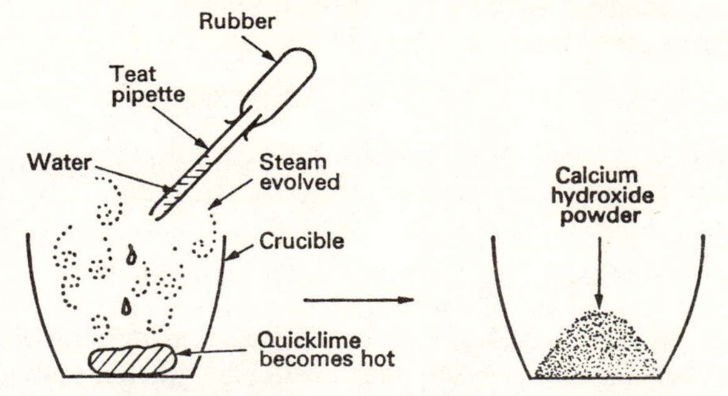

Fig. 24.2 Action of water on calcium oxide (quicklime)

Calcium oxide absorbs moisture and carbon dioxide when in air:

$$CaO + H_2O = Ca(OH)_2; \quad CaO + CO_2 = CaCO_3$$

It is used as a drying agent to dry ammonia and ethanol in desiccators.

Calcium hydroxide or *slaked lime* is a white powder. It absorbs carbon dioxide from air and forms the carbonate. It loses water when heated strongly. It is slightly soluble in water and is less soluble in hot water than in cold. The aqueous solution is called *lime water*, which is used as a test for carbon dioxide. A suspension of slaked lime in water is called *milk of lime*. Cold dry slaked lime absorbs chlorine and forms bleaching powder.

$$Ca(OH)_2(s) + Cl_2(g) = \underset{\text{Bleaching powder}}{CaOCl_2(s)} + H_2O$$

Uses of lime (quicklime) and slaked lime

Whitewash is a suspension of slaked lime in water. It is used to colour the walls of houses and buildings and to mark fields used for games and sports. The calcium hydroxide is changed by atmospheric carbon dioxide to calcium carbonate after several days.

Mortar is a mixture of slaked lime and sand which is made into a paste with water. The mixture sets to a hard mass which is able to stick bricks together. The setting is caused by the evaporation of water and the slow conversion of the slaked lime to calcium carbonate. The sand is a diluent and merely prevents the mortar cracking as it sets; it does not react chemically.

Cement is a mixture of calcium silicate and calcium aluminate, made by heating limestone and clay, which contains silicon(IV) oxide and aluminium oxide.

Concrete is a mixture of cement with sand and gravel. When mixed with water it sets to a hard, rock-like mass. The silicates and aluminates are hydrated and form long thread-like crystals. Concrete sets better when dampened while hardening because the water helps the formation of hydrates.

Calcium carbide is made from lime and coke heated in an electric furnace at a temperature of about 2000 °C:

$$CaO + 3C = CaC_2 + CO$$

It is used to prepare ethyne, C_2H_2 (acetylene), p. 158.

Glass is a mixture of silicates made by heating lime or limestone with sand and a metallic carbonate such as sodium or potassium carbonate.

Lime is used to soften water (see p. 146), to neutralize acidity in some soils, and to make clay soils more porous by making their particles larger.

Manufacture of lime in a lime kiln

Lime is made by heating lumps of limestone or chalk with producer gas:

$$CaCO_3 \rightleftharpoons CaO + CO_2$$

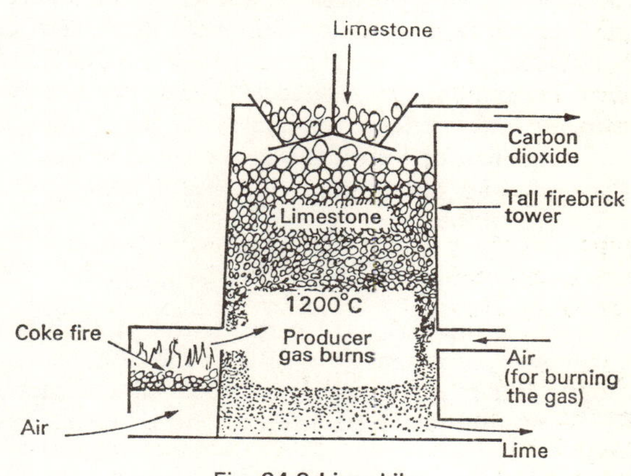

Fig. 24.3 Lime kiln

The action is reversible, but the current of gas and air sweeps away the carbon dioxide as soon as it is formed and prevents the reverse reaction. Lumps of lime are removed from the bottom of the kiln; see Fig. 24.3.

HARDNESS OF WATER

Distilled water, rain-water, and *all* soft waters readily form a stable lather of films and bubbles with soap. The water can then more easily wet any article to be washed because its surface tension is lower, see p. 164.

However, some waters do not readily form a lather with soap. Instead they first form a scum or precipitate with the soap; a lather cannot be obtained until this reaction is complete. Any water that forms a scum with soap is said to be *hard*. The hardness of water is caused by the presence of soluble salts of calcium or magnesium in the water. Hard water wastes soap.

Temporary hardness of water is caused by hydrogen carbonates of calcium or magnesium dissolved in the water. Underground water contains carbon dioxide absorbed from the air when the water fell as rain. If the water passes through rocks of limestone, chalk or magnesium carbonate, it soon dissolves a little of these rocks:

$$\underset{\text{insoluble}}{CaCO_3} + H_2O + CO_2 \rightleftharpoons \underset{\text{soluble}}{Ca(HCO_3)_2}$$

$$MgCO_3 + H_2O + CO_2 \rightleftharpoons Mg(HCO_3)_2$$

These soluble salts are easily removed by boiling the water which reverses the chemical reactions and precipitates the insoluble carbonates again.

Permanent hardness of water is caused by sulphates of calcium or magnesium dissolved in the water. If the water passes through rocks such as gypsum, $CaSO_4.2H_2O$, it becomes 'permanently' hard. The sulphates dissolved in the water are best removed by chemicals: they cannot be removed by boiling.

SOAP

Ordinary soaps are the sodium and potassium salts of some complicated organic acids such as stearic acid $(C_{17}H_{35}COO)^-H^+$, palmitic acid $(C_{15}H_{31}COO)^-H^+$, and oleic acid $(C_{17}H_{33}COO)^-H^+$. We will represent the chemical formula of soap by NaSt, where St is the negative radical or ion of the molecule of soap.

Calcium and magnesium ions react with soap to form a precipitate:

$$Ca^{2+} + \underset{\text{soap}}{2NaSt} = 2Na^+ + \underset{\text{scum}}{CaSt_2(s)}$$

The scum formed by hard water is the precipitate of the calcium and magnesium salts of the organic acids. A soapy lather is formed only when all the calcium and magnesium ions have been converted to insoluble salts.

Manufacture of soap

Vegetable oils and animal fats contain many different but similar chemical compounds known as esters, see p. 156; they consist of an organic acid such as stearic acid which is chemically combined with an alcohol called glycerol. A typical fat is glyceryl stearate.

Soap is made by heating a mixture of a fat and an aqueous caustic alkali with steam. The glycerol is released from the fat and soap is formed:

$$\underset{\text{fat}}{\text{Glyceryl stearate}} + NaOH = \underset{\text{soap}}{NaSt} + \text{glycerol}$$

Strong brine is then added and the soap separates as curds on the top; the glycerol remains at the bottom of the watery mixture.

To prepare soap in the laboratory

Dissolve a caustic alkali in methylated spirit or ethanol. Add lard or olive oil and heat gently on a water bath. When the reaction is complete, add a saturated solution of common salt to precipitate the soap. Filter off the soap.

WATER SOFTENING

As we have seen, only temporary hardness can be removed simply by boiling the water. Distillation of water removes both types of hardness because only water can evaporate from the solution and condense; the dissolved salts remain in the distillation flask.

Slaked lime removes temporary hardness only. It is added to water in reservoirs; the exact amount must be added because more than this quantity would itself cause hardness:

$$Ca(HCO_3)_2 + Ca(OH)_2 = 2CaCO_3(s) + 2H_2O$$

Washing soda, hydrated sodium carbonate, removes both types of hardness. Its use is the easiest way to soften small volumes of hard water in the home:

$$Ca^{2+}(aq) + CO_3^{2-}(aq) = CaCO_3(s)$$

Zeolite, sodium aluminium silicate, is a mineral that occurs naturally. Artificial zeolite is called *permutit*; its chemical formula can be represented by Na_2Z in which Z is the aluminium silicate radical, $Al_2O_3.xSiO_2$.

The permutit is able to exchange its sodium ions for calcium or magnesium ions. The sodium salts remain in solution in the water and do not interfere with the formation of lather by soap:

$$Ca^{2+} + Na_2Z = 2Na^+ + CaZ(s), \text{ calcium permutit}$$

The calcium permutit can be converted back to the sodium compound by soaking in brine; it can then be used again.

EFFECTS OF HARD WATER

Advantages The calcium salts in hard water are used by all animals in the formation of bones and teeth. The shells of birds' eggs also contain calcium salts.

Lead pipes cannot be used for drinking water if the water is soft because the lead dissolves as the poisonous $Pb(OH)_2$. Hard water quickly forms a coating of insoluble lead carbonate or lead sulphate on the inside of the pipes.

Disadvantages Hard water wastes soap. The scum of calcium and magnesium stearates sticks to clothes and retains the dirt. Laundries therefore prefer soft water.

The precipitate or 'fur' of calcium and magnesium carbonates that forms from hard water inside kettles and cooking pans is a bad conductor of heat. More gas or electricity is needed to heat the water.

Hard water also forms carbonate deposits called 'scale' inside boilers and boiler tubes. The boiler 'scale' reduces their heating efficiency and may block up the narrow tubes so much that steam cannot pass and the boiler bursts.

Stalactites and stalagmites Underground water in limestone areas contains calcium and magnesium hydrogen carbonates. When water drops from the roof of a cave, some of it evaporates and a tiny particle of calcium carbonate is deposited. After many years or even hundreds of years a column of calcium carbonate and magnesium carbonate, called a stalactite, is formed that hangs down from the roof of the cave. A similar column, called a stalagmite, may eventually form from the drops that fall on the floor of the cave. In time the stalactite and stalagmite sometimes meet and form one large column. Impurities in the water, such as iron salts, colour the columns.

ALUMINIUM

Extraction About 7 per cent of the earth's crust is aluminium. It is present in clay but cannot be extracted easily. Extraction by electrolysis of aluminium chloride is not possible because the chloride is too volatile. Aluminium is extracted from its hydrated oxide, $Al_2O_3.2H_2O$, called *bauxite*.

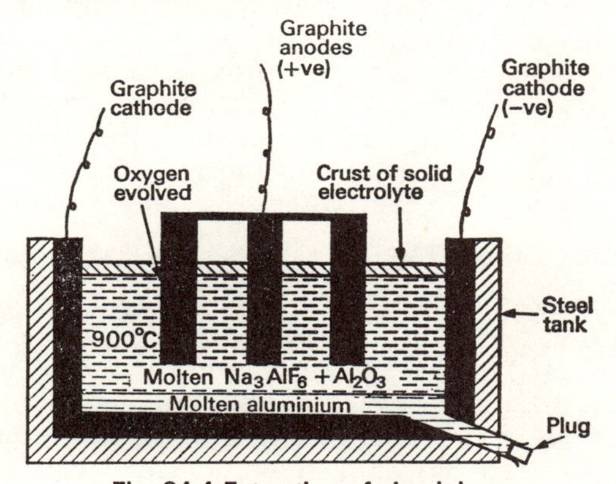

Fig. 24.4 Extraction of aluminium

The bauxite is first purified to remove silica, SiO_2, and iron oxide, Fe_2O_3. The pure aluminium oxide is dissolved in fused cryolite, Na_3AlF_6, at about 900 °C in a steel tank lined with graphite. This lining is made the cathode; the positive anodes are also graphite and they dip into the cryolite. Molten aluminium (m.p. 660 °C) collects on the floor of the cell and is run off at intervals.

The heating effect of the electric current keeps the temperature at the required value; the process is economical only where electric power is cheap.

More purified aluminium oxide is added as required. The graphite anodes are slowly oxidized to oxides of carbon:

At the cathode: $Al^{3+} + 3e = Al$

At the anode: $2O^{2-} - 4e = O_2$

i.e. $2Al_2O_3 + \text{electrical energy} = 4Al + 3O_2$.

Properties Aluminium is a silvery-white metal of low density (2·7 g/cm³) and with the usual physical properties of metals. A thin film of oxide, Al_2O_3, forms on it in air and prevents further corrosion. When heated strongly, it burns to form the oxide and nitride, AlN. It does not react with water because the oxide film prevents reaction; rubbing aluminium with mercury removes the oxide and the metal then reacts rapidly with air and with water. The oxide film also prevents the expected reactions with dilute acids Aluminium corrodes near the sea because sodium chloride present in the air attacks the oxide film.

Uses Aluminium is a bright metal of low density; it is not poisonous and it is a good conductor of heat. These properties make it suitable for cooking utensils. However, alkaline solutions corrode it and the utensils cannot be cleaned with washing soda.

Its strength and low density make it very useful in the construction of aircraft, trains, buses, lorries, and cars. The alloy *duralumin*, which contains 95 per cent aluminium together with copper and magnesium, is normally used because it is much stronger than pure aluminium.

Aluminium is used in overhead electric cables because its electrical conductivity is twice that of an equal mass of copper; the aluminium cable usually has a steel core to increase its tensile strength.

Thin aluminium foil is used as a protective wrapping for sweets, cigarettes, and foodstuffs, and it has replaced tin foil for these purposes.

Aluminium powder is used to make silver-coloured paint. The paint is used on petroleum and petrol storage tanks because it reflects heat radiation from the sun and keeps the liquids cool on hot, sunny days. The powder is also used in the *thermite process* to prepare liquid iron for welding and to prepare metals such as chromium and titanium. In this process, a mixture of aluminium powder and iron oxide is ignited by means of a strip of magnesium; the molten iron formed at the bottom of the mixture is used to weld broken railway lines:

$$Fe_2O_3 + 2Al = Al_2O_3 + 2Fe(l) + \text{heat}$$

Chromium is prepared by using chromium oxide and aluminium in the thermite process; the oxide cannot be reduced easily by carbon:

$$Cr_2O_3 + 2Al = Al_2O_3 + 2Cr + \text{heat}$$

TRANSITION METALS

Zinc, iron, and copper

The electronic structure of a calcium atom is 2, 8, 8, 2. In the next ten elements the additional electrons are not added to the outer *N* shell; they are added to the next inner shell, the *M* shell, which increases from 8 to 18 electrons. For example, the electronic structure of iron is 2, 8, 14, 2 and that of copper is 2, 8, 18, 1 and their ions are:

Fe^{2+}	2, 8, 14	Cu^{+}	2, 8, 18
Fe^{3+}	2, 8, 13	Cu^{2+}	2, 8, 17

The ten elements were originally called transition or transitional elements because they show some resemblances both to the elements preceding them and to those following them in the Periodic Table (refer to pp. 53–4).

These ten transition elements are all metals. They are lustrous and silvery white (except copper), good conductors of heat and electricity, and have high densities, high melting points, and high boiling points. Most form alloys with one another. They adsorb gases and for this reason they are good catalysts in gaseous reactions, for example, vanadium oxide in the contact process or nickel in hydrogenations (pp. 30 and 157).

The transition elements have two or more valencies, usually by losing two or three electrons and forming ions such as Fe^{2+} and Fe^{3+}. The copper atom loses either one or two electrons, forming Cu^{+} and Cu^{2+}. The ions are both hydrated and coloured in solution, for example, $Cu^{2+}(aq)$ or $Cu(H_2O)_4{}^{2+}$ is blue and $Fe^{3+}(aq)$ is yellow. The ions also form coloured complex ions such as tetraaminecopper(II) ion $Cu(NH_3)_4{}^{2+}$, which is deep blue.

Zinc differs from the other transition metals because its *M* shell contains 18 electrons and is complete and stable both in the zinc atom (2, 8, 18, 2) and ion (2, 8, 18).

ZINC

Extraction The metal occurs as zinc blende, ZnS, and as calamine, $ZnCO_3$. The ores are roasted in air to produce the oxide:

$$2ZnS + 3O_2 = 2ZnO + 2SO_2$$

$$ZnCO_3 = ZnO + CO_2$$

Two different methods are available for extracting the metal from the oxide.

In the *chemical process*, the oxide is mixed with coke, and heated to about 1400 °C by producer gas. Zinc boils at 907 °C and distils off as vapour which is condensed:

$$ZnO + C = CO + Zn(g)$$

In the *electrolytic extraction*, the oxide is converted to zinc sulphate by sulphuric acid. The sulphate solution is then electrolysed and zinc is deposited on the cathode:

$$Zn^{2+} + 2e = Zn(s)$$

Properties Zinc is a bluish-white hard metal. It forms the oxide when it burns in air or steam and it reacts with dilute acids and concentrated alkalis to yield hydrogen:

$$Zn + 2H^{+} = H_2 + Zn^{2+}$$

$$Zn + 2OH^{-} = H_2 + ZnO_2{}^{2-}, \text{ zincate ion}$$

Uses Galvanized iron is manufactured either by dipping sheets of steel into molten zinc or by electroplating them with zinc by making the steel the cathode in an electrolytic cell. Zinc metal can also be sprayed on to steel. The zinc coat prevents the iron from rusting because zinc is higher in the electrochemical series and forms ions in preference to iron:

$$Zn - 2e = Zn^{2+} \text{ rather than } Fe - 2e = Fe^{2+}$$

Galvanized iron is used on roofs and gutters, and to make buckets, dust bins and water tanks.

Brass is an alloy of zinc and copper; it is extensively used because of its pleasing appearance and its resistance to corrosion.

The metal case of dry batteries is made of zinc, which is the negative electrode.

Zinc oxide is used in paints; unlike lead paints, they are not turned black by hydrogen sulphide because zinc sulphide is white. Zinc carbonate and zinc oxide are used in lotions and ointments. Zinc chloride is used as a fungicide and to preserve wood.

IRON

Extraction The earth's crust contains 4 per cent iron; this metal is therefore the next most abundant metal after aluminium. However, iron is much easier to extract. It occurs as haematite, Fe_2O_3, magnetite, Fe_3O_4, spathic ore, $FeCO_3$, and iron pyrites, FeS_2.

The ores are roasted in air to remove water from the oxides and to convert the other ores to oxides:

$$FeCO_3 = FeO + CO_2; \quad 4FeO + O_2 = 2Fe_2O_3$$

The roasted ore, together with coke and limestone, is added to the top of a blast furnace. Air is mixed with oxygen and fuel oil and preheated to about 800 °C by waste gases leaving the furnace. The mixture of hot gases is blasted into the furnace through pipes near the bottom. The oxygen and oil greatly increase the efficiency of the furnace.

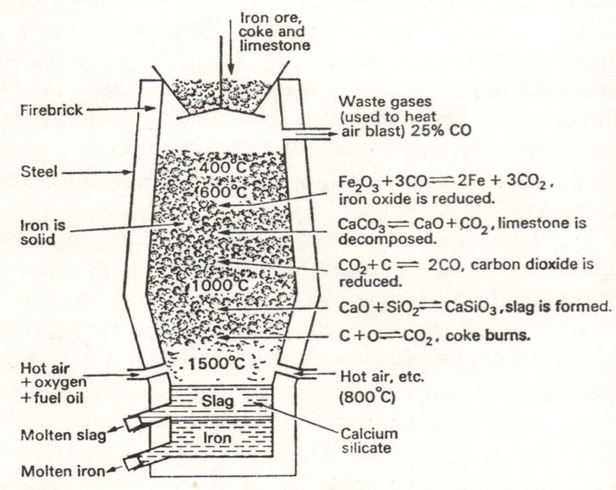

Fig. 24.5 Blast furnace for the extraction of iron

The simpler reactions which occur in the furnace are given on the diagram in Fig. 24.5. They include starting at the hottest part of the furnace near the bottom:

the combustion of coke to carbon dioxide;

the formation of a waste slag of calcium silicate from silica present in the ore and from lime;

the reduction of carbon dioxide to carbon monoxide by the white hot coke;

the decomposition of limestone to form calcium oxide;

the reduction of the iron oxide to solid iron near the top of the furnace mainly by carbon monoxide, but also by coke:

$$Fe_2O_3 + 3CO = 2Fe + 3CO_2$$

$$Fe_2O_3 + 3C = 2Fe + 3CO$$

The iron is produced as a solid at first but it melts as it passes down to the hotter parts of the furnace. It collects at the bottom as a liquid. The slag is less dense than the iron and floats on top of it. The iron and slag are removed as liquids. The other products are the waste gases (CO, CO_2, and N_2) which leave at the top of the furnace. A furnace can therefore operate continuously for months. The gases leaving the furnace contain about 25 per cent carbon monoxide and are burned to heat the air blast.

The molten iron is run into moulds and is called *pig iron* or *cast iron*; it expands slightly when it solidifies. Cast iron contains about 3 per cent of carbon, together with phosphorus, silicon, manganese, and sulphur as impurities, and the impurities make it brittle. It melts sharply and therefore cannot be welded by hammering. It is used to make gas burners, railings, and pipes which do not undergo great strain.

Manufacture of steel

Steel is an alloy of iron which contains between 0·15 and 1·5 per cent carbon, together with metals such as nickel, cobalt, and manganese. Its manufacture involves the removal of most of the carbon and silicon from iron and almost all of the sulphur and phosphorus which would make it brittle. We shall consider here two of the methods for making steel.

The Bessemer process The Bessemer converter, shown on the left of Fig. 24.6, is a steel vessel lined with basic oxides, MgO and CaO. It can be tilted on an axis. Molten iron is added when the converter is in a horizontal position so that the air-holes in the base are clear. A blast of oxygen and superheated steam is blown in and the converter is turned until it is vertical. The impurities burn off in the order silicon, manganese, and carbon; the carbon monoxide burns at the mouth with a roaring flame. Phosphorus forms its oxide which then combines with the lining to form a basic slag of calcium and magnesium phosphates:

$$3CaO + P_2O_5 = Ca_3(PO_4)_2, \text{ basic slag}$$

The slag is sold as fertilizer.

The basic oxygen furnace Formerly a blast of air was used, but the steel it produced contained dissolved nitrogen and was not suitable for many processes. In some converters the oxygen blast is merely passed over the surface of the molten iron and not blown through it. This is shown on the right of Fig. 24.6. About 300 tons of steel can be made every 45 minutes in this type of furnace.

Special steels are made by adding other elements to the final product obtained in the converter. Chromium and nickel are added to form stainless steel and steel which resists the corrosive action of sea water; tungsten and manganese are added to form very tough steels for stamping and crushing machinery.

Properties Pure iron is a white, soft metal, and ordinary iron is a grey metal. It rusts in moist air to form a hydrated oxide, $Fe_2O_3.xH_2O$. It burns in air and reacts with steam to form triiron tetraoxide,

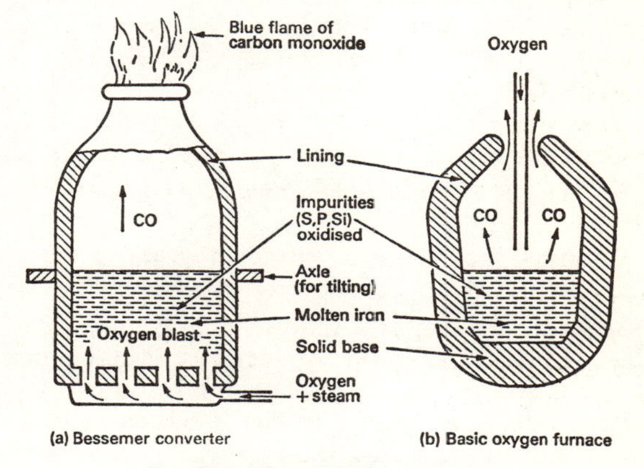

Fig. 24.6 Manufacture of steel

Fe_3O_4, a bluish-black solid. Iron reacts with dilute sulphuric acid or hydrochloric acid to form hydrogen and iron(II) salts. The hot metal reacts with hydrogen chloride gas to form the iron(II) chloride and with chlorine to form the iron(III) chloride:

$$Fe + 2HCl = H_2 + FeCl_2; \quad 2Fe + 3Cl_2 = 2FeCl_3$$

The metal adsorbs hydrogen, and this accounts for its catalytic action in the Haber process.

COPPER

Extraction The most important ore is copper pyrites, $CuFeS_2$. Others are copper glance, Cu_2S, and the oxide and carbonate.

The ore is roasted in air to form a molten mixture of copper(I) sulphide and iron(II) oxide:

$$2CuFeS_2 + 4O_2 = Cu_2S + 2FeO + 3SO_2$$

The product is heated with sand (SiO_2) and the iron oxide forms a silicate slag:

$$FeO + SiO_2 = FeSiO_3, \text{ iron(II) silicate}$$

Some of the copper sulphide forms oxide, which reacts with the rest of the sulphide to produce molten copper:

$$2Cu_2S + 3O_2 = 2Cu_2O + 2SO_2$$

$$Cu_2S + 2Cu_2O = 6Cu + SO_2$$

Sulphur dioxide comes out of the copper as it cools and makes the metal surface rough and blistered; the metal is called blister copper.

Pure copper is obtained by electrolytic refining. The impure copper is made the anode in a bath of copper sulphate solution with sheets of pure copper as the cathodes.

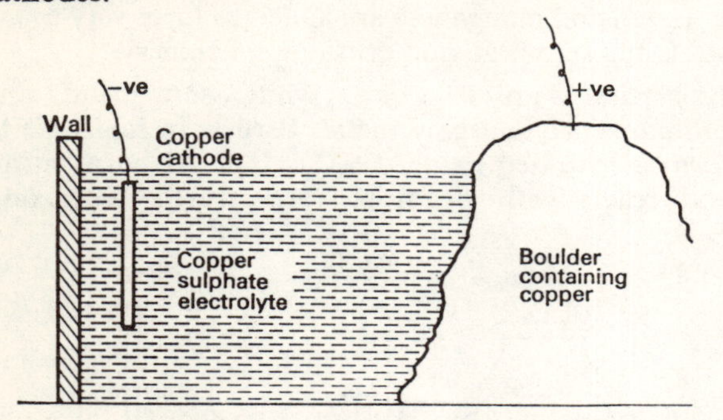

Fig. 24.7 Extraction of copper from a boulder

Copper is sometimes extracted from a boulder which contains the free element by using electrolysis.

Properties Copper is a reddish-brown metal, malleable, ductile, and a good conductor. It is stable in dry air, and forms a basic carbonate in moist air. It forms black oxide when heated in air. It has no action on steam and dilute acids, as is expected from its low position in the electrochemical series. It reduces concentrated nitric acid to oxides of nitrogen (NO_2 and NO, p. 128) and hot concentrated sulphuric acid to sulphur dioxide:

$$Cu + 2H_2SO_4 = SO_2 + CuSO_4 + 2H_2O$$

Like iron, it forms two series of compounds:

Copper(I) oxide Cu_2O; Copper(II) oxide CuO

Copper(I) chloride $CuCl$; Copper(II) chloride $CuCl_2$

Uses Copper is used in electric cables and flex, electric motors, dynamos, and other equipment, because of its very low electrical resistance. Its high thermal conductivity accounts for its use in boilers, steam pipes, and kettles.

Brass is an alloy of copper with up to 40 per cent zinc; the alloy is harder and more resistant to corrosion than copper. Duralumin contains a little copper.

Copper oxides CuO and Cu_2O

Copper(II) oxide is prepared by the action of heat on the solid carbonate, nitrate or hydroxide. The hydroxide is obtained as a jelly-like precipitate when aqueous caustic alkali is added to copper sulphate solution:

$$\underset{\text{green}}{CuCO_3} = CO_2 + CuO$$

$$\underset{\text{green}}{2Cu(NO_3)_2} = 4NO_2 + O_2 + 2CuO$$

$$Cu(OH)_2 = H_2O + \underset{\text{black}}{CuO(s)}$$

$$Cu^{2+}(aq) + 2OH^- = \underset{\text{blue}}{Cu(OH)_2(s)}$$

Copper(I) oxide is obtained as a red powder when Fehling's solution is warmed with glucose. Fehling's solution is a mixture of copper(II) sulphate, sodium hydroxide, and sodium potassium tartrate. The glucose reduces the Cu^{2+} ion in the solution:

$$Cu^{2+} + e = Cu^+$$

Copper(II) chloride is reduced to white, solid copper(I) chloride by boiling it with copper:

$$CuCl_2 + Cu = 2CuCl \quad \text{or} \quad Cu^{2+} + Cu = 2Cu^+$$

LEAD

Extraction The main ore is the sulphide, PbS, called galena. It is roasted in air to form the monoxide:

$$2PbS + 3O_2 = 2PbO + 2SO_2$$

The oxide is reduced with coke in a small blast furnace similar to that used for iron:

$$PbO + C = Pb + CO$$

Limestone is added to produce calcium oxide which removes silica as a slag:

$$CaO + SiO_2 = CO_2 + CaSiO_3, \text{ slag}$$

Properties Lead is a bright, lustrous, soft metal. It soon becomes dull in air owing to the formation of a film of hydroxide and carbonate. When heated in air it slowly forms the monoxide, PbO, and at about 450 °C it forms Pb_3O_4, which is called trilead tetraoxide, or red lead oxide:

$$2Pb + O_2 = 2PbO; \quad 3Pb + 2O_2 \rightleftharpoons Pb_3O_4,$$

Lead has no action on water, steam or dilute acids.

Uses Sheet lead is used on the roofs of buildings to keep out water and as a damp course to prevent water rising up a brick wall from the ground. Lead piping is common because it is soft, easily bent, does not corrode, is not attacked by hard water although soft water dissolves it, and broken parts are easily repaired by melting or soldering. Electricity and telephone cables and cables under the sea are covered with lead to keep out water.

The plates of accumulators contain lead. In the manufacture of sulphuric acid, the chambers are made of lead which resists the action of the acid. Thick lead sheets are used to protect people from radioactive radiation, for example in nuclear energy plants.

Solder contains lead and tin; it has a low softening point; and it does not melt sharply. When solder cools, the excess alloy can be wiped off a joint. *Type-metal* for printing contains lead, tin, and antimony; it expands when it solidifies and therefore it makes sharp type. It is tough and can be used to print thousands of pages. After use, the type-metal is readily melted and can be used over again.

Red lead oxide is a rust-inhibiting substance present in paint used on iron and steel. Lead(II) hydroxide carbonate, $2PbCO_3 . Pb(OH)_2$, is 'white lead' used in paints. The other white pigments, titanium(IV) oxide, TiO_2, and zinc oxide, are gradually replacing white lead because it is blackened by hydrogen sulphide in the atmosphere and it is poisonous when absorbed through the skin.

SUMMARY

Metals: readily lose electrons and form cations. They are bright, malleable, ductile, tough, good conductors, and usually have high melting points, high boiling points and high densities.
Non-metals, except hydrogen, readily gain electrons and form anions. Metals usually form basic oxides, displace hydrogen from dilute acids, form ionic halides, rarely form stable hydrides, and are electron donors or reducing agents.

Extraction of metals: by reduction, using carbon, carbon monoxide or electrolysis.
Aluminium: electrolysis of Al_2O_3 dissolved in cryolite, Na_3AlF_6.
Zinc: ZnS and $ZnCO_3$ are roasted to the oxide, which is reduced by coke or converted to the sulphate and electrolysed.
Iron: hot oxygen and fuel oil are blasted through the oxide mixed with coke and limestone. The reducing agent is CO.
Copper: pyrites, $CuFeS_2$, is roasted with sand, SiO_2. Cu_2S forms first and then copper. The slag is $FeSiO_3$. Copper is refined by electrolysis.
Steel: oxygen and steam are blown through molten iron in a Bessemer converter. The impurities are removed as oxides. In the *basic oxygen furnace* oxygen is blown over the surface of molten iron.

Alloy: a mixture or solution of two or more metals or carbon.
Duralumin: Al, with some Cu and Mg. Light and strong.
Brass: Cu and Zn. Hard and resists corrosion.
Steel: Fe with C, Cr, Ni, and other metals. Tough and ductile.
Solder: Sn and Pb. Low m.p., solidifies over a wide temperature range.
Type metal: Pb and Sn and antimony. Melts easily, expands when it solidifies, tough.

Uses of metals
Magnesium: light alloys in engines and aircraft; fireworks, flares, and incendiary bombs.
Aluminium: cooking utensils; light alloys; electric cables; protective wrapping; paint; thermite process for welding iron and for extraction of chromium.
Zinc: galvanizing iron; brass; dry batteries; paints; lotions and ointments.
Iron and steel: railings, pipes, girders, bridges, cars, buses, and engines.
Copper: electric cable, motors, dynamos; boilers, steam pipes, and kettles.

Hardness of water
Temporary hardness is caused by $Ca(HCO_3)_2$ and $Mg(HCO_3)_2$.
Permanent hardness is caused by $CaSO_4$ and $MgSO_4$. The Ca^{2+} and Mg^{2+} ions in hard water react with soap, NaSt, to form a scum, $CaSt_2$ and $MgSt_2$; St is the stearate, palmitate or oleate radical.
Removal of hardness: distillation, washing soda, and permutit remove all hardness; boiling, slaked lime, and ammonia remove temporary hardness only.
Soap: prepared from fats and hot caustic alkali.
Stalactites and *stalagmites:* columns of calcium carbonate and magnesium carbonate in some caves in limestone or chalk areas.
Boiler scale and *kettle fur:* calcium carbonate and magnesium carbonate.

Limestone: calcium carbonate; on heating it forms lime or quicklime, CaO, which forms slaked lime, $Ca(OH)_2$, with water.

QUESTIONS

1. Briefly explain each of the following, giving equations where appropriate. (*a*) 'Fur' is deposited inside kettles in some districts; (*b*) slaked lime removes temporary hardness of water but not permanent hardness; (*c*) washing soda softens hard water and forms a white precipitate; (*d*) limestone must be mixed with the iron ore added to a blast furnace; (*e*) caves are usually formed in limestone areas and stalactites and stalagmites form inside these caves.
2. Outline the method for the extraction of aluminium from purified bauxite. How does aluminium react with sodium hydroxide? State two important uses of aluminium as a metal. (C.)
3. Given soap solution and a sample of water, describe the experiments you would carry out to compare the amounts of temporary and permanent hardness present in the water. Describe what you observe when water is added slowly to a fairly large piece of quicklime. (C.)
4. Draw and label a simple diagram of a blast furnace used in the extraction of iron. State briefly the reactions which occur in the furnace. Suggest reasons why iron is not extracted by an electrolytic process such as that used to extract aluminium.
5. By reference to aluminium, iron, and copper only, discuss the principles underlying the extraction of metals. Indicate how the method of their extraction is linked to their position in the electrochemical series.
6. How is steel obtained from iron by the converter process (using oxygen) or by the basic oxygen process? Give two methods by which the quality of steel may be varied. (C.)
7. State three ways in which the metal zinc differs from the non-metal carbon. One at least of the differences must be a chemical difference. Give two uses for zinc. How and under what conditions does zinc react with (*a*) sodium hydroxide, (*b*) copper sulphate? (C.)
8. Using copper and iron as examples, discuss the properties of the transition elements. Use the facts that the electronic structure of iron is 2, 8, 14, 2 and of copper is 2, 8, 18, 1 to account for some of the properties you mention. (Among other things, your answer should mention catalysis and coloured ions.)

25 Organic Chemistry

MEANING OF ORGANIC CHEMISTRY

In early times, chemists had made most progress in studying the extraction of metals from their ores. The compounds that occurred in living plants and animals are much more complicated. Although it seemed that most of these chemicals contained carbon, hydrogen, and oxygen, it was generally thought that they obeyed different laws.

The Swedish chemist, Berzelius, named all chemicals that were derived from living matter 'organic' chemicals. In 1808 he suggested that they contain some 'vital force' of unknown character which was not present in 'inorganic' chemicals of mineral origin.

Carbamide, CON_2H_4, had been isolated from urine in the year 1778. In 1828 a German chemist prepared it 'without requiring a kidney of an animal, either man or dog'. His preparation disproved the 'vital force' theory. In 1848 organic chemistry was redefined as *the chemistry of the compounds of carbon.*

Some of the processes of organic chemistry have been used for thousands of years. The fermentation of beer from barley and of wine from the sugar in grapes continues to be practised in the same way today. Lavoisier discovered at the end of the eighteenth century that the souring of beer and wine by fermentation results from the formation of ethanoic (acetic) acid from alcohol, which in turn is formed from the sugar. Ethanoic acid was the first of the organic acids to be identified. Distillation has been used for about 1200 years to concentrate alcohol from fermented liquors and ethanoic acid from sour wines.

Nowadays several million organic compounds are known and organic compounds such as petrol, petroleum products, and plastics dominate civilized life and modern chemical industry. At one time the state of development of a country was judged by the quantity of sulphuric acid it produced; now it can be judged by the quantity of petroleum products it processes, changes, and uses.

Hydrocarbons

These are compounds of carbon and hydrogen only. Three simple ones and their graphical formulae are:

Methane CH_4	Ethene C_2H_4 (ethylene)	Ethyne C_2H_2 (acetylene)
H–C(H)(H)–H	$H_2C{=}CH_2$	H–C≡C–H
(tetrahedral)	(planar)	(linear)

In methane, each of the four valency bonds of carbon is joined to a hydrogen atom. The methane molecule is said to be saturated because each valency bond of the carbon atom is joined to a separate atom (a hydrogen atom) and the greatest combining power is exerted. Make a model of a methane molecule.

In ethene, two covalent bonds link two carbon atoms; in ethyne, three covalent bonds link the two carbon atoms. Ethene and ethyne are unsaturated compounds because the greatest combining power of carbon is not exerted. An ethene molecule can combine with two more hydrogen atoms (or similar atoms) and an ethyne molecule can combine with four more hydrogen atoms. Make models of the molecules.

Methane is the first member of a series of compounds, called the alkanes or paraffins. Ethene and ethyne are the first members of two different series called respectively the alkenes (olefines) and the alkynes (acetylenes).

ALKANES (PARAFFINS)

The first six members of the alkane series of hydrocarbons are:

methane	CH_4	ethane	C_2H_6	propane	C_3H_8
butane	C_4H_{10}	pentane	C_5H_{12}	hexane	C_6H_{14}

The first part of the name indicates the number of carbon atoms in one molecule. Thus, meth- indicates one carbon atom, but- indicates 4, pent- indicates 5, hex- indicates 6, and octadec- indicates 18. Octane has the formula C_8H_{18}. It is one of the alkanes in petrol.

The reasons for the use of the first four prefixes are:

methane is related to methanol;

ethane is related to ether which means burn or glow;

propane contains the same group as propanoic acid, which means first acid, because it was believed, wrongly, to be the first of the organic acids (p. 159);

butane contains the same group as the acid obtained from butter (the acid is butanoic acid, C_3H_7COOH).

The general formula of the alkanes is $C_xH_{(2x+2)}$. Thus, when x is 9, $(2x + 2)$ is 20 and the alkane is C_9H_{20}; it is called nonane.

To prepare methane CH_4

Anhydrous sodium ethanoate (acetate) is heated with soda-lime in a test-tube, see Fig. 25.1. The gas is

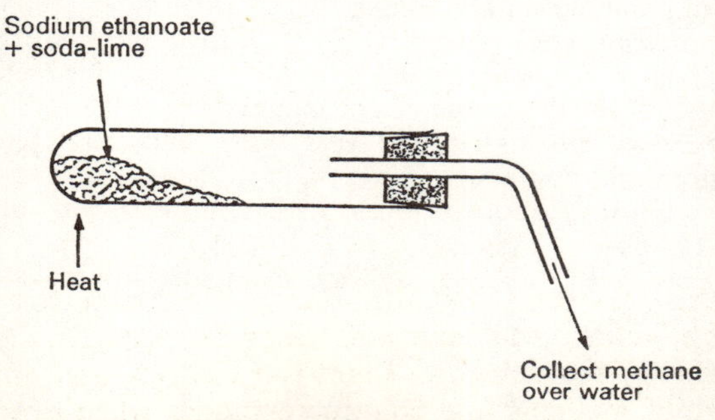

Fig. 25.1 Preparation of methane

collected over water. Soda-lime is lime that has been slaked with caustic soda solution, and its formula is usually written NaOH:

$$CH_3COONa + NaOH = CH_4(g) + Na_2CO_3$$

Methane is produced naturally by the bacterial decomposition of animal and vegetable matter in the absence of air. It occurs as *fire-damp* in coal mines, as *marsh gas* in marshy districts, and as the main constituent of *natural gas* under the sea and over petroleum or crude oil deposits. It is formed during the putrefaction of sewage and during the decomposition of cellulose in our intestines.

PROPERTIES OF THE ALKANES

Physical properties Methane is a colourless, tasteless, odourless gas. It is slightly soluble in water and is neutral to indicators. The melting points, boiling points, and densities of the alkanes increase with the number of carbon atoms in the molecule. The increase is due to the greater attraction between molecules as their mass increases, and therefore greater energy is required to separate them:

	Density (g/cm³)	*b.p.*	*m.p.*
Methane, CH_4	0·42 (at −162 °C)	−162 °C	−182 °C
Ethane, C_2H_6	0·45	−88 °C	−170 °C
Propane, C_3H_8	0·54	−39 °C	−184 °C
Butane, C_4H_{10}	0·60	1 °C	−138 °C
Pentane, C_5H_{12}	0·63	36 °C	−130 °C

Do not try to memorize these or any other figures. The melting points of methane and ethane are abnormal because the carbon atoms in their molecules are not in a zig-zag chain, as in other alkanes (p. 155).

Chemical properties The old name for the alkane series, paraffin, means 'little affinity'. The alkanes are not reactive compounds, and they do not normally react with acids, alkalis or oxidizing agents. Methane and ethane are very stable, but propane and higher alkanes (those with more carbon atoms per molecule) can decompose at a high temperature into simpler hydrocarbons; the process is called *cracking*. For example:

$$C_3H_8 = CH_4 + C_2H_4$$

Alkanes burn in a plentiful supply of oxygen or air to form carbon dioxide and water. In an insufficient supply, for example in the luminous Bunsen flame, carbon monoxide and carbon can also be formed. Alkanes that are gases or volatile liquids, such as petrol, form explosive mixtures with air:

$$CH_4 + 2O_2 = CO_2 + 2H_2O$$

$$CH_4 + O_2 = C + 2H_2O$$

Alkanes react with chlorine and bromine but not with iodine. Methane and chlorine react slowly at ordinary temperatures. Light is a catalyst, and a mixture of methane and chlorine explodes when exposed to bright sunlight or when it is sparked. In diffused sunlight the reaction is steady and hydrogen atoms are substituted by chlorine atoms; the reactions are examples of *substitution* reactions:

$$CH_4 + Cl_2 = HCl + CH_3Cl \quad \text{(monochloromethane)}$$

$$CH_3Cl + Cl_2 = HCl + CH_2Cl_2 \quad \text{(dichloromethane)}$$

$$CH_2Cl_2 + Cl_2 = HCl + CHCl_3 \quad \text{(trichloromethane)}$$

$$CHCl_3 + Cl_2 = HCl + CCl_4 \quad \text{(tetrachloromethane)}$$

Trichloromethane is also called chloroform; tetrachloromethane is carbon tetrachloride.

Concentrated bromine water is slowly decolorized by methane and other alkanes at room temperatures. Add about five drops of bromine water to a gas-jar of methane and leave for a few minutes. Remove the cover of the jar and blow gently—misty fumes show that hydrogen bromide has been formed.

USES OF ALKANES

The alkanes are the most widely used organic compounds. Crude oil or petroleum usually consists of up to eighty alkanes (from CH_4 to $C_{43}H_{88}$) and it contains petrol, kerosine, fuel oil, diesel oil, paraffin wax, and asphalt. Between 80 and 99 per cent of natural gas is methane; it is used to provide heat, light, and power. Natural gas from the Sahara is liquefied and shipped in refrigerator ships to several countries and is added to the gas supply; much of the gas called coal gas has never been near a lump of coal. Calor gas, bottled gas, and LPG (liquefied petroleum gas) are mixtures of propane and two kinds of butane, together with a little pentane. They are stored in steel cylinders under pressure, and are in liquid form; release of the pressure allows gas to escape. This 'liquid' gas is used as a fuel in homes, factories, and caravans because it is portable. It is sometimes used instead of acetylene (ethyne) for cutting metals. Alkanes which are liquid at ordinary pressures are used as solvents in varnishes and lacquers, as fuels for furnaces and kilns, as lubricating oils and in the manufacture of plastics, fats, and soaps.

SUBSTITUTION

All the atoms of an alkane molecule exert their full combining power with other atoms and the alkanes are *saturated.* New atoms cannot be added to an alkane molecule. An atom or radical can enter only if it replaces or *substitutes* a hydrogen atom; for example:

$$\text{(methane)}\ CH_3.H + XY = CH_3X + HY$$

$$\text{(ethane)}\ C_2H_5.H + XY = C_2H_5X + HY$$

These equations represent two *substitution reactions.*

The univalent group CH_3- is a methyl group, C_2H_5- is an ethyl group and C_2H_7- is a propyl group. The groups are alkyl groups or radicals. CH_3Br can be called methyl bromide or monobromoethane.

Ethane and chlorine can form six substitution compounds:

$$C_2H_5Cl \quad C_2H_4Cl_2 \quad C_2H_3Cl_3$$
$$C_2H_2Cl_4 \quad C_2HCl_5 \quad C_2Cl_6$$

HOMOLOGOUS SERIES

Organic compounds which possess similar groups of atoms have similar properties. They form a homologous series of compounds. By knowing the properties of one member of a series, the properties of other members can be stated with reasonable certainty and accuracy. The characteristics of a homologous series are:

1. Each member differs from the one before or the one after by a $-CH_2$ group:

$CH_3.CH_3$

```
   H  H
   |  |
H—C—C—H
   |  |
   H  H
```

ethane

$CH_3.CH_2.CH_3$

```
   H  H  H
   |  |  |
H—C—C—C—H
   |  |  |
   H  H  H
```

propane

$CH_3.CH_2.CH_2.CH_3$

```
   H  H  H  H
   |  |  |  |
H—C—C—C—C—H
   |  |  |  |
   H  H  H  H
```

butane

2. All members can be represented by a general formula, for example:

$$\text{alkanes } C_xH_{(2x+2)}; \quad \text{alkenes } C_xH_{2x}$$

3. The physical properties change gradually. The m.p., b.p., and density rise as molecular mass increases. Methane is a gas, pentane is a liquid, and octadecane, $C_{18}H_{38}$, is a solid. Increasing intermolecular forces account for the change in physical properties.

4. The chemical properties are similar. However, chemical reactivity decreases slightly with increasing molecular mass. For example, alkanes of high molecular mass do not burn as readily as methane and ethane.

ISOMERS

In inorganic chemistry one formula represents one compound. In organic chemistry there are many examples of several compounds which have the same empirical formula and the same molecular formula. The compounds are *isomers* and the phenomenon is called *isomerism*.

C_4H_{10} exists in two forms:

```
   H  H  H  H
   |  |  |  |
H—C—C—C—C—H
   |  |  |  |
   H  H  H  H
```

butane

```
   H  H  H
   |  |  |
H—C—C—C—H
   |  |  |
   H  |  H
      |
   H—C—H
      |
      H
```

2-methylpropane

Butane contains a chain of four carbon atoms; its isomer contains a chain of only three carbon atoms. Because the methyl group in the isomer is on the second carbon atom, its systematic name is 2-methylpropane.

There are three pentanes, C_5H_{12}. One has a chain of five carbon atoms; the second has a chain of four carbon atoms and is a 2-methylbutane; the third has a chain of only three carbon atoms and is dimethylpropane. These formulae, without the hydrogen atoms, show the difference:

```
                         C              C
                         |              |
C—C—C—C—C             C—C—C—C        C—C—C
                                        |
                                        C
```

Two dichloroethanes, $C_2H_4Cl_2$, exist. Their formulae can be written $CH_2Cl.CH_2Cl$ and $CH_3.CHCl_2$. The two chlorine atoms are combined either with separate carbon atoms or with one atom.

Make molecular models of all these isomers.

TETRAHEDRAL DISTRIBUTION OF BONDS IN METHANE CH_4

It is usual to write the four covalent bonds of a carbon atom at right angles to each other in one plane. As a result, the following two formulae look different:

```
   H             H
   |             |
X—C—X         H—C—X
   |             |
   H             X
```

However, there is only one compound CH_2X_2, for example, only one dichloromethane, CH_2Cl_2 and one dibromomethane, CH_2Br_2.

In 1874 it was suggested that the four covalent bonds are arranged symmetrically in three dimensions. The carbon atom is at the centre of a regular tetrahedron, and the four bonds are directed towards the four corners. This is the tetrahedral theory of the carbon atom. The angle between bonds is 109° 28′.

Make models of the following and show that the second and third diagrams represent the same substance. Rotate the models to make them coincide.

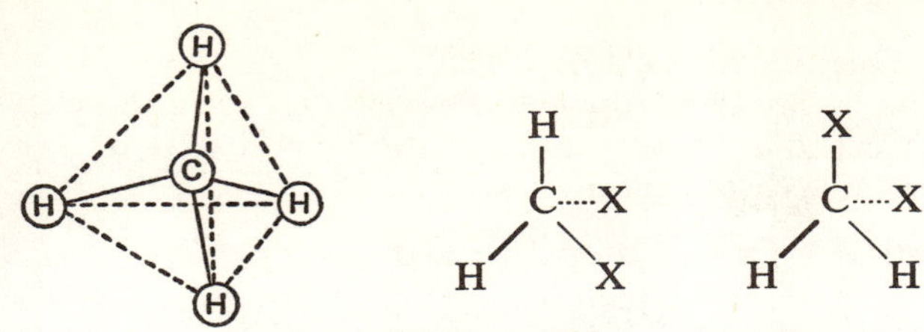

Make a model containing three or more carbon atoms. The carbon atoms are zig-zag and not straight:

ALCOHOLS

Alcohols have the general formula $C_xH_{(2x+1)}OH$ in which the —OH group is joined to a hydrocarbon radical:

Name	*Formula*	*b.p.*	*Density* (g/cm^3)
Methanol (methyl alcohol)	CH_3OH	65 °C	0·81
Ethanol (ethyl alcohol)	C_2H_5OH	78 °C	0·81
Propanol (propyl alcohol)	C_3H_7OH	97 °C	0·82
Butanol (butyl alcohol)	C_4H_9OH	117 °C	0·82

To prepare ethanol from glucose

1. Dissolve glucose in warm water, and add the warm solution to a large volume of cold water in a bottle.
2. Make brewer's yeast into a cream with a little water and add it to the glucose solution. Add a crystal of ammonium phosphate and of potassium nitrate, which are food for the yeast cells.
3. Leave the mixture at about 35 °C for three days. Fermentation occurs and carbon dioxide is evolved:

$$C_6H_{12}O_6 = 2C_2H_5OH + 2CO_2$$

4. Use fractional distillation, collecting the fraction that boils over below 95 °C, to remove most of the water. Re-distil the product, and collect the fraction that comes over between 78 °C and 82 °C. It contains about 96 per cent ethanol. If necessary add some calcium oxide to remove the small amount of water still left; it cannot be removed by distillation.

The enzyme called zymase, present in yeast, catalyses the decomposition of the glucose.

Reactions of ethanol

Burning Place a little ethanol on a clock-glass and light it. Observe the flame. Hold a dry gas-jar over the flame and then test the gas inside it with lime water.
Sodium Add small pieces of sodium, one at a time, to pure ethanol in a test-tube. Test any gas formed with a flame.
Phosphorus pentachloride Add a little of the chloride to pure ethanol in a test-tube. Identify any gas evolved.

Oxidation
(*a*) Add acidified potassium dichromate solution to ethanol and warm gently. Observe any colour change and smell the product.

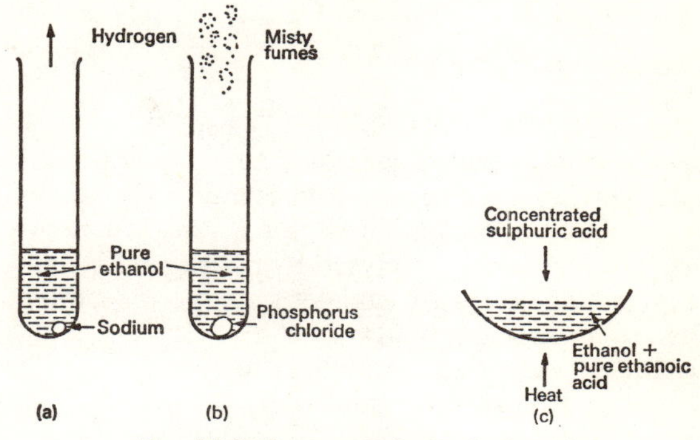

Fig. 25.2 Three reactions of ethanol

(*b*) Add acidified potassium manganate(VII) solution to ethanol and warm gently. Observe any colour change and smell the product.
Ethanoic acid Add concentrated sulphuric acid to a mixture of equal volumes of ethanol and pure ethanoic acid, formerly called glacial acetic acid, in a dish. Warm gently. Note any smell produced.

PROPERTIES OF ETHANOL

Ethanol is a colourless, volatile liquid, which is soluble in water. It has a burning taste; pure ethanol can cause unconsciousness and death.

Higher alcohols in the series are oily liquids and those with twelve or more carbon atoms per molecule are solids; higher alcohols mix less readily with water.
Chemical properties The reactions of ethanol can be classified as follows:

1. Properties of the —OH group (with sodium, phosphorus chlorides, and organic acids).
2. Oxidation, first to aldehydes and then to acids.
3. Dehydration, to ethoxyethane and to ethene.

Sodium (and potassium). Effervescence occurs and hydrogen is evolved. Evaporation of the solution leaves a white solid:

$$2C_2H_5OH + 2Na = H_2 + 2C_2H_5ONa,$$
sodium ethoxide

Compare this reaction:

$$2HOH + 2Na = H_2 + 2HONa,$$
sodium hydroxide

Phosphorus pentachloride There is a violent reaction and misty fumes of hydrogen chloride are evolved. This is a test for the —OH group in an organic compound:

$$C_2H_5OH + PCl_5 = C_2H_5Cl + HCl + POCl_3$$

$POCl_3$ is the formula of phosphorus trichloride oxide.

Organic acids: esterification Pure ethanoic acid (glacial acetic acid) reacts to form a sweet-smelling *ester*, ethyl ethanoate (ethyl acetate):

$$\text{Acid} + \text{Alcohol} = \text{Ester} + \text{Water}$$

$$CH_3COOH + HOC_2H_5 = CH_3COOC_2H_5 + H_2O$$

Compare this reaction:

$$ClH + HONa = ClNa + H_2O$$

The reaction is called *esterification* and it resembles neutralization. However, esterification is a much slower reaction because it is not a reaction between ions. The reaction is catalysed by hydrogen ions and it is reversible. Esters are covalent compounds whereas salts are ionic compounds.

Oxidation Acidified potassium dichromate oxidizes ethanol to ethanal (acetaldehyde):

$$C_2H_5OH + [O] \rightarrow CH_3CHO + H_2O, \text{ or}$$

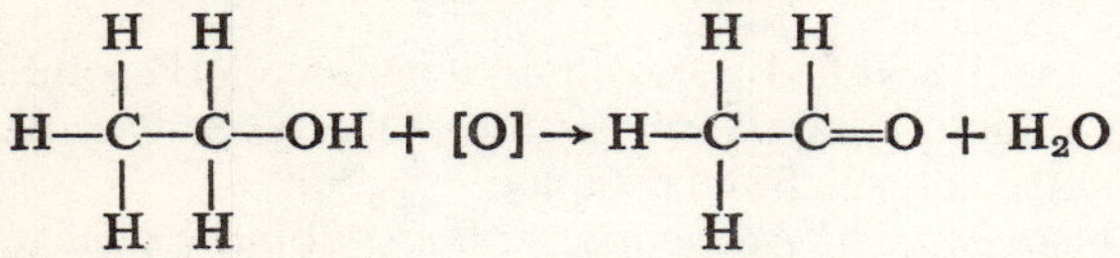

$$\begin{array}{ccccc} & H & H & & \\ & | & | & & \\ H- & C & - C & -OH & \\ & | & | & & \\ & H & H & & \end{array} + [O] \rightarrow \begin{array}{ccccc} & H & H & & \\ & | & | & & \\ H- & C & - C & =O & \\ & | & & & \\ & H & & & \end{array} + H_2O$$

The dichromate is changed from yellow to green. Acidified potassium manganate(VII) is a stronger oxidizing agent and the ethanal is further oxidized to ethanoic acid:

$$CH_3CHO + [O] \rightarrow CH_3COOH, \text{ or}$$

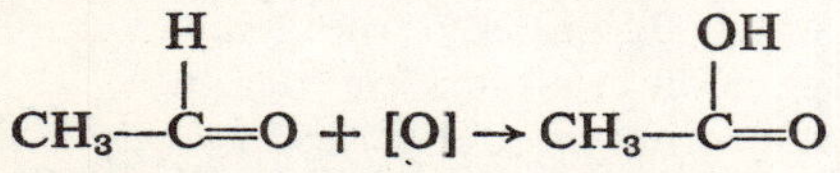

$$\begin{array}{ccc} & H & \\ & | & \\ CH_3- & C & =O \end{array} + [O] \rightarrow \begin{array}{ccc} & OH & \\ & | & \\ CH_3- & C & =O \end{array}$$

When ethanol vapour is passed over heated copper at about 300 °C, it is *dehydrogenated*; that is, hydrogen is removed:

$$C_2H_5OH = H_2 + CH_3CHO, \text{ ethanal}$$

Dehydration Concentrated sulphuric acid forms an ester when it reacts with ethanol under mild conditions. Ethanol is *dehydrated* (water is removed):

$$C_2H_5OH + H_2SO_4 = H_2O + C_2H_5HSO_4,$$
ethyl hydrogensulphate

At about 140 °C, with excess ethanol, the alcohol is dehydrated to ethoxyethane (diethyl ether):

$$2C_2H_5OH = H_2O + C_2H_5OC_2H_5$$

At about 180 °C, with excess acid, ethene is produced:

$$C_2H_5OH = H_2O + C_2H_4$$

Ethene is also formed when ethanol vapour is passed over heated aluminium oxide.

Note that the actions of copper and of aluminium oxide catalysts on ethanol vapour are different. Probably ethanol molecules are adsorbed on the surface of the catalysts in different ways before decomposition occurs.

ISOMERS OF ALCOHOLS

There are two propanols. One has the —OH group on the end carbon atom, and the other has it on the middle carbon atom:

a) $CH_3 . CH_2 . CH_2—OH$

b) $$\begin{array}{c} OH \\ | \\ CH_3 . CH . CH_3 \end{array} \text{ or } \begin{array}{ccccc} & H & OH & H & \\ & | & | & | & \\ H- & C & - C & - C & -H \\ & | & | & | & \\ & H & H & H & \end{array}$$

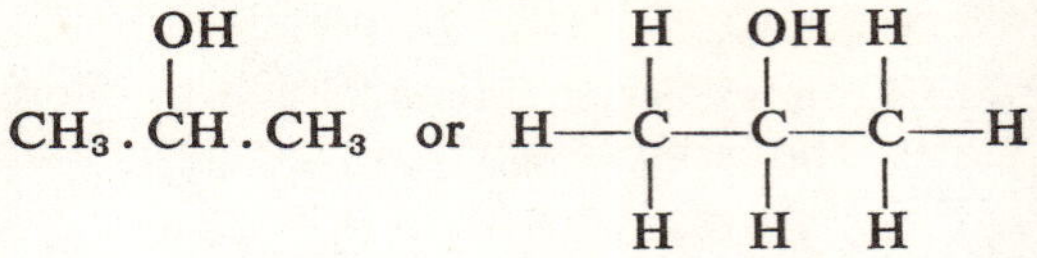

They are called propan-1-ol and propan-2-ol respectively.

Two compounds have the formula C_2H_6O.

a) $$\begin{array}{ccccc} & H & H & & \\ & | & | & & \\ H- & C & - C & -O- & H \\ & | & | & & \\ & H & H & & \end{array}$$ ethanol

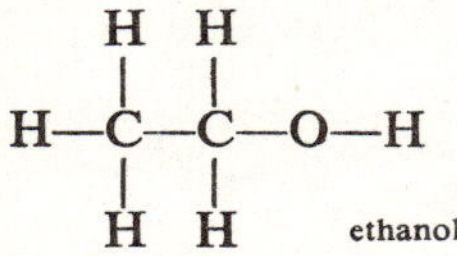

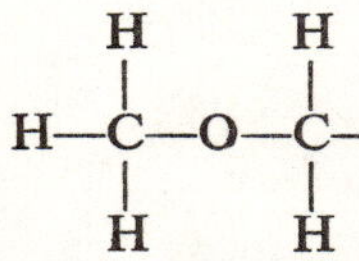

b) $$\begin{array}{ccccc} & H & & H & \\ & | & & | & \\ H- & C & -O- & C & -H \\ & | & & | & \\ & H & & H & \end{array}$$ (no —OH group) methoxymethane

Commercial alcohol and its uses

Alcohol is made by fermentation of carbohydrates, such as rice, potatoes, and cellulose. Beer, whisky, and wines contain ethanol.

Rectified spirit contains about 96 per cent ethanol. Absolute alcohol is 100 per cent ethanol. Methylated spirit is industrial ethanol to which poisons have been added to make it unfit for drinking.

Most industrial ethanol is now prepared from ethene. Methanol is made from water gas; see p. 119.

ALKENES (OLEFINES)

The general formula of the alkenes is C_xH_{2x} in which $x = 2$ or more. An alkene has two hydrogen atoms fewer than the corresponding alkane with the same number of carbon atoms.

Name	*Formula*	*b.p.*
Ethene (ethylene)	C_2H_4	−104 °C
Propene (propylene)	C_3H_6	−48 °C
Butene (butylene)	C_4H_8	−6 °C
Pentene	C_5H_{10}	30 °C

The alkenes contain a double bond of four electrons between two carbon atoms and are *unsaturated* compounds. This makes them chemically reactive:

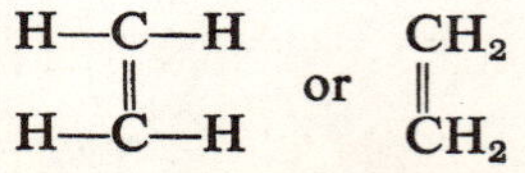

$$\begin{array}{c} H-C-H \\ \| \\ H-C-H \end{array} \text{ or } \begin{array}{c} CH_2 \\ \| \\ CH_2 \end{array} \text{ Ethene}$$

Two butene isomers are $CH_2 = CH . CH_2 . CH_3$ and $CH_2 = C(CH_3)_2$.

To prepare ethene C_2H_4 from ethanol

1. Soak some glass wool in ethanol and place it at the bottom of a boiling tube as in Fig. 25.3. Place pieces of broken porcelain on top of the glass wool.
2. Heat the broken porcelain strongly. Heat is conducted through the glass of the boiling tube and vaporizes the ethanol. Ethanol vapour passes through the hot porcelain pot and thermal decomposition occurs:

$$C_2H_5OH \rightarrow C_2H_4 + H_2O$$

or

$$CH_3CH_2OH \rightarrow H_2C{=}CH_2 + H_2O$$

3. Collect the ethene over water. Burn one jar of the gas. Add a few drops of bromine water to a second jar, and a little acidified potassium manganate(VII) solution to a third jar.

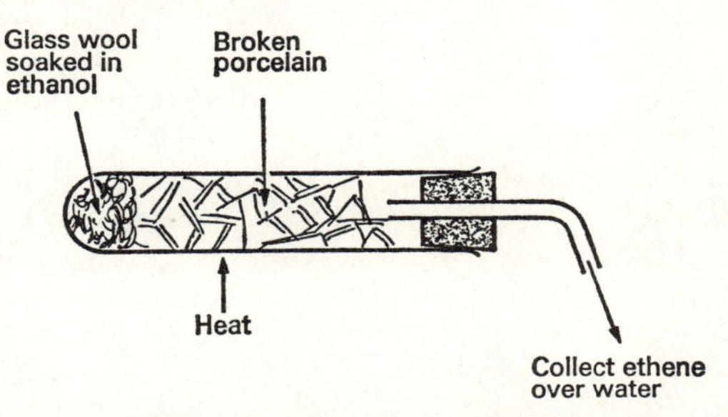

Fig. 25.3 Preparation of ethene (ethylene)

PROPERTIES OF ALKENES

Physical properties Ethene is a colourless gas with a sweetish smell. It is only slightly soluble in water but is fairly soluble in ethanol. It is neutral to indicators.

Chemical properties Alkenes burn in air with a luminous, smoky flame because of their high carbon content. Compare with the flame of methane which is *not* smoky:

$$C_2H_4 + 3O_2 = 2CO_2 + 2H_2O$$

Ethene burns in chlorine and deposits carbon:

$$C_2H_4 + 2Cl_2 = 4HCl + 2C \text{ (combustion)}$$

Addition reactions of ethene Most reactions of ethene are addition reactions in which one molecule of ethene combines with one molecule of a second substance to form a single compound. The addition occurs because of the double bond:

$$\begin{matrix} CH_2 \\ \| \\ CH_2 \end{matrix} + XY = \begin{matrix} CH_2X \\ | \\ CH_2Y \end{matrix} \quad \text{(X and Y are univalent)}$$

Hydrogen combines in the presence of nickel or platinum catalysts:

$$H_2C{=}CH_2 + H_2 \rightarrow H_3C{-}CH_3 \text{ or } C_2H_6\text{, ethane}$$

This is a *hydrogenation reaction* or the saturation of a double bond by hydrogen. Vegetable and animal oils were first hydrogenated in 1910 by heating to about 200 °C and bubbling hydrogen under pressure through them in the presence of finely-divided nickel. The oils are changed to fats, which are converted into margarine and other products.

Halogens Fluorine reacts violently with ethane. Chlorine and bromine react readily, and a solution of iodine in ethanol also reacts by addition. The products are colourless, and therefore the halogens are decolorized:

$$CH_2{=}CH_2 + Cl_2 \rightarrow CH_2Cl.CH_2Cl$$

1, 2-dichloroethane

All unsaturated compounds decolorize a few drops of a solution of bromine in tetrachloromethane. This is a test for unsaturation:

$$\begin{matrix} CH_2 \\ \| \\ CH_2 \end{matrix} + \begin{matrix} Br \\ | \\ Br \end{matrix} = \begin{matrix} CH_2Br \\ | \\ CH_2Br \end{matrix} \quad \text{1,2-dibromoethane}$$

Halogen acids Hydrogen iodide gas reacts readily, hydrogen bromide reacts slowly, and hydrogen chloride hardly reacts:

$$\begin{matrix} CH_2 \\ \| \\ CH_2 \end{matrix} + \begin{matrix} H \\ | \\ I \end{matrix} = \begin{matrix} CH_3 \\ | \\ CH_2I \end{matrix} \quad \text{iodoethane}$$

Sulphuric acid The concentrated acid adds on slowly but the fuming acid reacts quickly:

$$\begin{matrix} CH_2 \\ \| \\ CH_2 \end{matrix} + \begin{matrix} H \\ | \\ HSO_4 \end{matrix} \rightarrow \begin{matrix} CH_3 \\ | \\ CH_2HSO_4 \end{matrix} \text{ or } \begin{matrix} CH_3 \\ | \\ CH_2{-}O{-}SO_2{-}OH \end{matrix}$$

ethyl hydrogensulphate

This reaction is used in industry to separate ethene from saturated gases produced when petroleum is cracked, see p. 163. It is used in laboratories to absorb unsaturated alkenes from mixtures containing saturated alkanes, which are not absorbed.

Potassium manganate(VII) The purple solution, acidified with dilute sulphuric acid, is turned colourless:

$$\begin{matrix} CH_2 \\ \| \\ CH_2 \end{matrix} + H_2O + \underset{\text{from } KMnO_4}{[O]} \rightarrow \begin{matrix} CH_2OH \\ | \\ CH_2OH \end{matrix} \quad \text{Glycol or ethane-1,2-diol}$$

Alkaline manganate(VII) solution is turned green, and brown manganese(IV) oxide is precipitated. This reaction is used as a test for unsaturation; it is not given by alkanes.

Uses of ethene and other alkenes are given on p. 163.

ALKYNES (ACETYLENES)

The general formula of the alkynes is $C_xH_{(2x-2)}$ in which $x = 2$ or more. An alkyne molecule contains a triple bond of six electrons between two carbon atoms; this makes alkynes even more reactive than alkenes. Ethyne (acetylene) is H—C≡C—H.

Among other alkynes are propyne C_3H_4 or CH_3—C≡C—H, butyne C_4H_6, and so on.

To prepare ethyne (acetylene) C_2H_2

Place lumps of calcium carbide on sand in a flask. Add water drop by drop. So much heat is evolved that a layer of sand must be placed on the bottom of the flask to prevent the glass cracking. Collect the ethyne over water. It contains phosphine, PH_3, as an impurity because calcium phosphide, Ca_3P_2, is present in the carbide:

$$CaC_2 + 2H_2O = Ca(OH)_2 + C_2H_2(g)$$

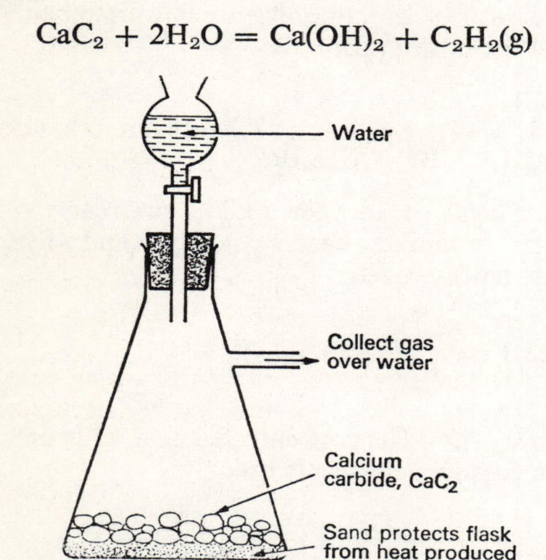

Fig. 25.4 Preparation of ethyne (acetylene)

PROPERTIES OF ETHYNE

Physical properties It is a colourless gas, with a sweetish smell when pure and an unpleasant smell when impure. It is almost insoluble in water. It explodes violently when under pressure and liquid ethyne is very dangerous. Cylinders of acetylene used in engineering shops contain the gas dissolved in propanone spread on a porous form of silica under a pressure of about twelve atmospheres.

Chemical properties Ethyne burns with a luminous and very sooty flame because of the high carbon content:

$$2C_2H_2 + 5O_2 = 4CO_2 + 2H_2O$$

It reacts explosively with chlorine, forming carbon:

$$C_2H_2 + Cl_2 = 2HCl + C$$

Addition reactions of ethyne One molecule of ethyne combines with either one or two molecules of a second substance to form single compounds. The additions occur because of the triple bond:

$$\begin{array}{c}CH\\ |||\\ CH\end{array} + XY \rightarrow \begin{array}{c}CHX\\ ||\\ CHY\end{array} \quad \text{(X and Y are univalent)}$$

$$\begin{array}{c}CH\\ |||\\ CH\end{array} + 2XY \rightarrow \begin{array}{c}CHX_2\\ |\\ CHY_2\end{array} \quad \text{(X and Y are univalent)}$$

Note that the two X atoms or groups combine with the same carbon atom. Addition compounds of the formula CHXY.CHXY rarely form. The structural formulae of the addition compounds are:

$$\begin{array}{ccccc} & X & & Y & \\ & | & & | & \\ H— & C & = & C & —H\end{array} \qquad \begin{array}{ccccc} & X & & Y & \\ & | & & | & \\ H— & C & — & C & —H \\ & | & & | & \\ & X & & Y & \end{array}$$

Hydrogen combines in the presence of nickel:

$$\begin{array}{c}CH\\ |||\\ CH\end{array} + H_2 = \begin{array}{c}CH_2\\ ||\\ CH_2\end{array} \qquad \begin{array}{c}CH\\ |||\\ CH\end{array} + 2H_2 = \begin{array}{c}CH_3\\ |\\ CH_3\end{array}$$

Halogens Under special conditions ethyne and chlorine can combine to form $CHCl_2 . CHCl_2$, 1, 1, 2, 2-tetrachloroethane, which is used as a solvent for oils and resins.

Reddish-brown bromine vapour, bromine water, and liquid bromine are rapidly decolorized:

$$\begin{array}{c}CH\\ |||\\ CH\end{array} + Br_2 = \begin{array}{c}CHBr\\ ||\\ CHBr\end{array}; \qquad \begin{array}{c}CH\\ |||\\ CH\end{array} + 2Br_2 = \begin{array}{c}CHBr_2\\ |\\ CHBr_2\end{array}$$

Halogen acids Hydrogen iodide reacts readily at room temperature, hydrogen bromide reacts when warmed, and hydrogen chloride hardly reacts:

$$\begin{array}{c}CH\\ |||\\ CH\end{array} + \begin{array}{c}H\\ |\\ Br\end{array} = \begin{array}{c}CH_2\\ ||\\ CHBr\end{array}$$

bromoethene

$$\begin{array}{c}CH\\ |||\\ CH\end{array} + 2\begin{array}{c}H\\ |\\ Br\end{array} = \begin{array}{c}CH_3\\ |\\ CHBr_2\end{array}$$

1,1-dibromoethane

Water Water adds to ethyne in the presence of dilute sulphuric acid and with mercury(II) sulphate as a catalyst:

$$CH{\equiv}CH + H_2O = CH_3CHO \text{ ethanal}$$

Potassium manganate(VII) The purple solution, of acidified manganate(VII) is decolorized. The equation is too difficult at this stage. This test and the decolorization of bromine are two tests for unsaturated compounds.

USES OF ETHYNE

The oxyacetylene flame produces a temperature well over 2000 °C. It is used for welding metals and for cutting up steel. Chlorinated products, for example, CH_2=CHCl, are used to produce solvents and plastics.

ORGANIC ACIDS

These acids all contain the carboxyl group—COOH or

$$—CO—OH \text{ or } —\underset{\underset{O}{\|}}{C}—OH$$

Their general formula is $C_xH_{(2x+1)}COOH$.

Name	*Formula*	*b.p.*
Methanoic acid (formic acid)	$H.COOH$	101 °C
Ethanoic acid (acetic acid)	$CH_3.COOH$	118 °C
Propanoic acid (propionic acid)	$C_2H_5.COOH$	141 °C

PROPERTIES OF ETHANOIC ACID

Physical properties It is a colourless liquid which changes to a crystalline solid, sometimes called glacial acetic acid, below 17 °C. It is hygroscopic and readily soluble in water. (Acids with more carbon atoms are less soluble; stearic acid is almost insoluble.) Ethanoic acid has a burning taste, is poisonous, and blisters the skin. The dilute acid is not poisonous, and vinegar contains about 4 per cent ethanoic acid.

Chemical properties It turns indicators to their acidic colour, neutralizes bases to form ethanoates (acetates), reacts with some metals to form hydrogen, and liberates carbon dioxide from carbonates:

$$CH_3COOH + HONa = H_2O + \underset{\text{sodium ethanoate}}{CH_3COONa},$$

The acid forms esters with alcohols, see p. 156.

Phosphorus pentachloride reacts vigorously and forms misty fumes of hydrogen chloride. This is a test for the —OH group:

$$CH_3COOH + PCl_5 = HCl + POCl_3 + \underset{\text{ethanoyl chloride}}{CH_3COCl}$$

CARBOHYDRATES

These are compounds of carbon, hydrogen, and oxygen only. The hydrogen and oxygen atoms are present in the ratio 2:1, and the general formula of a carbohydrate is $C_xH_{2y}O_y$. This formula could be written $C_x(H_2O)_y$ but this is not correct because carbohydrates do not contain water. The carbohydrates are classified into three groups:

1. Simple sugars: glucose and fructose. Both have the formula $C_6H_{12}O_6$ and they are isomers.
2. Complex sugars: sucrose and maltose. Both have the formula $C_{12}H_{22}O_{11}$ and they are isomers.
3. Complex carbohydrates: starch, dextrins, and cellulose. Their formulae are $(C_6H_{10}O_5)_n$ and n is a large number; n is not the same in the three compounds, and they are not isomers.

GLUCOSE (grape sugar) $C_6H_{12}O_6$

This sugar is present in the juices of many fruits, in the roots and leaves of many plants, in honey, and in the blood of animals. It is sometimes called blood sugar.

Glucose is formed in plants by photosynthesis, during which carbon dioxide and water combine by using the energy of sunlight with chlorophyll as a catalyst:

$$6CO_2 + 6H_2O + \text{energy} \xrightarrow{\text{chlorophyll}} C_6H_{12}O_6 + 6O_2$$

Glucose is not so sweet as ordinary sugar (sucrose or cane-sugar). The enzyme zymase in yeast ferments it and forms ethanol. Concentrated sulphuric acid dehydrates it to carbon (p. 140). During respiration in the bodies of animals, glucose reacts with oxygen of the blood to form carbon dioxide, water, and energy; the reaction is the reverse of photosynthesis.

FRUCTOSE (fruit-sugar) $C_6H_{12}O_6$

This is the sweetest sugar. It occurs naturally in plant juices, many fruits and honey.

SUCROSE (cane-sugar or beet-sugar) $C_{12}H_{22}O_{11}$

This is obtained from sugar-cane, sugar-beet, and sugar-palm. They are crushed and the sugar is extracted by dissolving it in hot water.

Dilute acids or the enzyme invertase in yeast convert sucrose into the two simple sugars, glucose and fructose:

$$C_{12}H_{22}O_{11} + H_2O = C_6H_{12}O_6 + C_6H_{12}O_6$$

STARCH $(C_6H_{10}O_5)_n$

The number n is very large, a thousand or more, and therefore the starch molecule is a macromolecule (a very large molecule).

The enzyme ptyalin in saliva converts starch to maltose, and dilute acids convert starch to glucose. Simple equations are:

$$2C_6H_{10}O_5 + H_2O \xrightarrow{\text{ptyalin}} C_{12}H_{22}O_{11}, \text{ maltose}$$

$$C_6H_{10}O_5 + H_2O \xrightarrow{\text{acids}} C_6H_{12}O_6, \text{ glucose}$$

Starch is insoluble in cold water. In hot water, it forms a colloidal solution. When dry starch is heated until it just begins to turn brown, it changes to compounds called dextrins. These have the same empirical formula as starch but their molecules are smaller. Toasted bread contains some dextrins.

CELLULOSE $(C_6H_{10}O_5)_m$ (m is about 5000)

This is the main substance in the cell walls of plants. Wood, paper, cotton, and linen consist mainly of cellulose. Filter paper and cotton wool are pure forms of cellulose. The cellulose around starch particles prevents their dissolving in cold water; hot water bursts the cellulose and allows the starch to form a solution. Animals can digest cellulose, but man cannot and it passes through his alimentary canal without change.

Tests for glucose and starch

Glucose Mix equal volumes of Fehling's solutions* A and B. Add glucose and warm. Finally, a red precipitate of copper(I) oxide is formed. Fructose and maltose also give this reaction.

Starch Add iodine solution (iodine dissolved in potassium iodide solution, as it is insoluble in water) to starch solution or to a slice of potato or yam. A deep bluish-black colour forms. Dextrins form a red colour with iodine solution.

* *Note:* Fehling's solution is kept in two separate bottles because the mixed substance does not keep well. One solution is aqueous copper sulphate; the other is sodium hydroxide and sodium potassium tartrate. The mixed solution may be regarded as copper(II) oxide in solution.

SUMMARY

Organic chemistry: the study of the chemistry of carbon compounds.

Hydrocarbon: a compound of carbon and hydrogen only.

Homologous series: a family of organic compounds of similar structure in which each differs from the next by the presence of an additional $—CH_2$ group.

Isomers: compounds with the same molecular formula but different structural formula; for example, ethanol CH_3CH_2OH and methoxymethane CH_3OCH_3. Both are C_2H_6O.

Alkanes (paraffins) $C_xH_{(2x+2)}$. Series of *saturated* compounds: all their atoms exert their full combining power—there are no double or triple bonds.

Methane: colourless, tasteless gas; present in natural gas. It burns in air and in chlorine. Chlorination forms *substitution* products CH_3Cl, CH_2Cl_2, etc., in diffused light.

Alcohols $C_xH_{(2x+1)}OH$. An alkyl radical is joined to an —OH group.

Preparation of ethanol, C_2H_5OH, from glucose by catalytic action of enzyme zymase in yeast.

Properties of the —OH group: forms H_2 with Na or K, HCl with PCl_5, and *esters* with acids; for example, ethyl ethanoate $CH_3COOC_2H_5$.

Oxidation, first to aldehyde and then to organic acid; for example:

$$CH_3CH_2OH \rightarrow CH_3CHO \rightarrow CH_3COOH$$

Dehydration, either to ethoxyethane or to an alkene; for example:

$$2C_2H_5OH - H_2O \rightarrow C_2H_5OC_2H_5$$

or

$$C_2H_5OH - H_2O \rightarrow CH_2{=}CH_2$$

Alkenes (olefines) C_xH_{2x}, where $x > 1$.

Preparation of ethene, C_2H_4, from ethanol either by heating, or with conc. H_2SO_4 at 180 °C.

Properties: colourless, sweet-smelling, neutral gas; burns in air with smoky flame and in chlorine.

Addition reactions: double bond breaks to add H_2, Cl_2, Br_2, HI, HBr, H_2SO_4, and —OH from acidified $KMnO_4$.

Alkynes (acetylenes) C_xH_{2x-2}, where $x > 1$.

Preparation of ethyne C_2H_2 by adding water to calcium carbide, CaC_2.

Properties: burns in air, explodes with chlorine to give C + HCl.

Addition reactions: triple bond can break in stages:
With hydrogen and Ni catalyst, C_2H_4 then C_2H_6.
With bromine, $CHBr{=}CHBr$ then $CHBr_2.CHBr_2$.
With HI (or HBr), $CH_2{=}CHBr$ then $CH_3.CHBr_2$.
With water (dilute H_2SO_4) and mercury(II) sulphate catalyst, CH_3CHO.

Test for unsaturated hydrocarbons: acidified $KMnO_4$ is decolorized; reddish bromine water is decolorized.

Organic acids $C_xH_{2x+1}COOH$; they contain the carboxyl group —COOH.

Properties: acidic to indicators, neutralize bases to form salts, liberate CO_2 from carbonates and H_2 from some metals, form esters with alcohols.

Test for hydroxyl (—OH) group: phosphorus pentachloride forms misty HCl fumes.

Carbohydrates: compounds of carbon, hydrogen, and oxygen with atoms H:O in ratio 2:1.

Simple sugars $C_6H_{12}O_6$, for example, glucose (blood sugar), fructose.

Complex sugars $C_{12}H_{22}O_{11}$, for example, sucrose, maltose.

Complex carbohydrates $(C_6H_{10}O_5)_n$, for example, starch, cellulose.

QUESTIONS

1. Give the names and formulae of two saturated hydrocarbons and one unsaturated hydrocarbon. State what you consider to be the meanings of saturated and unsaturated and illustrate your answer by one appropriate chemical reaction in each case. Outline how you could obtain one of the hydrocarbons in a test-tube by a simple chemical reaction.
2. Explain clearly and concisely what is meant by the following statements: (*a*) ethanol and methoxymethane, $(CH_3)_2O$, are structural isomers; (*b*) ethene and propene are members of a homologous series; (*c*) ethanol is dehydrated by the action of concentrated sulphuric acid at 180 °C, and ethene is produced.
3. Account for the fact that the H—C—H bond angle in methane is greater than the H—N—H bond angle in ammonia, which is greater than the H—O—H bond angle in water. In what way does the arrangements of the bonds in methane correspond to those of the bonds in diamond?
4. Mention any four physical or chemical differences between sodium hydroxide and ethanol. What are the similarities and differences between the reaction of ethanol with ethanoic acid and the reaction between sodium hydroxide and hydrochloric acid? Outline how you would prepare one named ester in a test-tube.
5. Give the chemical name of one common fat or oil. What is the action on this fat or oil of sodium hydroxide

solution? What is the action on either ethanol or ethanoic acid of phosphorus pentachloride?

6. Suggest a method of using copper(II) oxide to show that starch contains carbon and hydrogen. How would you demonstrate that starch is present in potato or rice?

7. Give the graphical formulae (those showing all the chemical bonds present) for the alcohols of molecular formula C_3H_8O. What products are formed when ethanol is (*a*) mildly oxidized, (*b*) vigorously oxidized? (C.)

8. Mention how you would obtain vinegar from a dilute solution of ethanol. How would you obtain fairly pure ethanoic acid (acetic acid) from the vinegar. Write equations for the action of this acid on sodium hydroxide, ethanol, and phosphorus pentachloride, and name the important product in each reaction.

26 Chemicals from Petroleum. Polymers and Macromolecules

Formation of petroleum

Petroleum was probably formed by the decomposition of animal and vegetable substances under pressure and at high temperatures. It was usually formed in shallow seas that contained plenty of animal and vegetable living organisms. The dead matter was covered with a deep layer of sand and clay. Bacterial decomposition under these conditions produced the petroleum. It is a mixture of hydrocarbons, mainly alkanes or paraffins.

Crude petroleum is a black viscous liquid which sometimes has a greenish sheen. Natural gas always occurs above it. The gas is mainly methane but also contains a little ethane, propane, and other hydrocarbons.

REFINING OF CRUDE PETROLEUM

Fractional distillation is used to separate crude oil into fractions or mixtures which boil within certain ranges of temperature. The products and the temperature ranges vary with the composition of the crude oil. Typical substances and temperatures are:

Fraction	*Boiling range*	*Uses*
Hydrocarbon gases	Below 40 °C	Fuel: 'bottled gas'
Gasoline	50–175 °C	Petrol
Kerosine (paraffin oil or naphtha)	175–250 °C	Fuel: 'cracked' to form petrol
Gas oil and diesel oil	250–350 °C	Boiler fuel; diesel fuel
Heavy oil	350–400 °C	Lubricating oil; fuel oil for furnaces; vaseline; paraffin wax
Bitumen	Residue	Road surfaces; roofing material

The crude oil is passed through pipes at about 400 °C.

The vapours formed are pumped into the bottom of a tall fractionating tower in which the temperature decreases from 400 °C at the bottom to about 40 °C at the top. Inside the tower are trays and pipes arranged so that the vapour passing up mixes thoroughly with the liquid dropping down. Different fractions are taken out from compartments at different levels. The simplified diagram in Fig. 26.1 makes the process clear.

The solid residue is bitumen (asphalt). Some of the hydrocarbon gases formed are used to heat the crude petroleum before it enters the column. The various fractions are treated by distillation, crystallization, solvent extraction or adsorption on charcoal or silica gel in order to obtain commercial products.

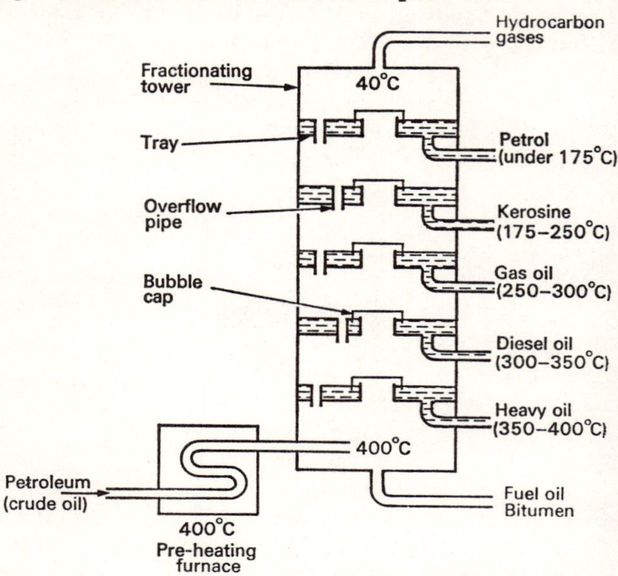

Fig. 26.1 Refining (fractionation) of petroleum

Natural gas

This is used as a fuel. Alkanes such as propane, butane, and pentane are liquefied by pressure and stored in metal 'bottles' or cylinders. Natural gas is often added to coal gas to produce town gas, and some town gas consists almost entirely of natural gas.

The methane in natural gas is partially oxidized by steam in the presence of a nickel catalyst or by oxygen:

$$CH_4 + H_2O = CO + 3H_2$$

$$2CH_4 + O_2 = 2CO + 4H_2$$

The mixture of hydrogen and carbon monoxide is used as a fuel. The carbon monoxide is also converted to carbon dioxide by treatment with steam in the presence of iron (see p. 30), and the carbon dioxide is removed. The hydrogen is used in the synthesis of ammonia.

If methane is burned in an insufficient supply of air, carbon black is formed:

$$CH_4 + O_2 = C + 2H_2O$$

The carbon black is used in the manufacture of printers' ink, shoe polishes and gramophone records; it is also added to rubber in order to make car tyres stronger and able to last longer.

Ethyne is now manufactured by heating natural gas to about 1500 °C; the products are then cooled rapidly to avoid decomposition of the ethyne:

$$2CH_4 = C_2H_2 + 3H_2$$

To demonstrate the cracking of heavy oils

Push a plug of glass wool to the bottom of a boiling-tube. Add medicinal paraffin to it drop by drop until the wool is saturated.

Fill the rest of the tube with broken porcelain. Arrange the tube as in Fig. 26.2. The delivery tube should lead into a gas-jar inverted in a trough of water.

Heat the broken porcelain strongly. Heat is conducted through the glass and porcelain to the paraffin, which boils. The vapour must pass through the hot porcelain. Collect the gaseous product in gas-jars.

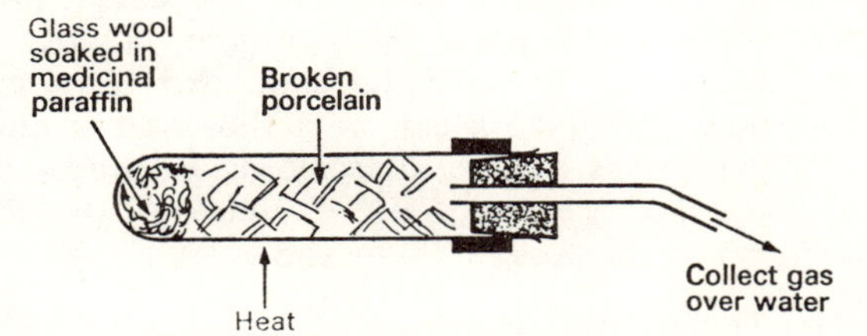

Fig. 26.2 The cracking of heavy oil

4. Light one jar of the gas and observe the flame.
5. Add two drops of bromine water to a second jar of gas. If the liquid is decolorized, the gas contains an unsaturated hydrocarbon.

COMMERCIAL CRACKING OF OILS

Cracking means the thermal decomposition of a hydrocarbon molecule into two or more hydrocarbon molecules each with a smaller number of carbon atoms; for example:

$$\underset{\text{propane}}{C_3H_8} = \underset{\text{methane}}{CH_4} + \underset{\text{ethene}}{C_2H_4}$$

$$\underset{\text{decane}}{C_{10}H_{22}} = \underset{\text{heptane}}{C_7H_{16}} + \underset{\text{propene}}{C_3H_6}$$

The first use of petroleum was to provide kerosine for lighting and heating purposes, and its petrol content was both useless and a danger. The development of the motor car reversed the position, and today over 80 per cent of petroleum is used as petrol. Since fractionation does not yield sufficient petrol, the less volatile hydrocarbons are cracked to form petrol and gases, which are mainly alkenes. Petrol hydrocarbons contain between five and nine carbon atoms and their formulae are C_5H_{12} to C_9H_{20}.

Thermal cracking is done by heating the gas oil from petroleum to about 500 °C under pressure. Decomposition occurs. The final product is fractionated to obtain petrol. The alkenes are collected and any unchanged oil is used again. The alkenes are produced by dehydrogenation of alkanes:

$$\underset{\text{alkane}}{—CH_2—CH_2—} \rightarrow \underset{\text{alkene}}{—CH{=}CH—} + H_2 \text{ dehydrogenation}$$

Catalytic cracking is a similar process done at a lower temperature with silicon(IV) and aluminium oxide catalysts. The products are different from those obtained in thermal cracking. The catalysts cause the formation of molecules in which one carbon atom is joined to three or four other carbon atoms, e.g.

$$\begin{array}{c} —C—\underset{\substack{|\\ C}}{C}—C— \end{array} \quad \text{or} \quad \begin{array}{c} —C—\overset{\substack{C\\ |}}{\underset{\substack{|\\ C}}{C}}—C— \end{array}$$

and molecules containing rings of six carbon atoms. Petrol containing such substances is better than petrol with straight-chain molecules, e.g.

$$—C—C—C—C—C—$$

PETROLEUM IN INDUSTRIAL PROCESSES

Hydrogen Refer to p. 30.

Sulphur Refer to p. 134.

Alkenes The gases obtained by the cracking of heavy oils are used to produce plastics, detergents, solvents, and insecticides. Some details are given on pp. 165–6.

Ethanol Ethene is hydrated directly at a high temperature and under great pressure. Phosphoric acid is a catalyst:

$$CH_2{=}CH_2 + H_2O = CH_3CH_2OH$$

In an indirect process, ethene is absorbed in concentrated sulphuric acid and then water is added to hydrolyse the product:

$$C_2H_4 \xrightarrow{H_2SO_4} C_2H_5HSO_4 \xrightarrow{H_2O} C_2H_5OH + H_2SO_4$$

Most industrial ethanol is made by these two processes.

Glycol Ethene is mixed with excess air or oxygen under high pressure and with a silver catalyst:

$$2CH_2{=}CH_2 + O_2 = 2(CH_2)_2O, \text{ ethene oxide}$$

Water converts the ethene oxide to glycol:

$$(CH_2—CH_2)O + H_2O = (CH_2OH)_2, \text{ glycol}$$

Glycol is used to manufacture Terylene.

Detergents Hydrocarbons with ten to eighteen carbon atoms are treated with sulphuric acid; for example:

$$C_xH_{2x} + H_2SO_4 = C_xH_{(2x+1)} . HSO_4$$

The sodium salts of the acids are detergents.

To demonstrate the preparation of a detergent

1. Place about 1 cm³ of castor oil in a test-tube.
2. Carefully add about 2 cm³ of concentrated sulphuric acid and stir with a glass rod. Heat is evolved and it causes some of the oil to turn black, owing to formation of carbon.
3. Add the product to about 3 cm³ of water in a second tube (do not add the water to the acid mixture) and stir well.

4. Pour away the aqueous layer, which contains the excess sulphuric acid, and wash the solid product with water. Shake the solid with water to show that it produces a lather.

$$\text{Castor oil} + \text{sulphuric acid} = \text{Detergent}$$

Commercial detergents are sodium salts because the free acids would be corrosive and dangerous.

THE ACTION OF DETERGENTS ON GREASE AND DIRT

Soap is the commonest detergent. It reduces the surface tension of water and allows the water to moisten the cloth being washed. A detergent changes grease into tiny droplets which are easily removed from the cloth and it softens hard water.

In everyday life the term 'detergent' means any substance that is used as a substitute for soap. These detergents are made from petroleum and not from fats and oils. Their advantage over soap is that their calcium and magnesium salts are soluble in water and therefore they form no scum in hard water.

Add detergent to (*a*) a cloth soaked in olive oil or other oil, and (*b*) a cloth marked with soot. Shake well and then examine the result.

One constituent of soap is sodium octadecanoate (stearate), $C_{17}H_{35}COONa$. The —COONa group is at the end of a chain of seventeen carbon atoms in the group $C_{17}H_{35}$. The —COONa group readily mixes with water and is called *hydrophilic* (water-loving). The hydrocarbon group is not soluble in water and is *hydrophobic* (water-hating), but it is soluble in oils and fats. Detergents have similar water-soluble and fat-soluble groups at the ends of their molecules. For example, the —SO_4Na group is water-soluble and the hydrocarbon group is fat-soluble.

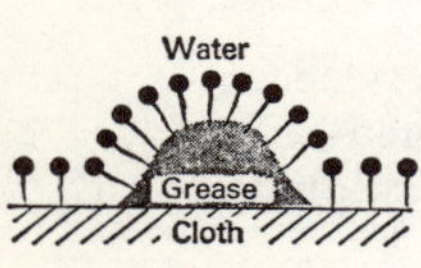

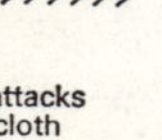

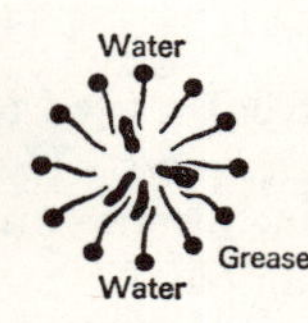

(a) Detergent attacks grease on cloth (b) Grease is partly emulsified (c) Grease is split up into small 'soluble' droplets

Fig. 26.3 How a detergent emulsifies grease

The fat-soluble or hydrocarbon part of detergent molecules penetrates grease on cloth and the water-soluble part remains in the water. Therefore the grease is split up into many tiny globules which is called an *emulsion* (of grease in water) and it leaves the cloth. A detergent molecule contains ten to eighteen carbon atoms. A detergent with less than ten carbon atoms in its molecules cannot dissolve grease sufficiently and a detergent with more than eighteen carbon atoms is only slightly soluble in water.

SYNTHETIC POLYMERS

ADDITION-POLYMERIZATION

Carbon readily forms long chains of atoms and differs in this respect from the other elements we study. Some chains contain thousands of carbon atoms. Rubber, cellulose, silk, starch, and proteins are examples of naturally occurring substances which contain very long chains of carbon atoms.

Nitrogen dioxide, NO_2, polymerizes to form dinitrogen tetroxide, N_2O_4 or $(NO_2)_2$. NO_2 is called the *monomer* and $(NO_2)_2$ is the *dimer*; the names mean 'one part' and 'two parts'. Some organic compounds form compounds in which the monomer X forms a polymer X_n in which n can be a thousand or more. The polymer has a large molecule or *macromolecule* and it is called a *high polymer*. Starch is $(C_6H_{10}O_5)_n$ in which n varies from 1000 to about 3000.

Fibres and plastics

Wool, silk, cotton, and linen form natural fibres. A fibre is a solid whose lengths are hundreds or thousands of times greater than their widths. Rayon was the first synthetic fibre and it was originally called artificial silk. Polythene, polystyrene, nylon, and Terylene are more common now. A plastic is a substance which can be shaped by heating and pressure at some stage in its manufacture to form a product that is stable at ordinary temperatures. Bakelite and Perspex are common plastics and Bakelite was the first synthetic plastic. Polythene, polystyrene, and nylon are also plastics. All are high polymers and consist of macromolecules.

CONDENSATION-POLYMERIZATION OR POLYCONDENSATION

In addition-polymerization, only single molecules combine and there is no loss or gain of material; only one substance is formed. In condensation-polymerization, sometimes called polycondensation, single molecules combine to form a macromolecule; water or another simple substance, such as ammonia, is also formed as a second product:

$$\text{Addition-polymerization:} \quad nX \rightarrow X_n$$
$$\text{Condensation-polymerization:} \quad nX \rightarrow Y_n + nH_2O$$

Possibly starch is formed from glucose by condensation-polymerization:

$$\underset{\text{glucose}}{nC_6H_{12}O_6} = nH_2O + \underset{\text{starch}}{(C_6H_{10}O_5)_n}$$

To study the action of heat on Perspex

Heat small pieces of Perspex in the apparatus shown in Fig. 26.4; do not smell the poisonous fumes. The distillate is the monomer which is the methyl ester of an unsaturated acid; it is called methyl methylpropenate, $H_2O{=}C(CH_3)COOCH_3$.

Polymerizing the monomer Dodecanoyl peroxide catalyses the polymerization. Add a little to the distillate and warm at about 50 °C in a beaker of water for about one hour.

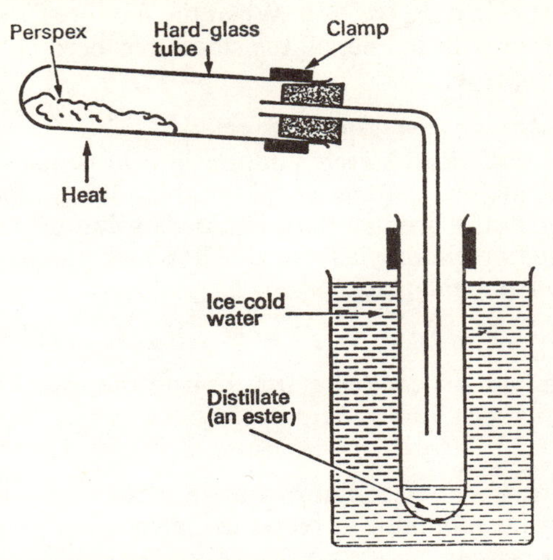

Fig. 26.4 Obtaining the monomer from Perspex

Polyethene (*Polythene or polyethylene*)
Addition-polymerization is an addition reaction and the monomer must contain a double bond. The simplest organic compound with a double bond is ethene:

$$2nC_2H_4 = (C_2H_4)_n$$

or

$$2n(H_2C{=}CH_2) = (\text{—}CH_2\text{—}CH_2\text{—}CH_2\text{—}CH_2\text{—})_n$$

There is a CH_3— group at each end of the chain, and there can be as many as 1400 carbon atoms in the chain.

Polyethene or Polythene is manufactured by two processes. In one, ethene is heated to about 200 °C under a pressure of about 1500 atmospheres in the presence of a trace of oxygen as a catalyst. In the second process, ethene is passed into a suspension of a catalyst in a hydrocarbon solvent at about 60 °C and 10 atmospheres pressure.

Polythene is a tough, white, waxy solid. It is not affected by acids, bases, or solvents, or oxidizing or reducing agents. It can bend easily without breaking and it is a good insulator. It is used as a wrapping material and for making cold-water pipes, ropes, and twines.

Polystyrene
Substituted ethenes, $XHC{=}CH_2$, also polymerize. The formula of styrene is $C_6H_5 . HC{=}CH_2$, in which X is the phenyl group, C_6H_5—. Styrene polymerizes readily:

$$2n(C_6H_5 . HC{=}CH_2) = (\text{—}C_6H_5 . HC\text{—}CH_2\text{—}C_6H_5 . HC\text{—}CH_2\text{—})_n, \quad \text{polystyrene}$$

The general equation for the polymerization of substituted ethenes is:

$$nXHC{=}CH_2 \rightarrow \text{—}\underset{H}{\overset{X}{C}}\text{—}\underset{H}{\overset{H}{C}}\text{—}\underset{H}{\overset{X}{C}}\text{—}\underset{H}{\overset{H}{C}}\text{— etc.}$$

Heat, light or catalysts change the monomer styrene to polystyrene, and the value of n in its formula can be as high as 5000. Polystyrene resembles polyethene, and it is resistant to water, acids, alkalis, and solvents.

To polymerize styrene
Mix styrene with an equal volume of kerosine, which catalyses the polymerization. Boil for about an hour at a temperature of 150 °C. Use a vertical condenser, called a reflux condenser, to condense the vapours and return them to the reaction flask. Allow the product to cool and pour it into a large volume of methanol. White solid polystyrene forms.

Perspex
This plastic is made from a disubstituted ethene $H_2C{=}CXY$ in which two of the hydrogen atoms of ethene are replaced by radicals. Perspex is prepared from methyl methylpropenate, $H_2C{=}C(CH)_3COOCH_3$, in which X is CH_3 and Y is $COOCH_3$. The formula for Perspex is:

$$\text{—}\underset{H}{\overset{H}{C}}\text{—}\underset{Y}{\overset{X}{C}}\text{—}\underset{H}{\overset{H}{C}}\text{—}\underset{Y}{\overset{X}{C}}\text{— etc.}$$

Compare this formula with that for polystyrene. Since a perspex macromolecule contains many ester groups, $COOCH_3$, the plastic is called a *polyester*.

Perspex is often used instead of glass. It has a lower density than glass and is not brittle, but it is easily scratched. It is used for aeroplane windows, lenses, and prisms.

Terylene
Terylene is made from glycol, $HO . CH_2 . CH_2 . OH$, and terephthalic acid, $HO . OC . C_6H_4 . CO . OH$, by condensation-polymerization. The molecules of both substances have two reactive groups, —OH and —COOH, one at each end; water is eliminated when they react. A simple equation for the polycondensation is:

$$n\underset{\text{glycol}}{H\text{—}O\text{—}X\text{—}O\text{—}H} + n\underset{\text{terephthalic acid}}{HO\text{—}CO\text{—}Y\text{—}CO\text{—}OH}$$

$$= \underset{\text{Terylene (a polyester)}}{(\text{—}O\text{—}X\text{—}O\text{—}CO\text{—}Y\text{—}CO\text{—})_n} + nH_2O$$

The texture of Terylene is similar to that of wool, and the polymer is therefore suitable for making suits and clothing. It dries quickly because it does not absorb water, and permanent creases can be put in it.

Nylon

This high polymer is tough, hard-wearing, and does not rot or deteriorate. It is used as a fibre for making hosiery, brushes, ropes, and cloth for furnishings; parachutes, carpets, and 'furs' are also made from it. It can also be moulded as a plastic and used to make gear wheels and small bearings and runners.

Nylon is made from a diamine, $H_2N . A . NH_2$, and a diacid, HO . CO . B . CO . OH. The diamine has two amine groups, NH_2, one at each end of its molecules, and the diacid has two acidic carboxyl groups. Water is eliminated when the two substances react by polycondensation. A simple equation is:

$$\underset{\text{diamine}}{H_2N\text{—}A\text{—}NH_2} + \underset{\text{diacid}}{HO\text{—}CO\text{—}B\text{—}CO\text{—}OH}$$
$$\rightarrow \underset{\text{nylon}}{\text{—}HN\text{—}A\text{—}NH\text{—}CO\text{—}B\text{—}CO\text{—}} + 2H_2O$$

The —NH—CO— group is called the amide group, and nylon is therefore an example of a *polyamide*.

Nylon 66 is made from a diamine $H_2N . (CH_2)_6 . NH_2$ and hexanedioic acid, $HOOC . (CH_2)_4 . COOH$. The 66 is used because both molecules contain 6 carbon atoms. Nylon 610 is made from a diamine with 6 carbon atoms and a diacid with 10 carbon atoms. Nylon 6 is made from the monomer $H_2N . (CH_2)_5COOH$, which is both an amine and an acid.

The structures of Terylene and nylon may be clearer with block diagrams:

```
                          O          O
                          ||         ||
Terylene  —O—[X]—O—C—[Y]—C— etc.

           H         H  O          O
           |         |  ||         ||
Nylon     —N—[A]—N—C—[B]—C— etc.
```

NATURAL MACROMOLECULES

Carbohydrates

Starch $(C_6H_{10}O_5)_n$, is a carbohydrate that is made by plants. It is the final product of photosynthesis in which carbon dioxide and water are probably changed to glucose which then changes by condensation-polymerization to starch. The macromolecules of starch have molecular masses between 10^5 and 10^6. Starch is readily converted back to glucose.

To obtain glucose from starch

1. Add a few drops of dilute hydrochloric acid to starch solution in a boiling tube. Boil the mixture for a few minutes.
2. Allow to cool and then add sodium hydroxide solution until the mixture turns litmus blue.
3. Test for the presence of glucose by warming the alkaline mixture with Fehling's solution (p. 160). A red precipitate of copper(I) oxide, Cu_2O, forms. Starch has no action on Fehling's solution; glucose reduces the solution to the red oxide.
4. Repeat the test, but this time boil the mixture of starch and acid for a long time. Remove a few drops of the mixture from time to time and add it to iodine solution, which turns blue if starch is still in the mixture (p. 160). When the mixture forms no colour with iodine, all the starch has been changed to glucose.

To study the action of saliva on starch

1. Add saliva to starch solution in a test-tube. Keep the mixture at about 35 °C for about thirty minutes.
2. Warm the product with Fehling's solution.

A red precipitate is formed. The sugar produced by the saliva is maltose:

$$2n(C_6H_{10}O_5) + nH_2O = nC_{12}H_{22}O_{11}, \text{ maltose}$$

Saliva acts on cooked starch during the digestion of food.

Proteins

All living matter contains proteins. They are present in eggs as albumin, in blood as haemoglobin, and in milk as casein.

Protein molecules are macromolecules formed by condensation-polymerization of one or more compounds called amino-acids. These organic compounds have an amino-group $—NH_2$ at one end of a chain of carbon atoms and an acidic carboxyl group —COOH at the other end. Water is eliminated between opposite ends of two molecules:

$$nHOOC\text{—}X\text{—}NH_2 + nHOOC\text{—}X\text{—}NH_2$$
$$= (\text{—}OC\text{—}X\text{—}NH\text{—}CO\text{—}X\text{—}NH\text{—})_n + 2nH_2O$$

Two different amino-acids can condense, and X and Y represent different groups in the two molecules. The smallest molecular mass of a protein is about 6000; that of haemoglobin is about 66 000 and that of tobacco virus is twenty million. The —NH—CO— group in proteins is called a *peptide linkage*; some proteins contain more than 2000 of these linkages in a molecule.

Boiling acids and alkalis break the peptide linkages and addition of water occurs. That is, the change produced is the reverse of the condensation reaction. The catalytic action of enzymes in digestive juices causes the same change during digestion of proteins in food.

SUMMARY

Petroleum: refining produces hydrocarbon gases, gasoline, kerosine, gas oil, diesel oil, heavy oil, and bitumen.

Natural gas contains mainly methane and other alkanes.

Cracking is the thermal decomposition of hydrocarbon molecules into simpler molecules. A molecule may be decomposed and/or dehydrogenated and chains of carbon atoms may form rings.

Thermal cracking: done by heat and pressure.

Catalytic cracking: done at lower temperature with silicon(IV) oxide and aluminium oxide catalysts.
Products from petroleum: hydrogen, sulphur, alkenes, ethanol, detergents, fibres, and plastics.
Detergents: a molecule contains a water-soluble group, $—SO_4Na$, and a fat-soluble group with 10 to 18 carbon atoms. Detergents emulsify grease.

Synthetic polymers: polyethene (Polythene), polystyrene, Perspex, Terylene, and nylon.
Polyethene: made from ethene heated under great pressure or warmed under moderate pressure with a catalyst.
Addition-polymerization: the combination of several molecules of one substance, called the monomer:

$$nC_2H_4 = (C_2H_4)_n.$$

Polystyrene: made by polymerization of styrene,

$$C_6H_5 . HC{=}CH_2,$$

a monosubstituted ethene.
Perspex: made from methyl methylpropenate,

$$H_2C{=}C(CH_3)COOCH_3,$$

a disubstituted ethene and the methyl ester of an unsaturated acid. Perspex is a polyester.
Terylene: made from glycol, $HO . CH_2 . CH_2 . OH$, and terephthalic acid, $HO . OC . C_6H_4 . CO . OH$. Water is eliminated. Terylene is a polyester.
Condensation-polymerization or *polycondensation:* the combination of many molecules, either of one compound or of two compounds, with the elimination of water or ammonia.

$$nX—OH + nHO—CO—Y = (—X—O—CO—Y—)_n \text{ (polyester)} + nH_2O$$

$$nA—NH_2 + nHO—CO—B = (—A—NH—CO—B—)_n \text{ (polyamide)} + nH_2O$$

Nylon: made from a diacid, $HOOC . (CH_2)_4 . COOH$ and a diamine, $H_2N . (CH_2)_6 . NH_2$ to form nylon 66. Nylon is a polyamide.

Natural macromolecules
Starch, $(C_6H_{10}O_5)_n$, is converted by acids to glucose and by saliva to maltose, both of which reduce Fehling's solution.
Proteins: formed by condensation-polymerization of amino-acids, a molecule of which contains a NH_2 and a COOH group. Proteins contain peptide linkages, —NH—CO—, which are broken by hot acids and alkalis and by enzymes in digestive juices.

QUESTIONS

1. What chemical changes are possible during the cracking of an oil? How would you demonstrate cracking in the laboratory?
2. Petroleum is the raw material used in the manufacture of many chemicals. Outline the fractional distillation of crude petroleum, and also outline how hydrogen, sulphur, and ethanol are obtained from petroleum.
3. Describe briefly the chemical structure of a molecule of a detergent (not soap) and explain how the detergent emulsifies grease. In what way is soap less suitable than other detergents for many purposes?
4. Explain the difference between addition-polymerization and condensation-polymerization. In your answer refer to the formation of Polythene (or polystyrene), nylon, and Terylene. (The structure of the linkages in these polymers should be given, but no further details of their structure.)
5. Indicate concisely how the macromolecules of starch are probably built up from glucose molecules. Explain how protein macromolecules are produced from amino-acids. How would you break down starch to glucose in the laboratory, and by what test would you prove the formation of glucose?
6. Describe one test which indicates the presence of a double bond in the styrene molecule (styrene is a liquid). By reference to ethane, polyethene (Polythene), methyl methylpropenate, and Perspex, explain the meanings of monomer, polymer, and macromolecule.
7. Describe, with illustrative examples, different kinds of polymerization. How does the chemist exploit these processes to make materials having specified properties which are in common everyday use? (L.)
8. Certain substances produced by living things consist of giant molecules. Write a concise account of such substances. Name and discuss the structure of three other compounds which consist of giant molecules.

APPENDIX

Simple Tests for Chemicals

IDENTIFICATION OF GASES

Four simple tests enable you to identify the common gases.

1. *Note the colour* Chlorine, hydrogen halides, nitrogen dioxide and bromine and iodine vapours are coloured. Hydrogen halides are colourless but form misty fumes in damp air.
2. *Note the smell* Chlorine, hydrogen halides, nitrogen dioxide, sulphur dioxide, hydrogen sulphide, and ammonia have characteristic smells.
3. *Note the action of the gas on a burning splint* Hydrogen and hydrogen sulphide burn. Oxygen and dinitrogen monoxide (N_2O) re-light a glowing splint and make a burning splint burn brightly.
4. *Put damp indicator papers in the gas*; for example, red and blue litmus papers. Ammonia is the only common gas that is alkaline. Chlorine bleaches litmus, and it first turns blue litmus red. Hydrogen halides, nitrogen dioxide, sulphur dioxide, carbon dioxide, and hydrogen sulphide are acidic gases.

CONFIRMATORY TESTS A stopper dipped in concentrated hydrochloric acid forms white fumes with ammonia.

A stopper dipped in concentrated aqueous ammonia forms white fumes with hydrogen halides and chlorine.

Carbon dioxide turns lime water milky.

Sulphur dioxide decolorizes potassium manganate(VII) solution which has been acidified with dilute sulphuric acid, and turns acidified potassium dichromate solution from orange to green.

Hydrogen sulphide turns filter paper dipped in lead ethanoate or lead nitrate solution to black lead sulphide. Sometimes the sulphide on the paper is coloured a shiny, brownish black.

Water vapour condenses to a colourless liquid, which turns white copper sulphate blue.

ACTION OF HEAT ON SUBSTANCES

Heat a solid in a test-tube. Boil a solution to dryness in a boiling tube and then heat the solid residue:

Observation	*Inference*
Oxygen	NO_3^-; O_2^{2-} (peroxide ion) or dioxide
Nitrogen dioxide	NO_3^-
Carbon dioxide	CO_3^{2-}; HCO_3^-
Sulphur dioxide	SO_3^{2-}; HSO_3^-

Ammonium salts sometimes form a white sublimate, and water is formed by hydrates, hydrogen carbonates, and some hydroxides.

FLAME TEST

Add one or two drops of concentrated hydrochloric acid to the solid on a clock-glass. Dip the end of a clean nichrome or platinum wire in the mixture. Place the mixture in a flame:

Observation	*Inference*
Golden yellow (brilliant)	Na^+
Red	Ca^{2+} or Li^+
Lilac	K^+
Bluish-green	Cu^{2+}

Sodium is present in most ordinary compounds and therefore a yellow flame that is not brilliant should be ignored. Potassium salts contain sodium, and its yellow flame makes the lilac flame difficult to see. Look at the flame through a blue glass, which removes the yellow light and the lilac flame appears crimson.

ACTION OF REAGENTS ON SUBSTANCES

1. *Dilute hydrochloric acid*

Add the cold acid to the solid or solution. Warm gently if necessary:

Observation	*Inference*
Carbon dioxide	CO_3^{2-}; HCO_3^-
Sulphur dioxide	SO_3^{2-}; HSO_3^-
Hydrogen sulphide	S^{2-}

2. *Concentrated sulphuric acid*

Add the cold acid to the solid. Warm gently if necessary, but never boil because the very hot acid forms white sulphur trioxide fumes which may cause incorrect observations. Carbon dioxide, sulphur dioxide, and hydrogen sulphide are formed as with hydrochloric acid. In addition, the following ions can be identified:

Observation	*Inference*
Hydrogen chloride	Cl^-
Hydrogen bromide (and bromine)	Br^-
Hydrogen iodide (and iodine)	I^-

If the test indicates the presence of a halide radical, mix the original solid with manganese(IV) oxide and warm the mixture with concentrated sulphuric acid. Chlorides form chlorine, bromides form bromine as a reddish gas, and iodides form iodine as a violet vapour.

3. *Barium nitrate or chloride*

To a solution of the original substance, add aqueous barium nitrate or chloride. Add excess nitric acid to any precipitate.

Sulphates form a white precipitate which is not changed by the acid. Carbonates and sulphites form white precipitates which dissolve in the excess acid:

$$Ba^{2+} + SO_4^{2-} = BaSO_4\text{, white}$$

4. *Silver nitrate solution*
To a solution of the original substance, add excess dilute nitric acid to decompose any carbonate or sulphite and then add silver nitrate solution.

Chlorides form a white precipitate, soluble in aqueous ammonia. Bromides form a pale yellow precipitate, slightly soluble in dilute ammonia solution. Iodides form a yellow precipitate, insoluble in ammonia solution:

$$Ag^{+}(aq) + Cl^{-}(aq) = AgCl(s)\text{, white}$$

5. *Chlorine water*
To a solution of the original substance, add chlorine water or pass chlorine gas.

Bromides form bromine as a reddish solution and iodides form iodine as a brown solution and then as a black solid.

Add tetrachloromethane, CCl_4, to the coloured solution. Bromine dissolves in it to form a reddish solution and iodine forms a violet solution:

$$Cl_2 + 2Br^{-} = 2Cl^{-} + Br_2(aq\text{ or }l)$$

6. *Sodium hydroxide solution*
To a solution of the original substance, add sodium hydroxide slowly. Note if a precipitate forms and if it dissolves in excess of the alkali. Smell the product to detect ammonia:

Observation	*Inference*
Ammonia	NH_4^+
White ppt., insoluble in excess	Ca^{2+}
White ppt., soluble in excess	Zn^{2+}, Pb^{2+}, Al^{3+}
Blue ppt., insoluble in excess	Cu^{2+}
Green ppt., insoluble in excess	Fe^{2+} iron(II)
Reddish-brown ppt., insoluble in excess	Fe^{3+} iron(III)

The precipitates are hydroxides. Those of zinc, lead, and aluminium are amphoteric and therefore dissolve in excess alkali.

7. *Aqueous ammonia*
Repeat the previous test with ammonia solution. The precipitates are formed in the same way.

In excess ammonia, blue copper hydroxide dissolves to form a deep blue solution, containing the tetraamine copper(II) ion, $Cu(NH_3)_4^{2+}$. Lead and aluminium hydroxides do not dissolve in excess ammonia but zinc hydroxide does, forming the $Zn(NH_3)_4^{2+}$ ion.

8. *Potassium hexacyanoferrate(II) and (III) solutions*
These are also called potassium ferrocyanide and potassium ferricyanide solution.

The (II) compound in solution forms a dark blue precipitate with solutions of iron(III) salts.

The (III) compound in solution forms a dark blue precipitate with solutions of iron(II) salts. Formulae and equations are not required here.

QUESTIONS

1. By what reactions and observations would you detect chloride ions in potassium chloride and lead ions in lead nitrate?
2. Describe two chemical reactions which would serve to distinguish between copper(II) oxide and manganese(IV) oxide.
3. Give an account of the action of concentrated sulphuric acid on a carbonate, a chloride, a sulphite, and a nitrate.
4. Give one example of the use in the laboratory of any five of the following reagents: potassium permanganate, ammonia solution, iron(II) suphate, solid lead nitrate, iodine solution, barium nitrate. Write equations for any reactions you mention.
5. A specimen of sodium sulphite has partially oxidized during storage. How would you detect the presence of both sulphite ion and sulphate ion in the specimen? How would you test for the presence of potassium ions in the specimen?
6. Identify the substances *A* and *B* from the information given, and give equations where appropriate:
(*a*) *A* was a white crystalline solid. When it was heated brown fumes were evolved, and a glowing splint held in these fumes burst into flames. A yellow residue was obtained, which fused with the glass. *A* was soluble in water and the solution gave a dense white precipitate with dilute sulphuric acid.
(*b*) *B* was a crystalline solid readily soluble in water to give a pale green solution. When the solution was treated with ammonia solution a light-green coloured precipitate was obtained which darkened to olive-green, and then to a brownish colour. When *B* was heated a fine red powder was left in the tube and acid gases were evolved which (i) decolorized potassium manganate(VII) and (ii) turned a drop of acidified barium chloride solution cloudy. (L.)
7. *C*, *D*, and *E* were three crystalline compounds of sodium. (i) When dissolved in water, *C* turned blue litmus red, *D* had no effect on red or blue litmus, and *E* turned red litmus blue. (ii) *D* and *E* were heated separately in dry test-tubes. *D* gave off a gas which relit a glowing splint and steam alone was evolved from *E*. (iii) A solution of *C* was added to *E*. There was an immediate effervescence of a gas which gave a white precipitate when passed into lime water. (iv) When barium chloride solution was added to a solution of *C*, a white precipitate, insoluble in hydrochloric acid, was formed. (v) When a mixture of *D* and copper turnings was warmed with concentrated sulphuric acid, a brown gas, which turned moist blue litmus red, was evolved.
Identify the compounds *C*, *D*, and *E* and explain the observations. (C.)

Answers to Numerical Questions

Chapter 1, p. 8

9. 18 g
11. 33·3 g, 11·33 g
12. 12 g NaCl; 1 g NaCl, 25 g KNO_3; 37 g NaCl, 25 g KNO_3

Chapter 9, p. 51

3. 3:4
4. 160·2 g, 160·2 g
5. 2
7. 56
8. 20 per cent, $SnCl_4$

Chapter 13, p. 74

2. 2 F/mole of hydrogen, 4 F/mole of oxygen
6. 0·57 g copper, 1·84 g silver
7. *Y, Z, W, X*
8. 2·0 g
9. 16 800 cm^3

Chapter 14, p. 81

3. 3:1
9. (*a*) *Q*, (*b*) *T*, (*c*) *U*, (*d*) *Q*, (*g*) *R*, *P*

Chapter 15, p. 85

1. (*a*) $2x$, (*b*) $0{\cdot}5x$
2. 473/293
3. 2·016, 32·032
4. 5 g, 10 000 cm^3, 35·7 g
6. PH_3
9. 16

Chapter 16, p. 88

2. 560 cm^3
3. 1 g
4. (*a*) 29 900 cm^3, (*b*) 77·3 g
5. (*a*) 10 cm^3 CO_2, 5 cm^3 O_2; (*b*) 2·5 cm^3 O_2; (*c*) 4 cm^3 CO_2, 12 cm^3 O_2
6. 32
7. Ag_2O
8. $Cu(OH)_2 . 2CuCO_3$
9. Li_3N, 1

Chapter 17, p. 96

2. 6 kcal mol^{-1}. 25·2 kJ mol^{-1}, 6 °C
6. 3218 kJ mol^{-1}, 2046 kJ mol^{-1}
7. 413 kJ mol^{-1}
8. (*a*) 100 cm^3, (*b*) 360 kJ, (*c*) 80 g, (*d*) 104 kJ

Chapter 18, p. 100

6. (*b*) about 1250 s

Chapter 20, p. 112

1. (*a*) 4·9 g; (*b*) 73 g
2. (*a*) 0·1 M *or* M/10; (*b*) 0·4 M; (*c*) 0·25 M *or* M/4; (*d*) 0·125 M *or* M/8
3. (*a*) 1·1 dm^3 *or* 1100 cm^3; (*b*) 120 cm^3; (*c*) 500 cm^3
4. (*a*) 20 cm^3; (*b*) 25 cm^3; (*c*) 10 cm^3; (*d*) 10 cm^3
5. (*a*) 50 cm^3; (*b*) 100 cm^3; (*c*) 500 cm^3; (*d*) 1000 cm^3
6. 13·25 g per dm^3
7. 0·25 M *or* 10 g per dm^3; 40 cm^3
8. 16 per cent
9. 2
10. (*b*) 6·69 g
11. 2·125 g; 2·8 dm^3 *or* 2800 cm^3
12. 0·5 M *or* M/2; 560 cm^3

Chapter 23, p. 142

8. 1·2 g, 4480 cm^3

Index

If there are several page references, the main reference or references are given first.

LOGARITHMS

											Mean differences								
	0	1	2	3	4	5	6	7	8	9	1	2	3	4	5	6	7	8	9
10	0000	0043	0086	0128	0170	0212	0253	0294	0334	0374	4	8	12	17	21	25	29	33	37
11	0414	0453	0492	0531	0569	0607	0645	0682	0719	0755	4	8	11	15	19	23	26	30	34
12	0792	0828	0864	0899	0934	0969	1004	1038	1072	1106	3	7	10	14	17	21	24	28	31
13	1139	1173	1206	1239	1271	1303	1335	1367	1399	1430	3	6	10	13	16	19	23	26	29
14	1461	1492	1523	1553	1584	1614	1644	1673	1703	1732	3	6	9	12	15	18	21	24	27
15	1761	1790	1818	1847	1875	1903	1931	1959	1987	2014	3	6	8	11	14	17	20	22	25
16	2041	2068	2095	2122	2148	2175	2201	2227	2253	2279	3	5	8	11	13	16	18	21	24
17	2304	2330	2355	2380	2405	2430	2455	2480	2504	2529	2	5	7	10	12	15	17	20	22
18	2553	2577	2601	2625	2648	2672	2695	2718	2742	2765	2	5	7	9	12	14	16	19	21
19	2788	2810	2833	2856	2878	2900	2923	2945	2967	2989	2	4	7	9	11	13	16	18	20
20	3010	3032	3054	3075	3096	3118	3139	3160	3181	3201	2	4	6	8	11	13	15	17	19
21	3222	3243	3263	3284	3304	3324	3345	3365	3385	3404	2	4	6	8	10	12	14	16	18
22	3424	3444	3464	3483	3502	3522	3541	3560	3579	3598	2	4	6	8	10	12	14	15	17
23	3617	3636	3655	3674	3692	3711	3729	3747	3766	3784	2	4	6	7	9	11	13	15	17
24	3802	3820	3838	3856	3874	3892	3909	3927	3945	3962	2	4	5	7	9	11	12	14	16
25	3979	3997	4014	4031	4048	4065	4082	4099	4116	4133	2	3	5	7	9	10	12	14	15
26	4150	4166	4183	4200	4216	4232	4249	4265	4281	4298	2	3	5	7	8	10	11	13	15
27	4314	4330	4346	4362	4378	4393	4409	4425	4440	4456	2	3	5	6	8	9	11	13	14
28	4472	4487	4502	4518	4533	4548	4564	4579	4594	4609	2	3	5	6	8	9	11	12	14
29	4624	4639	4654	4669	4683	4698	4713	4728	4742	4757	1	3	4	6	7	9	10	12	13
30	4771	4786	4800	4814	4829	4843	4857	4871	4886	4900	1	3	4	6	7	9	10	11	13
31	4914	4928	4942	4955	4969	4983	4997	5011	5024	5038	1	3	4	6	7	8	10	11	12
32	5051	5065	5079	5092	5105	5119	5132	5145	5159	5172	1	3	4	5	7	8	9	11	12
33	5185	5198	5211	5224	5237	5250	5263	5276	5289	5302	1	3	4	5	6	8	9	10	12
34	5315	5328	5340	5353	5366	5378	5391	5403	5416	5428	1	3	4	5	6	8	9	10	11
35	5441	5453	5465	5478	5490	5502	5514	5527	5539	5551	1	2	4	5	6	7	9	10	11
36	5563	5575	5587	5599	5611	5623	5635	5647	5658	5670	1	2	4	5	6	7	8	10	11
37	5682	5694	5705	5717	5729	5740	5752	5763	5775	5786	1	2	3	5	6	7	8	9	10
38	5798	5809	5821	5832	5843	5855	5866	5877	5888	5899	1	2	3	5	6	7	8	9	10
39	5911	5922	5933	5944	5955	5966	5977	5988	5999	6010	1	2	3	4	5	7	8	9	10
40	6021	6031	6042	6053	6064	6075	6085	6096	6107	6117	1	2	3	4	5	6	8	9	10
41	6128	6138	6149	6160	6170	6180	6191	6201	6212	6222	1	2	3	4	5	6	7	8	9
42	6232	6243	6253	6263	6274	6284	6294	6304	6314	6325	1	2	3	4	5	6	7	8	9
43	6335	6345	6355	6365	6375	6385	6395	6405	6415	6425	1	2	3	4	5	6	7	8	9
44	6435	6444	6454	6464	6474	6484	6493	6503	6513	6522	1	2	3	4	5	6	7	8	9
45	6532	6542	6551	6561	6571	6580	6590	6599	6609	6618	1	2	3	4	5	6	7	8	9
46	6628	6637	6646	6656	6665	6675	6684	6693	6702	6712	1	2	3	4	5	6	7	7	8
47	6721	6730	6739	6749	6758	6767	6776	6785	6794	6803	1	2	3	4	5	5	6	7	8
48	6812	6821	6830	6839	6848	6857	6866	6875	6884	6893	1	2	3	4	4	5	6	7	8
49	6902	6911	6920	6928	6937	6946	6955	6964	6972	6981	1	2	3	4	4	5	6	7	8
50	6990	6998	7007	7016	7024	7033	7042	7050	7059	7067	1	2	3	3	4	5	6	7	8
51	7076	7084	7093	7101	7110	7118	7126	7135	7143	7152	1	2	3	3	4	5	6	7	8
52	7160	7168	7177	7185	7193	7202	7210	7218	7226	7235	1	2	2	3	4	5	6	7	7
53	7243	7251	7259	7267	7275	7284	7292	7300	7308	7316	1	2	2	3	4	5	6	6	7
54	7324	7332	7340	7348	7356	7364	7372	7380	7388	7396	1	2	2	3	4	5	6	6	7
55	7404	7412	7419	7427	7435	7443	7451	7459	7466	7474	1	2	2	3	4	5	5	6	7